Springer Series in
OPTICAL SCIENCES 150

founded by H.K.V. Lotsch

Editor-in-Chief: W. T. Rhodes, Atlanta

Editorial Board: A. Adibi, Atlanta
T. Asakura, Sapporo
T. W. Hänsch, Garching
T. Kamiya, Tokyo
F. Krausz, Garching
B. Monemar, Linköping
H. Venghaus, Berlin
H. Weber, Berlin
H. Weinfurter, München

Springer Series in
OPTICAL SCIENCES

The Springer Series in Optical Sciences, under the leadership of Editor-in-Chief *William T. Rhodes*, Georgia Institute of Technology, USA, provides an expanding selection of research monographs in all major areas of optics: lasers and quantum optics, ultrafast phenomena, optical spectroscopy techniques, optoelectronics, quantum information, information optics, applied laser technology, industrial applications, and other topics of contemporary interest.
With this broad coverage of topics, the series is of use to all research scientists and engineers who need up-to-date reference books.

The editors encourage prospective authors to correspond with them in advance of submitting a manuscript. Submission of manuscripts should be made to the Editor-in-Chief or one of the Editors. See also www.springer.com/series/624

Editor-in-Chief
William T. Rhodes
Georgia Institute of Technology
School of Electrical and Computer Engineering
Atlanta, GA 30332-0250, USA
E-mail: bill.rhodes@ece.gatech.edu

Editorial Board
Ali Adibi
Georgia Institute of Technology
School of Electrical and Computer Engineering
Atlanta, GA 30332-0250, USA
E-mail: adibi@ee.gatech.edu

Toshimitsu Asakura
Hokkai-Gakuen University
Faculty of Engineering
1-1, Minami-26, Nishi 11, Chuo-ku
Sapporo, Hokkaido 064-0926, Japan
E-mail: asakura@eli.hokkai-s-u.ac.jp

Theodor W. Hänsch
Max-Planck-Institut für Quantenoptik
Hans-Kopfermann-Straße 1
85748 Garching, Germany
E-mail: t.w.haensch@physik.uni-muenchen.de

Takeshi Kamiya
Ministry of Education, Culture, Sports
Science and Technology
National Institution for Academic Degrees
3-29-1 Otsuka, Bunkyo-ku
Tokyo 112-0012, Japan
E-mail: kamiyatk@niad.ac.jp

Ferenc Krausz
Ludwig-Maximilians-Universität München
Lehrstuhl für Experimentelle Physik
Am Coulombwall 1
85748 Garching, Germany   *and*
Max-Planck-Institut für Quantenoptik
Hans-Kopfermann-Straße 1
85748 Garching, Germany
E-mail: ferenc.krausz@mpq.mpg.de

Bo Monemar
Department of Physics
and Measurement Technology
Materials Science Division
Linköping University
58183 Linköping, Sweden
E-mail: bom@ifm.liu.se

Herbert Venghaus
Fraunhofer Institut für Nachrichtentechnik
Heinrich-Hertz-Institut
Einsteinufer 37
10587 Berlin, Germany
E-mail: venghaus@hhi.de

Horst Weber
Technische Universität Berlin
Optisches Institut
Straße des 17. Juni 135
10623 Berlin, Germany
E-mail: weber@physik.tu-berlin.de

Harald Weinfurter
Ludwig-Maximilians-Universität München
Sektion Physik
Schellingstraße 4/III
80799 München, Germany
E-mail: harald.weinfurter@physik.uni-muenchen.de

Please view available titles in *Springer Series in Optical Sciences*
on series homepage http://www.springer.com/series/624

Cornelia Denz
Sergej Flach
Yuri S. Kivshar

Editors

Nonlinearities in Periodic Structures and Metamaterials

With 132 Figures

Professor Dr. Cornelia Denz
Westfälische Wilhelms-Universität
Institut für Angewandte Physik
Corrensstr. 2–4, 48149 Münster, Germany
E-mail: denz@uni-muenster.de

Dr. Sergej Flach
Max-Planck-Institut für Physik Komplexer Systeme
Nöthnitzer Str. 38, 01187 Dresden, Germany
E-mail: flach@mpipks-dresden.mpg.de

Professor Yuri S. Kivshar
Australian National University
Research School of Physics and Engineering
0200 Canberra, Australia
E-mail: ysk@internode.on.net

ISSN 0342-4111
ISBN 978-3-642-02065-0   e-ISBN 978-3-642-02066-7
DOI 10.1007/978-3-642-02066-7
Springer Heidelberg Dordrecht London New York

Library of Congress Control Number: 2009928303

© Springer-Verlag Berlin Heidelberg 2010
This work is subject to copyright. All rights are reserved, whether the whole or part of the material is concerned, specifically the rights of translation, reprinting, reuse of illustrations, recitation, broadcasting, reproduction on microfilm or in any other way, and storage in data banks. Duplication of this publication or parts thereof is permitted only under the provisions of the German Copyright Law of September 9, 1965, in its current version, and permission for use must always be obtained from Springer-Verlag. Violations are liable to prosecution under the German Copyright Law.
The use of general descriptive names, registered names, trademarks, etc. in this publication does not imply, even in the absence of a specific statement, that such names are exempt from the relevant protective laws and regulations and therefore free for general use.

Cover design: SPi Publisher Services

Printed on acid-free paper

Springer is part of Springer Science+Business Media (www.springer.com)

Preface

In nature, numerous systems with periodically modulated spatial parameters are known. As a result, *linear waves* that propagate in such system exhibit many common properties. For example, the same symmetry properties of periodic potentials that restrict electrons and phonons in crystalline lattices to specific *energy bands*, also limit light waves of a given frequency in the photonic structures with periodic variations of the refractive index (*photonic crystals*) to specific propagation directions. For a given propagation direction, the frequency regimes of transparency and total reflection are analogous to the energy bands and *energy gaps* of electrons in a lattice. The same is true for the dynamics of Bose-Einstein condensates loaded on optical lattices where matter waves experience Bragg reflection being characterized by the *bandgap spectrum*, and for various plasmons in networks of Josephson junctions or metal surfaces.

As the amplitude of the wave excitations is increased and the response of the material becomes *nonlinear*, wave propagation in periodic and spatially modulated structures becomes far more complex. Features as nonlinear resonances, inelastic scattering, self-trapping, and dynamical localization influence the transport properties significantly. In particular, stable monochromatic and spatially localized excitations, which exist in nonlinear spatially periodic systems, are known as "intrinsic localized modes", "discrete breathers" or "lattice solitons", depending on the particular implementation. These strongly nonlinear modes have no analogs in the linear theory and they arise as generic solutions of the dynamics of nonlinear, spatially periodic continuous or discrete Hamiltonian systems. They have been observed in various systems, such as surface and bulk lattice vibrations of solids, excitations in layered anti-ferromagnetic structures, Josephson junction networks, coupled nonlinear optical waveguides, driven micro-mechanical cantilever arrays, and carbon nanotubes grown in periodically organized arrays. More recent developments and successful experimental observations involve one- and two-dimensional photonic lattices and photonic crystals, localized matter waves, and vortices in Bose-Einstein condensates loaded onto optical lattices. Quantization and subsequent tunnelling of gap solitons in Bose-Einstein condensates on optical lattices is connecting recent experimental emphasis with current theoretical research in the field of nonlinear localized excitations.

The effects mentioned above are mostly associated with the bandgap spectra and resonant wave interaction with spatially periodic media. Another specific feature of these systems can be revealed in the limit when the wavelength is much larger than the periodic structure, when the averaged macroscopic properties of such a microstructured material may differ dramatically from the properties of the components it consists of. Fascinating examples are microstructured metamaterials that demonstrate highly unusual properties.

In *metamaterials*, often associated with left-handed materials, both magnetic permeability and electric permittivity become simultaneously negative, thus allowing the propagation of electromagnetic waves with the Poynting vector anti-parallel to the wave vector and, therefore, with the basic feature of light reversal. This leads to some very interesting effects such as the reversal of the Doppler shift for radiation, and the reversal of Cherenkov radiation. In addition, one of the most basic principles of optics, Snell's law, is "reversed" at the interface of a left-handed medium with a normal right-handed material, so that the electromagnetic waves experience *negative refraction*. One of the possible applications of left-handed materials is a flat lens made of a slab of negatively refracting material: materials with negative refractive indices would amplify evanescent waves thus retaining the information they contain and achieving unprecedented resolution which overcomes the diffraction limit of conventional imaging. The study of nonlinear effects in metamaterials generates many surprises and unpredicted results that are expected to reveal many novel and fascinating phenomena.

The purpose of this book is to present theoretical, numerical, and experimental expertise in the study of nonlinear effects in seemingly different types of periodic systems, and thereby to unite the fundamental concepts of the different, but actually vastyl expanding fields. The book presents novel theoretical and experimental approaches as well as techniques for analyzing stable nonlinear excitations. Moreover, nonlinear wave propagation and interaction in highly anisotropic and periodic structures are given, which are characterized by the simultaneous existence of continuous and discrete modes, including photonic lattices and photonic crystals in optics, optically trapped Bose-Einstein condensates in atomic physics, and left-handed metamaterials in the physics of microwaves.

The authors of the chapters are leading experts in these fields, they advance the nonlinear physics of periodic systems and facilitate key experimental observations of many nonlinear effects predicted theoretically.

The book is aiming at an audience that is already familiar with the basics of light propagation in optical systems. Especially, we address young, advanced scientists as well as scientists in research groups with experimental as well as theoretical expertise. It will offer insights into the basic mathematical principles of localization, wave scattering, transport, quantization methods as well as computational aspects and - most importantly - work out the common grounds of the above listed seemingly different physical situations where the apparatus is to be applied.

In particular, the chapters collected in this book discuss and study properties as well as control of nonlinear waves and localized excitations in all systems mentioned, including nonlinear effects in photonic lattices, Bose-Einstein condensates in optical lattices, and the possibility of similar nonlinear excitations in microstructural metamaterials that exhibit negative refraction. Special attention is paid to the interplay between nonlinear and linear modes, which result in a number of interesting resonant scattering and trapping effects, and on quantum effects which are relevant for Bose-Einstein condensates. We are confident that the joint gathering of contributions of experimentalists and theorists from each subfield who are working on these aspects so far quite independently, will boost the implementation of ideas and experimental techniques in all three fields of research.

The work on this volume was additionally promoted by the International Seminar and Workshop "Nonlinear Physics in Periodic Structures and Metamaterials", held at the Max-Planck-Institute for the Physics of Complex Systems (Dresden) March 19–30, 2007. We also like to thank the Australian Research Council, the German Academic Exchange Service, the German Research Foundation, as well as the University of Münster for their continous support.

A book like this one that compiles contributions of different authors is as good as the authors' contributions. Therefore, we thank all authors for their excellent articles as well as their support of the edition process. We also like to thank Patrick Rose for the perfect compilation of all articles.

Münster, Dresden, Canberra *Cornelia Denz*
December 2008 *Sergej Flach*
 Yuri S. Kivshar

Contents

Part I Nonlinear Effects in Reduced-Dimensional Structures: From Wave Guide Arrays to Slow Light

1 Nonlinear Effects in One-Dimensional Photonic Lattices
Detlef Kip, Milutin Stepić 3
1.1 Introduction 3
1.2 Linear properties and Waveguide Array Formation 4
    1.2.1 Band-gap Structure and Floquet-Bloch Modes of One-Dimensional Lattices 4
    1.2.2 Fabrication of Nonlinear Waveguide Arrays 6
1.3 Light Localization and Lattice Solitons 7
    1.3.1 Lattice Solitons 7
    1.3.2 Discrete Modulational Instability 8
    1.3.3 Discrete Vector Solitons 9
    1.3.4 Higher Order Lattice Solitons 10
    1.3.5 Discrete Dark Solitons 11
1.4 Interactions of Light Beams in One-Dimensional Photonic Lattices 12
    1.4.1 Interactions with Defects 12
    1.4.2 Blocker Interaction 13
    1.4.3 Collinear Interaction 13
References 15

2 Nonlinear Optical Waves in Liquid Crystalline Lattices
Gaetano Assanto, Andrea Fratalocchi 21
2.1 Introduction 21
2.2 Photonic Lattices in Nematic Liquid Crystals 22
2.3 Discrete Dynamics 24
    2.3.1 Discrete Solitons in NLC 24
    2.3.2 Nonlinear Steering in NLC Lattices 26
2.4 Lattice Dynamics 29
    2.4.1 Multi-gap Lattice Solitons 29

	2.4.2	Light-driven Landau-Zener Tunneling	31
2.5	Summary		34
References	....		35

3 Nonlinear Optics and Solitons in Photonic Crystal Fibres
Dmitry V. Skryabin, William J. Wadsworth 37

3.1	Introduction	37
3.2	Supercontinuum Generation and Frequency Conversion: Techniques and Applications	38
	3.2.1 Femtosecond Supercontinua	38
	3.2.2 Long-pulse Supercontinua	40
3.3	Solitons in Solid-core PCFs and Their Role in Supercontinuum Generation	42
	3.3.1 Soliton Fission and Intrapulse Raman Scattering	42
	3.3.2 Resonant Radiation from Solitons	43
	3.3.3 Mixing of Solitons with Dispersive Radiation, Radiation Trapping and Short-wavelength Edge of Supercontinuum ..	44
	3.3.4 Red Shifted Radiation and Soliton Self-frequency Shift Cancelation	46
	3.3.5 Other Soliton Effects in Solid-core PCFs	48
3.4	Pulse Compression in PCFs	48
3.5	Nonlinear and Quantum Optics in Hollow-core PCFs	49
3.6	Summary	50
References	....	51

4 Spatial Switching of Slow Light in Periodic Photonic Structures
Andrey A. Sukhorukov 55

4.1	Introduction	55
4.2	Dispersion and Tuning of the Speed of Light in Nonlinear Periodic Structures	56
4.3	Slow-light Switching in Waveguide Couplers	59
	4.3.1 All-optical Switching in Bragg-grating Couplers	59
	4.3.2 Tunneling of Slow Light in Photonic-crystal Couplers	63
4.4	Slow Optical Bullets	66
4.5	Summary	68
References	....	68

Part II Nonlinear Effects in Multidimensional Lattices: Solitons and Light Localization

---

5 Introduction to Solitons in Photonic Lattices
Nikolaos K. Efremidis, Jason W. Fleischer, Guy Bartal, Oren Cohen, Hrvoje Buljan, Demetrios N. Christodoulides, Mordechai Segev 73

| 5.1 | Introduction to Optical Periodic Systems .... | 73 |

5.2	Optically Induced Lattices	74
5.3	Coupled-mode Theory	78
5.4	Linear Properties	79
5.5	One-dimensional Lattice Solitons	84
5.6	Two-dimensional Lattice Solitons	89
5.7	Vortex Solitons in Lattices	92
5.8	Random-phase lattice solitons	94
References		95

6 Complex Nonlinear Photonic Lattices: From Instabilities to Control
Jörg Imbrock, Bernd Terhalle, Patrick Rose, Philip Jander, Sebastian Koke, Cornelia Denz 101

6.1	Introduction	101
6.2	Optically Induced Lattices in Photorefractive Media	102
6.3	Anisotropy in Nonlinear Photonic Lattices	107
	6.3.1 Orientation Anisotropy	107
	6.3.2 Polarization Anisotropy	108
6.4	Two-dimensional Discrete Solitons in Nonlinear Photonic Lattices	108
6.5	Hybrid Lattices	112
6.6	Multiperiodic Lattices	113
6.7	Complex Beam Propagation in Complex Lattices	115
6.8	Controlling Instabilities of Counterpropagating Solitons by Optically Induced Photonic Lattices	118
6.9	Summary	123
References		124

7 Light Localization by Defects in Optically Induced Photonic Structures
Jianke Yang, Xiaosheng Wang, Jiandong Wang, and Zhigang Chen 127

7.1	Introduction	127
7.2	Optically Induced Lattices and Defects	128
7.3	Linear Defect Modes in 1D Lattices	130
7.4	Linear Defect Modes in 2D Square Lattices	132
7.5	Linear Defect Modes in 2D Ring Lattices	135
7.6	Nonlinear Defect Modes	137
7.7	Summary	140
References		140

8 Polychromatic Light Localisation in Periodic Structures
Dragomir N. Neshev, Andrey A. Sukhorukov, Wieslaw Z. Krolikowski, Yuri S. Kivshar 145

8.1	Introduction	145
8.2	Polychromatic Light in Periodic Structures	147
8.3	Nonlinear Localisation of Polychromatic Light	148

	8.3.1	Collective Nonlinear Interactions in Media with Slow Nonlinearity 149
	8.3.2	Polychromatic Gap Solitons 149
8.4	Experimental Studies of Polychromatic Self-trapping 151	
	8.4.1	Experimental Setup................................... 151
	8.4.2	Nonlinear Spectral-spatial Reshaping 153
	8.4.3	Generation of Polychromatic Gap Solitons 154
	8.4.4	Interaction with an Induced Defect 156
	8.4.5	Spatial-spectral Reshaping by Interaction with a Surface . . 157
8.5	Summary 159	
References 159		

Part III Periodic Structures for Matter Waves: From Lattices to Ratchets

9 Bose-Einstein Condensates in 1D Optical Lattices: Nonlinearity and Wannier-Stark Spectra

Ennio Arimondo, Donatella Ciampini, Oliver Morsch 165
9.1 Introduction.. 165
9.2 Optical Lattice 166
9.3 Analysis of the Interference Pattern 170
9.4 Nonlinear Optical Lattice 171
9.5 Bloch Oscillations... 173
9.6 Landau-Zener Tunneling 174
9.7 Resonantly Enhanced Quantum Tunnelling..................... 176
9.8 Summary 178
References 178

10 Transporting Cold Atoms in Optical Lattices with Ratchets: Mechanisms and Symmetries

Sergey Denisov, Sergej Flach, Peter Hänggi 181
10.1 Introduction... 181
10.2 Single Particle Dynamics..................................... 182
10.3 Symmetries 183
      10.3.1 Symmetries of a Periodic Function with Zero Mean 183
      10.3.2 Symmetries of the Equations of Motion 184
      10.3.3 The Case of Quasiperiodic Functions..................... 185
10.4 Dynamical Mechanisms of Rectification: The Hamiltonian Limit 186
10.5 Resonant Enhancement of Transport with Quantum Ratchets 189
10.6 Summary 192
References 192

11 Atomic Bose-Einstein Condensates in Optical Lattices with Variable Spatial Symmetry
Sebastian Kling, Tobias Salger, Carsten Geckeler, Gunnar Ritt, Johannes Plumhof, Martin Weitz 195
11.1 Introduction 195
11.2 Principle of Optical Multiphoton Lattices 196
11.3 Experimental Approach 198
11.4 Measurements and Results 199
11.5 Quantum Ratchets 201
References 202

12 Symmetry and Transport in a Rocking Ratchet for Cold Atoms
Ferruccio Renzoni 205
12.1 Introduction 205
12.2 Symmetries of a Rocking Ratchet 206
    12.2.1 The Dissipationless Case 206
    12.2.2 Weak dissipation 207
    12.2.3 Quasiperiodic Driving 207
12.3 Dissipative Optical Lattices 208
12.4 Rocking Ratchet for Cold Atoms 209
    12.4.1 Biharmonic Driving 210
    12.4.2 Multifrequency Driving 212
12.5 Summary 214
References 214

Part IV Metamaterials: From Linear to Nonlinear Features

13 Optical Metamaterials: Invisibility in Visible and Nonlinearities in Reverse
Natalia M. Litchinitser, Vladimir M. Shalaev 217
13.1 Introduction 217
13.2 Optical Metamaterials: New Degrees of Freedom 219
13.3 A Route to Invisibility 220
    13.3.1 Transformation Approach 222
    13.3.2 Cloaking Device: From Microwaves to Optics 222
13.4 Nonlinear Optics with Backward Waves in Negative Index Materials 227
    13.4.1 Second-harmonic Generation 229
    13.4.2 Optical Parametric Amplification: Loss Compensation in NIMs 230
    13.4.3 Bistability in Couplers 232
    13.4.4 Bistability in Layered Structures 234
    13.4.5 Solitons in Resonant Plasmonic Nanostructures 235
13.5 Summary 236
References 237

14 Nonlinear Metamaterials
Ilya V. Shadrivov 241
14.1 Introduction... 241
14.2 Nonlinear Response of Metamaterials: Theory 242
      14.2.1 Nonlinear Magnetic Permeability 243
      14.2.2 Nonlinear Dielectric Permittivity 246
14.3 Nonlinear Metamaterials: Experiments......................... 246
14.4 Nonlinearity-controlled Transmission 247
14.5 Electromagnetic Spatial Solitons in Metamaterials 250
14.6 Second-order Nonlinear Effects in Metamaterials 251
14.7 Conclusions 256
References 256

15 Circuit Model of Gain in Metamaterials
Allan D. Boardman, Neil King, Yuriy Rapoport 259
15.1 Introduction... 259
15.2 Negative Resistance Structures 260
15.3 Diode Inclusions 260
15.4 Discussion of Stability 264
15.5 Numerical Analysis... 267
15.6 Summary 270
References 271

16 Discrete Breathers and Solitons in Metamaterials
George P. Tsironis, Nikos Lazarides, Maria Eleftheriou 273
16.1 Introduction... 273
16.2 Magnetic Breathers 275
      16.2.1 Hamiltonian Discrete Breathers 277
      16.2.2 Dissipative Discrete Breathers......................... 278
16.3 Magnetic Solitons 280
16.4 Electromagnetic Solitons 282
16.5 Summary 285
References 286

Index 289

List of Contributors

Ennio Arimondo
CNR-INFM and CNISM
Dipartimento di Fisica E. Fermi
Università di Pisa
Via Buonarroti 2
56127 Pisa, Italy
arimondo@df.unipi.it

Gaetano Assanto
Nonlinear Optics and
OptoElectronics Lab (NooEL),
CNISM and INFN
University "Roma Tre"
Via della Vasca Navale 84
00146 Rome, Italy
assanto@uniroma3.it

Guy Bartal
Physics Department and
Solid State Institute
Technion-Israel
Institute of Technology
Haifa 32600, Israel

Allan D. Boardman
Institute for Materials Research
University of Salford
Salford, M6 5WT, United Kingdom
a.d.boardman@salford.ac.uk

Hrvoje Buljan
Department of Physics
University of Zagreb
PP 332, Zagreb, Croatia
hbuljan@phy.hr

Zhigang Chen
Department of Physics and
Astronomy
San Francisco State University
CA 94132, USA
zchen@stars.sfsu.edu

Demetrios N. Christodoulides
College of Optics and Photonics
University of Central Florida
Orlando, Florida 32813, USA
demetri@creol.ucf.edu

Donatella Ciampini
CNR-INFM and CNISM
Dipartimento di Fisica E. Fermi
Università di Pisa
Via Buonarroti 2
I-56127 Pisa, Italy
ciampini@df.unipi.it

Oren Cohen
Physics Department and
Solid State Institute
Technion-Israel
Institute of Technology
Haifa 32600, Israel

List of Contributors

Sergey Denisov
Institut für Physik
Universität Augsburg
Universitätsstr. 1
86135 Augsburg, Germany
sergey.denisov
  @physik.uni-augsburg.de

Cornelia Denz
Institut für Angewandte Physik and
Center for Nonlinear Science
Westfälische Wilhelms-Universität
Münster
Corrensstr. 2/4
48149 Münster, Germany
denz@uni-muenster.de

Nikolaos K. Efremidis
Department of Applied Mathematics
University of Crete
71409 Heraclion, Crete, Greece
nefrem@tem.uoc.gr

Maria Eleftheriou
Department of Physics
University of Crete and Institute of
Electronic Structure and Laser
Foundation of Research and
Technology Hellas (FORTH)
P.O. Box 2208
71003 Heraklion, Greece
marel@physics.uoc.gr

Sergej Flach
Max-Planck-Institut für Physik
Komplexer Systeme
Nöthnitzer Str. 38
01187 Dresden, Germany
flach@mpipks-dresden.mpg.de

Jason W. Fleischer
Electrical Engineering Department
Princeton University
Princeton, NJ 08544, USA
jasonf@princeton.edu

Andrea Fratalocchi
Research Centers CRS SOFT
INFM-CNR and Museo Storico per
la Fisica "Enrico Fermi"
University "Sapienza"
P.le Aldo Moro 2
00185 Rome, Italy
andrea.fratalocchi@roma1.
infn.it

Carsten Geckeler
Institut für Angewandte Physik
Universität Bonn
Wegelerstr. 8
53115 Bonn, Germany

Peter Hänggi
Institut für Physik
Universität Augsburg
Universitätsstr. 1
86135 Augsburg, Germany
hanggi@physik.uni-augsburg.de

Jörg Imbrock
Institut für Angewandte Physik and
Center for Nonlinear Science
Westfälische Wilhelms-Universität
Münster
Corrensstr. 2/4
48149 Münster, Germany
imbrock@uni-muenster.de

Philip Jander
Institut für Angewandte Physik and
Center for Nonlinear Science
Westfälische Wilhelms-Universität
Münster
Corrensstr. 2/4
48149 Münster, Germany
phj@uni-muenster.de

Neil King
Institute for Materials Research
University of Salford
Salford, M6 5WT, United Kingdom
dr.n.king@googlemail.com

Detlef Kip
Institute of Physics and Physical
Technologies
Clausthal University of Technology
Leibnizstr. 4
38678 Clausthal-Zellerfeld, Germany
dkip@pe.tu-clausthal.de

Yuri S. Kivshar
Nonlinear Physics Centre
Research School of Physics and
Engineering
Australian National University,
Canberra, 0200 ACT, Australia
ysk@intermode.on.net

Sebastian Kling
Institut für Angewandte Physik
Universität Bonn
Wegelerstr. 8
53115 Bonn, Germany
kling@iap.uni-bonn.de

Sebastian Koke
Max-Born-Institut
Max-Born-Str. 2A
12489 Berlin
koke@mbi-berlin.de

Wieslaw Z. Krolikowski
Laser Physics Centre
Research School of Physics and
Engineering
Australian National University,
Canberra, 0200 ACT, Australia
wzk111@rsphysse.anu.edu.au

Nikos Lazarides
Department of Physics
University of Crete and Institute of
Electronic Structure and Laser
Foundation of Research and
Technology Hellas (FORTH)
P.O. Box 2208
71003 Heraklion, Greece
nl@physics.uoc.gr

Natalia M. Litchinitser
Department of Electrical Engineering
State University of New York at
Buffalo
309 Bonner Hall
Buffalo, New York 14260, USA
natashal@buffalo.edu

Oliver Morsch
CNR-INFM and CNISM
Dipartimento di Fisica E. Fermi
Università di Pisa
Via Buonarroti 2
I-56127 Pisa, Italy
morsch@df.unipi.it

Dragomir N. Neshev
Nonlinear Physics Centre
Research School of Physics and
Engineering
Australian National University,
Canberra, 0200 ACT, Australia
dragomir.neshev@anu.edu.au

Johannes Plumhof
Institut für Angewandte Physik
Universität Bonn
Wegelerstr. 8
53115 Bonn, Germany

Yuriy Rapoport
Physics Faculty
Taras Shevchenko Kyiv National
University
Prospect Glushkov 6
22 Kyiv, Ukraine
laser@i.kiev.ua

Ferruccio Renzoni
Department of Physics and
Astronomy
University College London
Gower Street
London WC1E 6BT, United
Kingdom
f.renzoni@ucl.ac.uk

Gunnar Ritt
Institut für Angewandte Physik
Universität Bonn
Wegelerstr. 8
53115 Bonn, Germany

Patrick Rose
Institut für Angewandte Physik and
Center for Nonlinear Science
Westfälische Wilhelms-Universität
Münster
Corrensstr. 2/4
48149 Münster, Germany
patrick.rose@uni-muenster.de

Tobias Salger
Institut für Angewandte Physik
Universität Bonn
Wegelerstr. 8
53115 Bonn, Germany

Mordechai Segev
Physics Department and
Solid State Institute
Technion-Israel
Institute of Technology
Haifa 32600, Israel
msegev@techunix.technion.ac.il

Ilya V. Shadrivov
Nonlinear Physics Centre
Research School of Physics and
Engineering
Australian National University
Canberra ACT 0200, Australia
ivs124@rsphysse.anu.edu.au

Vladimir M. Shalaev
School of Electrical and Computer
Engineering and Birck
Nanotechnology Center
Purdue University
West Lafayette, Indiana 47907, USA
shalaev@purdue.edu

Dmitry V. Skryabin
Centre for Photonics and Photonic
Materials, Department of Physics
University of Bath
Bath BA2 7AY, United Kingdom
d.v.skryabin@bath.ac.uk

Milutin Stepić
Vinča Institute of Nuclear Sciences
P.O.B. 522
11001 Belgrade, Serbia
mstepic@vinca.rs

Andrey A. Sukhorukov
Nonlinear Physics Centre and Centre
for Ultrahigh-bandwidth Devices for
Optical Systems (CUDOS)
Research School of Physics and
Engineering
Australian National University
Canberra, ACT 0200, Australia
ans124@rsphysse.anu.edu.au

Bernd Terhalle
Institut für Angewandte Physik and
Center for Nonlinear Science
Westfälische Wilhelms-Universität
Münster
Corrensstr. 2/4
48149 Münster, Germany
bernd.terhalle@uni-muenster.de

George P. Tsironis
Department of Physics
University of Crete and Institute of
Electronic Structure and Laser
Foundation of Research and
Technology Hellas (FORTH)
P.O. Box 2208
71003 Heraklion, Greece
gts@physics.uoc.gr

William J. Wadsworth
Centre for Photonics and Photonic
Materials, Department of Physics
University of Bath
Bath BA2 7AY, United Kingdom
w.j.wadsworth@bath.ac.uk

Jiandong Wang
Department of Mathematics and
Statistics
University of Vermont
VT 05401, USA
jwang@cems.uvm.edu

Xiaosheng Wang
Department of Physics and
Astronomy
San Francisco State University
CA 94132, USA
gxkren@gmail.com

Martin Weitz
Institut für Angewandte Physik
Universität Bonn
Wegelerstr. 8
53115 Bonn, Germany

Jianke Yang
Department of Mathematics and
Statistics
University of Vermont
VT 05401, USA
jyang@cems.uvm.edu

Part I

Nonlinear Effects in Reduced-Dimensional Structures: From Wave Guide Arrays to Slow Light

1

Nonlinear Effects in One-Dimensional Photonic Lattices

Detlef Kip[1] and Milutin Stepić[1,2]

[1] Institute of Physics and Physical Technologies, Clausthal University of Technology, Leibnizstr. 4, 38678 Clausthal-Zellerfeld, Germany
  dkip@pe.tu-clausthal.de
[2] Vinča Institute of Nuclear Sciences, P.O.B. 522, 11001 Belgrade, Serbia
  mstepic@vinca.rs

1.1 Introduction

Optical waves propagating in photonic periodic structures are known to exhibit a fundamentally different behavior when compared to their homogeneous counterparts in bulk materials. In such systems the spatially periodic refractive index experienced by light waves is analogous to the situation in crystalline solids, where electrons travel in a periodic Coulomb potential [1]. Consequently, the propagating extended (Floquet Bloch) modes of a linear periodic optical system form a spectrum that is divided into allowed bands, separated by forbidden gaps, too, and the two different physical systems share most of their mathematical description. Photonic band-gap materials, which may be artificially fabricated to be periodic in three, two, or only one dimension, hold strong promise for future photonic applications like miniaturized all-optical switches, filters, or memories [2]. Here novel opportunities are offered when nonlinear material response to light intensity is taken into account. When studying such nonlinear photonic crystals it turns out that light propagation is governed by two competing processes: linear coupling among different lattice sites and energy localization due to nonlinearity. For an exact balance of these counteracting effects self-localized states can be obtained, which are called lattice solitons [3–6].

Uniform one-dimensional (1D) waveguide arrays (WAs) may be understood as a special case of 1D photonic crystals with a periodicity of the refractive index scaled to the wavelength of light. These arrays consist of equally spaced identical channel waveguides, where energy is transferred from one site to another through evanescent coupling or tunnelling of light. Although such arrays share many of their linear and nonlinear properties with other periodic systems in nature, for example excitons in molecular chains [7], charge density waves in electrical lattices [8], Josephson junctions [9], spin waves in antiferromagnets [10], or Bose-Einstein condensates in periodic optical traps [11], they have some advantages making them attractive candidates for studying general nonlinear lattice problems: Due to the larger wavelength of light when compared to, e.g., electrons, wave amplitudes can be directly imaged, thus allowing for a full experimental control of input and output signals. The relatively easy sample fabrication and compact experimental setups, together with suitable working environments at room temperature without the need for vacuum chambers, have put the optics domain at the forefront of research on nonlinear periodic systems.

In this chapter we will provide a brief overview on light propagation and soliton dynamics in 1D nonlinear WAs, and will discuss some recent experimental results on the example of arrays in photorefractive lithium niobate (LiNbO$_3$). In the following section, we discuss some basic linear properties of WAs like discrete diffraction, normal and anomalous diffraction, and methods to engineer tailored photonic band structures using different experimental techniques and material systems. The third part is devoted to nonlinear light propagation in 1D WAs. After discussing the instability regimes of extended Floquet-Bloch (FB) modes in 1D lattices, which coincide with the occurrence of discrete modulation instability, we give an overview of different types of localized nonlinear excitations, for example multi-hump, dark, or vector lattice solitons, that have been investigated in WAs. Finally, the last section is devoted to the interaction of light with lattice defects and other light beams, which may form the basic elements for novel applications in photonics.

1.2 Linear properties and Waveguide Array Formation

1.2.1 Band-gap Structure and Floquet-Bloch Modes of One-Dimensional Lattices

In absence of nonlinear effects optical beams will spread in space because of diffraction while pulses will experience temporal broadening due to dispersion. Although diffraction is an omnipresent geometrical effect and dispersion is material dependent and absent in vacuum, both effects occur because of different rates of phase accumulation for different spatial or temporal frequencies. In physics, the dispersion relation is the relation between the system's energy (or propagation constant) and its corresponding momentum (Bloch momentum). The dispersion relation of linear waves in bulk or continuous media has a parabolic form [12]. Consequently, in a 1D planar waveguide layer unlimited transverse propagation of modes results in a continuous dispersion spectrum with the same parabolic shape. A vivid example for a planar waveguide fabricated in LiNbO$_3$ is given in Fig. 1.1a. By a modified prism coupler setup [13] the effective indices $n_{\text{eff}} = \beta\lambda/2\pi$ have been measured (normalized to the substrate index n_{sub}) as a function of Bloch momentum, where β is the corresponding (longitudinal) propagation constant and λ is the light wavelength. Having in mind analogies drawn between dispersion and diffraction [12, 14],

1 Nonlinear Effects in One-Dimensional Photonic Lattices 5

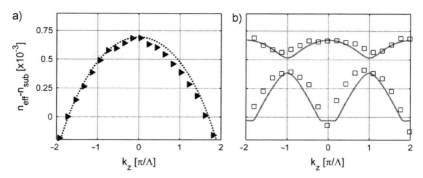

Fig. 1.1. Experimentally measured band structures of (**a**) a planar waveguide and (**b**) a 1D WA (grating period $\Lambda = 8\,\mu\text{m}$). Symbols are measured propagation constants. The dashed line in (**a**) is just a guide for the eye, whereas in (**b**) solid lines show the corresponding calculated band structure

diffraction is determined by the curvature at the corresponding point of the dispersion curve while the direction of propagation of light is normal to this curve. As can be seen, in this example the diffraction coefficient is negative (normal diffraction) for all propagating waves.

In media with a periodic index modulation a band structure arises with allowed bands separated by gaps where light propagation is forbidden [12,15]. The form of the band-gap structure depends on system parameters such as, for example, the distance between adjacent channels of the nonlinear WA and the strength of the refractive index modulation, which can be fully controlled in the fabrication process. To take up the previous example, an additional 1D periodic index modulation can be formed in the planar waveguide of Fig. 1.1a by two-beam holographic recording of an elementary grating [16]: Each refractive index maximum of the modulated pattern forms a single-mode channel waveguide which is evanescently coupled to its first neighbors. An example of the obtained band structure which shows the first two bands of a LiNbO_3 WA is given in Fig. 1.1b. While diffraction in bulk media is always normal, in periodic media diffraction can reverse its sign leading to regions of anomalous diffraction, for example, within the first band for $\pi/2 < k_z\Lambda < \pi$ and around the center of the first Brillouin zone (BZ) in the second band. Here, k_z stands for the transverse component of the wave number, and Λ denotes the grating period. Furthermore, diffraction may even vanish at certain points in the dispersion diagram (e.g., for $k_z\Lambda \approx \pi/2$ in the first band), allowing for almost diffraction-free propagation of light.

Another example of a measured band structure with four guided bands of a 1D WA with stronger modulation is given in Fig. 1.2a. Experimental values of propagation constants are denoted by squares, whereas solid lines correspond to numerically calculated bands. If the condition $n_{\text{eff}} - n_{\text{sub}} > 0$ is

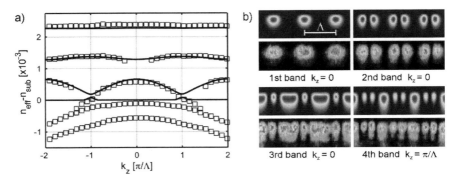

Fig. 1.2. (a) Band-gap structure of WA with period $\Lambda = 8\,\mu\text{m}$. (b) Intensity of FB modes from different bands: numerical results (*top*) and experimental data (*bottom*)

fulfilled modes are guided, otherwise they are radiative. The implementation of the prism coupling method [13] allows for the selective excitation of pure FB modes of the periodic structure. Some illustrative examples of excited modes are given in Fig. 1.2b. Numerical results shown in the upper rows correspond fairly well to the experimentally obtained images measured at the samples' output facet.

1.2.2 Fabrication of Nonlinear Waveguide Arrays

One-dimensional WAs have been fabricated in quite different materials ranging from semiconductors [4,17] and photorefractives [18,19], to polymers [20], glasses [21], and liquid crystals [22]. WAs in the semiconductor AlGaAs have been formed by, e.g., reactive-ion etching of adequate wafers with epitaxially fabricated layers. This semiconductor crystal possesses an instantaneous Kerr-like focusing nonlinearity for optical wavelengths in the infrared, and typical optical powers required to obtain suitable nonlinearites are in the range of 10^2–10^3 W. In silica-based glasses either ion exchange in molten salts or direct writing using femtosecond lasers has been used. WAs in polymers have been fabricated by UV lithography, whereas in liquid crystals a set of regularly spaced transparent electrodes has been used. In photorefractive crystals, where nonlinearities are based on light-induced space charge fields and the electrooptic effect, two different methods for WA formation have been used so far: induction of index gratings by illumination of the crystal with light [18], or permanent index changes due to indiffusion of titanium stripe patterns [19]. Light-induced lattices are based on the interference of two or more writing laser beams propagating inside the bulk sample. Such lattices are both rewritable and dynamically tunable. One may control the coupling between channels by adjusting the intensity of the recording light while Bragg reflection is defined by the angle between the interfering beams. However, the achievable refractive index modulations are rather limited and clumsy equipment is required to stabilize the interference patterns. On the other hand, there exist several methods to fabricate permanent waveguides and structures in photorefractive crystals [23]. In LiNbO$_3$ the method of in-diffusion of titanium has been used to form permanent WAs with lattice periods ranging from 2 to 20 microns. Furthermore, in-diffusion of impurities like iron or copper may be used to tailor the photorefractive properties of the material. Besides its wide use in nonlinear optics, for example for frequency conversion and fast optical modulation of light, LiNbO$_3$ possess a rather high nonlinear index change at very low light intensities. However, this material is also sensitive to holographic light scattering and has a rather long build-up time for nonlinear index changes in the range of seconds or even minutes.

1.3 Light Localization and Lattice Solitons

1.3.1 Lattice Solitons

Lattice solitons are localized structures which exist due to the exact balance between periodicity and nonlinear effects. They comprise both discrete and gap solitons. Discrete solitons exist in the first (semi-infinite) band-gap due to total internal reflection. Near the top of the first band, which is located at the center of the first BZ (see Fig. 1.1b), where beam diffraction is normal, unstaggered (adjacent elements are in-phase) discrete solitons may exist provided that a self-focusing or positive nonlinearity is present [4, 24–27]. The prediction of the existence of fundamental optical lattice solitons in WAs dates back to 1988 [3], and ten years later the group of Silberberg succeeded in the experimental observation of such solitons in a Kerr-like focusing medium [4], which has stimulated intense research in this field [28–30].

Gap solitons [5, 7, 31–33] are yet another type of stable nonlinear structures that can be observed in periodic media. Due to a nonlinear index change the propagation constant of these solitons is shifted inside the gap in-between two allowed bands. Fundamental gap solitons may be excited either from the top of the second band at the edge of the first BZ (normal diffraction) in lattices with self-focusing nonlinearity [34], or from the first band at the edge of the first BZ (anomalous diffraction) in lattices exhibiting self-defocusing nonlinearity [33]. In the latter case, soliton structures are of staggered form (adjacent elements are out-of-phase) [35–37].

A recent example of discrete gap soliton formation in a LiNbO$_3$ WA with defocusing nonlinearity is given in Fig. 1.3a. The top image of the output facet is taken immediately after light is coupled in and monitors linear discrete diffraction inside the array. With increasing recording time the nonlinearity builds up and finally the light is trapped predominantly in a single channel.

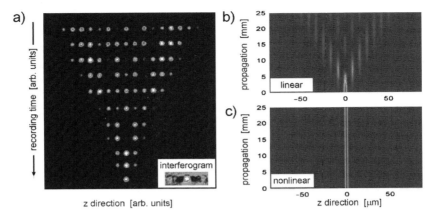

Fig. 1.3. Gap soliton formation in a LiNbO$_3$ WA with period $\Lambda = 7.6\,\mu\text{m}$ at the edge of the first BZ of the first band. (**a**) Output intensity for single-channel excitation with input power $P_{\text{in}} = 30\,\mu\text{W}$. (**b**), (**c**) Related BPM simulations for the linear (**b**) and nonlinear (**c**) case

The inset shows the corresponding interferogram of the output light with a superimposed plane wave, which represents an experimental proof for the staggered amplitude of the formed soliton. A numerical simulation (based on a beam propagation method (BPM)) which corresponds to the case of discrete diffraction is presented in Fig. 1.3b, while Fig. 1.3c shows the nonlinear case of stable soliton propagation inside the gap.

1.3.2 Discrete Modulational Instability

Experimentally, discrete and gap solitons may be obtained through the mechanism of modulational instability (MI) of a wide input beam. Discrete MI represents a nonlinear phenomenon in which initially smooth extended waves of the periodic system (FB modes) disintegrate into regular soliton trains under the combined effects of nonlinearity and diffraction. It has been predicted that FB modes exhibiting anomalous diffraction become unstable in the presence of self-defocusing nonlinearity while modes exhibiting normal diffraction break up under the effect of a self-focusing nonlinearity [3,35,38–40]. Experimentally, this has been proven for the first time in AlGaAs arrays exhibiting a focusing cubic nonlinearity [41], followed later by related experiments in both quadratic [42] and defocusing WAs [43].

An example of numerical and experimental evidence of discrete MI in LiNbO$_3$ in the first and second band is presented in Fig. 1.4 [44]. The experimental pictures on the top consist of 75 intensity line scans each, which have been taken from the output facet every minute, mimicing the time evolution

1 Nonlinear Effects in One-Dimensional Photonic Lattices 9

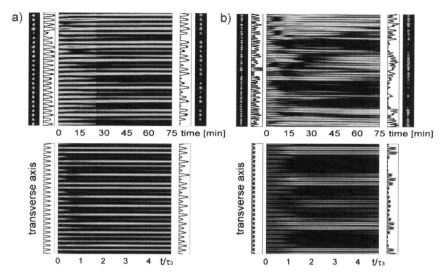

Fig. 1.4. Discrete MI in a defocusing WA: Comparison of experimentally measured and simulated light intensity at the output facet. (**a**) Edge of the first BZ in the first band for $P_{\text{in}} = 10\,\mu\text{W}$ (*top*) and related numerical simulation (*bottom*), and (**b**) at the center of the first BZ in the second band for $P_{\text{in}} = 21\,\mu\text{W}$ (*top*) and related numerical simulation (*bottom*)

of light intensity. Discrete MI may be observed only for a limited region of in-coupled light power in-between lower and upper MI thresholds [39]. Here the upper threshold arises from saturation of the nonlinearity, which stabilizes the system by decreasing the nonlinear gain and increases the threshold for the onset of MI.

1.3.3 Discrete Vector Solitons

Vector solitons [45] are composite structures that consist of two or more components which are individually incapable to form stable structures, but which mutually self-trap in a nonlinear medium. Discrete vector solitons (DVS) in 1D WAs are yet another, more complex class of vector solitons which have been investigated both theoretically [46–49] and experimentally [50,51]. Recently it has been recognized that both complex vector structures whose components stem from different bands [52–54] and composite band-gap solitons [55,56] may be found in nonlinear periodic systems, too.

First experiments on DVSs in 1D media have been performed by Stegeman's group using AlGaAs WAs with cubic nonlinearity [50], where both TE and TM components have a single-hump structure. Whereas in these media a separation of four-wave mixing processes and cross-phase modulation

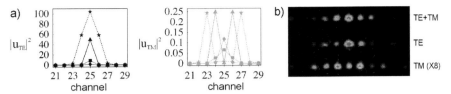

Fig. 1.5. Discrete vector soliton formation. (**a**) Stationary profiles of TE (lhs) and TM modes (rhs). Diamonds, squares, triangles and stars correspond to $\nu = -5, -1, 0$ and 1.1. (**b**) Measured stationary output of a DVS for mutually incoherent input beams with power ratio $P_{TE}/P_{TM} = 1.5$, both components together (*top*), TE (*middle*) and TM component alone (*bottom*, amplified 8 times)

is possible, these two terms are non-separable in arrays with saturable nonlinearity [51]. Here the power of the dominating TE mode grows in a similar fashion as the on-site mode from Ref. [57], giving rise to speculations that such iso-frequency DVSs could be moved and routed across the array. Interestingly, the TM mode exhibits a splitting into a two-hump structure. Fig. 1.5 shows results obtained for a LiNbO$_3$ WA with saturable defocusing nonlinearity. Numerically obtained stationary profiles of TE and TM modes for different values of soliton parameter ν are presented in Fig. 1.5a. The shape of the DVS slightly changes for different power ratios P_{TE}/P_{TM}, however, the center is mostly TE polarized while tails have dominant TM polarization. An experimental example for mutually incoherent input beams is given in Fig. 1.5b. As predicted, a dominating single-hump TE polarized component and a weaker double-humped TM component are observed [51].

1.3.4 Higher Order Lattice Solitons

It is well known that even 1D lattices support a wide spectrum of various strongly localized modes. Except the most often studied on-site and inter-site solitons (modes A and B, respectively) [58–64], various forms of lattice solitons such as twisted [36,61,65], quasi-rectangular [66], multi-hump solitons [67–70], and higher-order soliton trains [71] have been studied as well. Higher order lattice solitons are complex structures which may be intuitively viewed as a nonlinear combination of on-site solitons residing in adjacent channels. Such multi-hump structures are stable above a critical power threshold which can be estimated by linear stability analysis [68].

Recently, higher order lattice solitons have been observed experimentally in a Cu-doped LiNbO$_3$ WA using simultaneous in-phase excitation of two or three channels. Stationary profiles of such multi-humped solitons are presented in Fig. 1.6a. Experimentally observed images of an even two-hump soliton, which has been excited by two individual in-phase Gaussian beams, and a three-soliton train, which has been excited by a single super-Gaussian beam, are shown in Fig. 1.6b. The corresponding numerical results are given

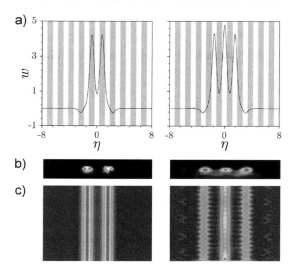

Fig. 1.6. Higher-order solitons in a WA. (**a**) Stationary profiles of an even two-hump soliton (lhs) and a three-soliton train (rhs). (**b**) Experimental images on the output facet for input power $P_{\text{in}} = 10\,\mu\text{W}$. (**c**) BPM results showing stable propagation of two- and three-channel input excitations

in Fig. 1.6c. Generally, the performed investigations indicate that the here used excitation of multi-humped solitons is quite efficient even in rather short arrays and confirm the possibility of dense soliton packing in form of soliton trains.

1.3.5 Discrete Dark Solitons

As noted in Ref. [59], the modes A and B can be seen as two dynamical states of a single mode moving across the array. The difference in their energy is related to the Peierls-Nabarro (PN) potential, which represents a barrier that has to be overcome in order to move a discrete soliton half of the lattice period aside. In media with cubic self-focusing nonlinearity the PN potential grows with increase of mode power, thus disabling stable propagation of mode B and free steering of large amplitude solitons [60,62]. On the other hand, in arrays with saturable nonlinearity it has been discovered that the PN potential can vanish and reverse its sign [36,63,72]. Therefore stable propagation of mode B becomes possible and solitons may be steered through the lattice. Numerical evidence of stable propagation of bright inter-site modes were presented for both saturable [63] and cubic-quintic nonlinearities [73].

Beside bright solitons 1D lattices may support also dark discrete solitons [74–77]. Such solitons have one or more dark elements on a constant bright background and possess a π phase jump across the center of the structure. In LiNbO$_3$ arrays it has been demonstrated both analytically and experimentally

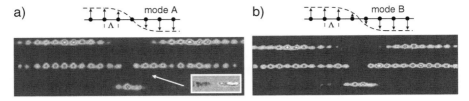

Fig. 1.7. Formation of discrete dark solitons. (**a**) Phase profile of unstaggered on-site dark soliton, formation of stable soliton state and guiding of a weak probe beam, respectively. The inset shows the corresponding interferogram. (**b**) The same for the unstaggered inter-site dark soliton

that the dark mode B can propagate in stable manner, too [64,76]. Experimental results on dark soliton formation in a $LiNbO_3$ WA are given in Fig. 1.7 [76]. On the lhs the situation for mode A with a phase jump located on-channel is monitored. The first row shows linear discrete diffraction of the dark notch, while in the nonlinear case (second row) a narrow dark soliton with staggered phase profile (see inset) is formed. The rhs shows the analogue situation for mode B, where the tailored input light pattern has been shifted by half a lattice period to locate the phase jump in-between channels. The lowest rows show the guiding of weak probe beams that are launched after the pump light was turned off. Here for mode A a single waveguide is formed while for mode B a two-channel-wide guiding structure is obtained.

1.4 Interactions of Light Beams in One-Dimensional Photonic Lattices

Among the most interesting properties of spatial optical solitons is the nonlinear interaction that takes place when solitons intersect or propagate close enough to each other within the nonlinear material [78]. Especially in discrete media like coupled WAs, a realization of all-optical functions would strongly benefit from the inherent multi-port structure of the array. Therefore, optical lattice solitons are prominent candidates to become main information carriers in future all-optical networks, and many new applications like all-optical switching [79–83], steering [6,7,21,63,84–87], and amplification [88] have been proposed.

1.4.1 Interactions with Defects

Having in mind that perfectly periodic media do not exist, several groups have investigated the interaction of lattice solitons with various structural defects. Generally, defects can be created by changing the spacing of two adjacent waveguides in an otherwise uniform array [89], by variation of the effective index or the width of a single channel [90, 91], or by optical induction techniques [92]. Defects can either attract or repel solitons, and soliton trapping has been investigated in the presence of both linear and nonlinear defects [93]. In modulated arrays additional defects can be used for Bloch wave filtering [91], and the number of bounded modes in an array can be dynamically controlled [90]. On the other hand, uniform linear WAs with nonlinear defects have been proposed as suitable candidates for the observation of Fano resonances [94].

1.4.2 Blocker Interaction

Weak probe beams launched into a lattice will spread quickly in transverse direction because of evanescent coupling of energy among adjacent sites. However, diffraction may be considerably reduced if the beam is launched at an angle corresponding to diffraction-less propagation [13]. Recently, interactions of such low-power (linear) probe beams with both coherent [95] and incoherent bright blocker solitons [96] have been studied in Kerr-like semiconductor WAs. In defocusing and saturable $LiNbO_3$ arrays both bright and dark blocker solitons were used for probe beam deflection [97]. It has been also realized that such nonlinear processes, of which an example is presented in Fig. 1.8, are suitable for the realization of all-optical beam splitters with adjustable splitting ratios.

1.4.3 Collinear Interaction

Interactions and collisions of discrete solitons have been investigated mainly numerically [7, 98–101]. Depending on the relative phase between the beams, their amplitude and the type of nonlinearity, soliton repulsion, fusion, and fission as well as energy transfer and oscillatory behavior have been observed. In arrays exhibiting a cubic nonlinearity and, in most experimental realizations, also in saturable arrays, strong soliton beams are pinned to a certain channel. Therefore, mostly interactions of co-propagating parallel beams have been investigated experimentally [102, 103]. Fig. 1.9 presents an example of co-propagating solitons launched in-phase into two channels of a $LiNbO_3$ WA [103]. Fig. 1.9a depicts a comparison between experimentally (top) and numerically (bottom) obtained results in the linear case of discrete diffraction. In the lower power regime (Fig. 1.9b) soliton fusion in the central channel is observed, a process that does not occur in cubic media [102]. In the region of higher power (Fig. 1.9c) an almost independent soliton-like propagation (pinning) of the two beams is found. Interestingly, in the case of out-of-phase beams in discrete media with self-defocusing nonlinearity, a pure oscillatory behavior of beams is found by means of numerical simulations [103].

Interactions of counter-propagating solitons in 1D WA have been experimentally investigated in both $LiNbO_3$ [104] and strontium-barium niobate crystals [105]. Main result is the experimental confirmation of the existence

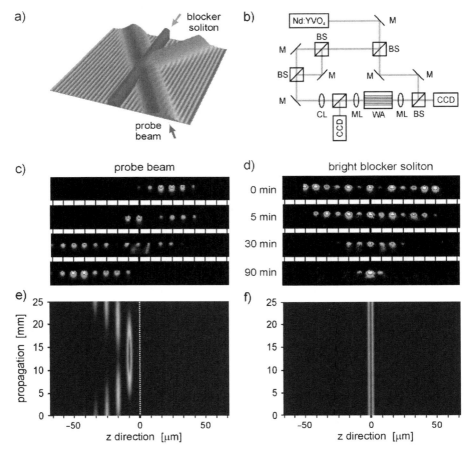

Fig. 1.8. (a) Interaction scheme of a weak probe beam with a counter-propagating bright blocker soliton. (b) Experimental setup (for notation see Ref. [97]). (c), (d) Temporal evolution of the intensity on the output facet when a low-power probe beam and a bright soliton beam of higher power intersect. (e), (f) BPM simulation of steady-state propagation of probe beam (propagation downwards) and bright soliton (propagation upwards), respectively

of three dynamical regimes predicted theoretically [106]. For low input power a regime of stable propagation of counter-propagating beams is found where vector solitons are formed. As this stable co-existence of counter-propagating beams does not exist in bulk media, this proves the stabilizing effect of the lattice on soliton propagation. However, when the input power is increased, instability occurs also in the lattice leading to discrete beam displacements, and finally a regime of high optical power is reached showing chaotic dynamics.

Beside in uniform WAs, various nonlinear effects have been investigated in engineered arrays [83, 107], binary arrays [108], double-periodic lattices [17],

1 Nonlinear Effects in One-Dimensional Photonic Lattices   15

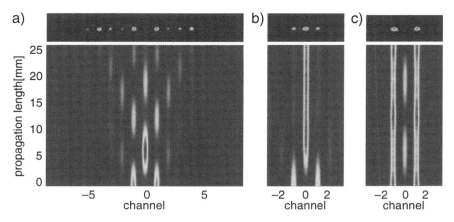

Fig. 1.9. Comparison of in-phase interaction of two collinearly propagating beams for different input powers in a defocusing lattice. Experimental output on endfacet (*top*) and BPM simulation (*bottom*). (**a**) Discrete diffraction, (**b**) fusion of solitons at low power, and (**c**) soliton-like propagation for higher input power

chirped arrays [109] and arrays of curved waveguides [110]. Some other types of lattice solitons such as incoherent solitons [111], random phase solitons [112], polychromatic solitons [113] and surface solitons [114] will be covered in detail in other chapters of this book.

References

1. J. Singleton, *Band Theory and Electronic Properties of Solids*, Oxford (2001)
2. J.D. Joannopoulos, R.D. Meade, and J.N. Winn, *Photonic Crystals: Molding the Flow of Light*, Princeton (1995)
3. D.N. Christodoulides and R.I. Joseph, Opt. Lett. **13**, 794 (1988)
4. H.S. Eisenberg, Y. Silberberg, R. Morandotti, A.R. Boyd, and J.S. Aitchison, Phys. Rev. Lett. **81**, 3383 (1998)
5. A.A. Sukhorukov, Y.S. Kivshar, H.S. Eisenberg, and Y. Silberberg, IEEE J. Quant. Electron. **39**, 31 (2003)
6. D.N. Neshev, A.A. Sukhorukov, B. Hanna, W. Królikowski, and Y.S. Kivshar, Phys. Rev. Lett. **93**, 083905 (2004)
7. A.S. Davydov and N.I. Kislukha, Phys. Stat. Sol. B **59**, 465 (1973)
8. P. Marquié, J.M. Bilbault, and M. Remoissenet, Phys. Rev. E **51**, 6127 (1995)
9. A.V. Ustinov, Phys. D **123**, 315 (1998)
10. B.I. Swanson, J.A. Brozik, S.P. Love, G.F. Strouse, A.P. Sreve, A.R. Bishop, W.Z. Wang, and M.I. Salkola, Phys. Rev. Lett. **82**, 3288 (1999)
11. A. Trombettoni and A. Smerzi, Phys. Rev. Lett. **86**, 2353 (2001)
12. D. Mandelik, H.S. Eisenberg, Y. Silberberg, R. Morandotti, and J.S. Aitchison, Phys. Rev. Lett. **90**, 053902 (2003)
13. C.E. Rüter, J. Wisniewski, and D. Kip, Opt. Lett. **31**, 2768 (2006)

14. H.S. Eisenberg, Y. Silberberg, R. Morandotti, and J.S. Aitchison, Phys. Rev. Lett. **85**, 1863 (2000)
15. P. Yeh, A. Yariv, and C.-S. Hong, J. Opt. Soc. Am. **67**, 423 (1977)
16. E. Smirnov, C.E. Rüter, D. Kip, K. Shandarova, and V. Shandarov, Appl. Phys. B **88**, 359 (2007)
17. P. Millar, J.S. Aitchison, J.U. Kang, G.I. Stegeman, A. Villeneuve, G.T. Kennedy, and W. Sibbet, J. Opt. Soc. Am. B **14**, 3224 (1997)
18. N.K. Efremidis, S. Sears, D.N. Christodoulides, J.W. Fleischer, and M. Segev, Phys. Rev. E **66**, 046602 (2002)
19. F. Chen, M. Stepić, C.E. Rüter, D. Runde, D. Kip, V. Shandarov, O. Manela, and M. Segev, Opt. Express **13**, 4314 (2005)
20. T. Pertsch, T. Zentgraf, U. Peschel, A. Brauer, and F. Lederer, Appl. Phys. Lett. **80**, 3247 (2002)
21. K.M. Davis, K. Miura, N. Sugimoto, and K. Hirao, Opt. Lett. **21**, 1729 (1996)
22. K.A. Brzdąkiewicz, M.A. Karpierz, A. Fratalocchi, G. Assanto, and E. Nowinowski-Kruszelnick, Opto-Electron. Rev. **13**, 107 (2005)
23. D. Kip, Appl. Phys. B **67**, 131 (1998)
24. A.B. Aceves, C. De Angelis, T. Peschel, R. Muschall, F. Lederer, S. Trillo, and S. Wabnitz, Phys. Rev. E **53**, 1172 (1996)
25. J.C. Eilbeck, P.S. Lomdahl, and A.C. Scot, Phys. Rev. B **30**, 4703 (1984)
26. R. Scharf and A.R. Bishop, Phys. Rev. A **43**, 6535 (1991)
27. E.W. Laedke, K.W. Spatschek, and S.K. Turitsyn, Phys. Rev. Lett. **73**, 1055 (1994)
28. R. Iwanow, R. Schiek, G.I. Stegeman, T. Pertsch, F. Lederer, Y. Min, and W. Sohler, Phys. Rev. Lett. **93**, 113902 (2004)
29. A. Fratalocchi, G. Assanto, K.A. Brzdąkiewicz, and M.A. Karpierz, Opt. Lett. **29**, 1530 (2004)
30. J.W. Fleischer, G. Bartal, O. Cohen, T. Schwartz, O. Manela, B. Freedman, M. Segev, H. Buljan, and N.K. Efremidis, Opt. Express **13**, 1780 (2005)
31. W.P. Su, J.R. Schrieffer, and A.J. Heeger, Phys. Rev. Lett. **42**, 1698 (1979)
32. C.M. de Sterke and J.E. Sipe, Opt. Lett. **14**, 871 (1989)
33. J.W. Fleischer, T. Carmon, M. Segev, N.K. Efremidis, and D.N. Christodoulides, Phys. Rev. Lett. **90**, 023902 (2003)
34. D. Mandelik, R. Morandotti, J.S. Aitchinson, and Y. Silberberg, Phys. Rev. Lett. **92**, 093904 (2004)
35. Y.S. Kivshar, Opt. Lett. **18**, 1147 (1993)
36. A. Maluckov, M. Stepić, D. Kip, and L. Hadžievski, Eur. Phys. J. B **45**, 539 (2005)
37. M. Matuszewski, C.R. Rosberg, D.N. Neshev, A.A. Sukhorukov, A. Mitchell, M. Trippenbach, M.W. Austin, W. Królikowski, and Y.S. Kivshar, Opt. Express **14**, 254 (2006)
38. T. Dauxois and M. Peyrard, Phys. Rev. Lett. **70**, 3935 (1993)
39. M. Stepić, C.E. Rüter, D. Kip, A. Maluckov, and L. Hadžievski, Opt. Commun. **267**, 229 (2006)
40. M. Jablan, H. Buljan, O. Manela, G. Bartal, and M. Segev, Opt. Express **15**, 4623 (2007)
41. J. Meier, G.I. Stegeman, D.N. Christodoulides, Y. Silberberg, R. Morandotti, H. Yang, G. Salamo, M. Sorel, and J.S. Aitchison, Phys. Rev. Lett. **92**, 163902 (2004)

1 Nonlinear Effects in One-Dimensional Photonic Lattices   17

42. R. Iwanow, G.I. Stegeman, R. Schiek, Y. Min, and W. Sohler, Opt. Express **13**, 7794 (2005)
43. M. Stepić, C. Wirth, C.E. Rüter, and D. Kip, Opt. Lett. **31**, 247 (2006)
44. C.E. Rüter, J. Wisniewski, M. Stepić, and D. Kip, Opt. Express **15**, 6320 (2007)
45. S.V. Manakov, Sov. Phys. JETP **38**, 248 (1974)
46. S. Darmanyan, A. Kobyakov, E. Schmidt, and F. Lederer, Phys. Rev. E **57**, 3520 (1998)
47. M.J. Ablowitz and Z.H. Musslimani, Phys. Rev. E **65**, 056618 (2002)
48. R.A. Vicencio, M.I. Molina, and Y.S. Kivshar, Phys. Rev. E **71**, 056613 (2005)
49. E.P. Fitrakis, P.G. Kevrekidis, B.A. Malomed, and D.J. Frantzeskakis, Phys. Rev. E **74**, 026605 (2006)
50. J. Meier, J. Hudock, D.N. Christodoulides, G. Stegeman, Y. Silberberg, R. Morandotti, and J.S. Aitchison, Phys. Rev. Lett. **91**, 143907 (2003)
51. R.A. Vicencio, E. Smirnov, C.E. Rüter, D. Kip, and M. Stepić, Phys. Rev. A **76**, 033816 (2007)
52. O. Cohen, T. Schwartz, J.W. Fleischer, M. Segev, and D.N. Christodoulides, Phys. Rev. Lett. **91**, 113901 (2003)
53. A.A. Sukhorukov and Y.S. Kivshar, Phys. Rev. Lett. **91**, 113902 (2003)
54. Y. Lahini, D. Mandelik, Y. Silberberg, R. Morandotti, Opt. Express **13**, 1762 (2005)
55. A.S. Desyatnikov, E.A. Ostrovskaya, Y.S. Kivshar, and C. Denz, Phys. Rev. Lett. **91**, 153902 (2003)
56. T. Song, S.M. Liu, R. Guo, Z.H. Liu, and Y.M. Gao, Opt. Express **14**, 1924 (2006)
57. M. Stepić, D. Kip, L. Hadžievski, and A. Maluckov, Phys. Rev. E **69**, 066618 (2004)
58. M. Peyrard and M.D. Kruskal, Phys. D **14**, 88 (1984)
59. Y.S. Kivshar and D.K. Campbell, Phys. Rev. E **48**, 3077 (1993)
60. Y.S. Kivshar, W. Królikowski, and O.A. Chubykalo, Phys. Rev. E **50**, 5020 (1994)
61. S. Darmanyan, A. Kobyakov, and F. Lederer, Sov. Phys. JETP **86**, 682 (1998)
62. R. Morandotti, U. Peschel, J.S. Aitchison, H.S. Eisenberg, and Y. Silberberg, Phys. Rev. Lett. **83**, 2726 (1999)
63. L. Hadžievski, A. Maluckov, M. Stepić, and D. Kip, Phys. Rev. Lett. **93**, 033901 (2004)
64. L. Hadžievski, A. Maluckov, and M. Stepić, Opt. Express **15**, 5687 (2007)
65. P.G. Kevrekidis, A.R. Bishop, and K.Ø. Rasmussen, Phys. Rev. E **63**, 036603 (2001)
66. S. Darmanyan, A. Kobyakov, F. Lederer, and L. Vazquez, Phys. Rev. B **59**, 5994 (1999)
67. T. Kapitula, P.G. Kevrekidis, and B.A. Malomed, Phys. Rev. E **63**, 036604 (2001)
68. Y.V. Kartashov, V.A. Vysloukh, and L. Torner, Opt. Express **12**, 2831 (2004)
69. G. Kalosakas, Phys. D **216**, 44 (2006)
70. E. Smirnov, C.E. Rüter, D. Kip, Y.V. Kartashov, and L. Torner, Opt. Lett. **32**, 1950 (2007)
71. P.J.Y. Louis, E.A. Ostrovskaya, C.M. Savage, and Y.S. Kivshar, Phys. Rev. A **67**, 013602 (2003)

72. T.R.O. Melvin, A.R. Champneys, P.G. Kevrekidis, and J. Cuevas, Phys. Rev. Lett. **97**, 124101 (2006)
73. R. Carretero-González, J.D. Talley, C. Chong, and B.A. Malomed, Phys. D **216**, 77 (2006)
74. M. Johansson and Y.S. Kivshar, Phys. Rev. Lett. **82**, 85 (1999)
75. R. Morandotti, H.S. Eisenberg, Y. Silberberg, M. Sorel, and J.S. Aitchison, Phys. Rev. Lett. **86**, 3296 (2001)
76. E. Smirnov, C.E. Rüter, M. Stepić, D. Kip, and V. Shandarov, Phys. Rev. E **74**, 065601 (R) (2006)
77. E.P. Fitrakis, P.G. Kevrekidis, H. Susanto, and D.J. Frantzeskakis, Phys. Rev. E **75**, 066608 (2007)
78. G.I. Stegeman and M. Segev, Science **286**, 1518 (1999)
79. W. Królikowski, U. Trutschel, M. Cronin-Golomb, and C. Schmidthattenberger, Opt. Lett. **19**, 320 (1994)
80. O. Bang and P.D. Miller, Opt. Lett. **21**, 1105 (1996)
81. W. Królikowski and Y.S. Kivshar, J. Opt. Soc. Am. B **13**, 876 (1996)
82. T. Pertsch, U. Peschel, and F. Lederer, Opt. Lett. **28**, 102 (2003)
83. R.A. Vicencio, M.I. Molina, and Y.S. Kivshar, Opt. Lett. **28**, 1942 (2003)
84. A.B. Aceves, C. De Angelis, S. Trillo, and S. Wabnitz, Opt. Lett. **19**, 332 (1994)
85. Y.V. Kartashov, L. Torner, and V.A. Vysloukh, Opt. Lett. **29**, 1102 (2004)
86. A. Fratalocchi, G. Assanto, K.A. Brzdąkiewicz, and M.A. Karpierz, Appl. Phys. Lett. **86**, 051112 (2005)
87. C.R. Rosberg, I.L. Garanovich, A.A. Sukhorukov, D.N. Neshev, W. Królikowski, and Y.S. Kivshar, Opt. Lett. **31**, 1498 (2006)
88. A.B. Aceves, G.G. Luther, C. De Angelis, A.M. Rubenchik, and S.K. Turitsyn, Phys. Rev. Lett. **75**, 73 (1995)
89. R. Morandotti, H.S. Eisenberg, D. Mandelik, Y. Silberberg, D. Modotto, M. Sorel, C.R. Stanley, and J.S. Aitchison, Opt. Lett. **28**, 834 (2003)
90. H. Trompeter, U. Peschel, T. Pertsch, F. Lederer, U. Streppel, D. Michaelis, and A. Brauer, Opt. Express **11**, 3404 (2003)
91. A.A. Sukhorukov and Y.S. Kivshar, Opt. Lett. **30**, 1849 (2005)
92. F. Fedele, J. Yang, and Z. Chen, Opt. Lett. **30**, 1506 (2005)
93. L. Morales-Molina and R.A. Vicencio, Opt. Lett. **31**, 966 (2006)
94. A.E. Miroshnichenko and Y.S. Kivshar, Phys. Rev. E **72**, 056611 (2005)
95. J. Meier, G.I. Stegeman, D.N. Christodoulides, Y. Silberberg, R. Morandotti, H. Yang, G. Salamo, M. Sorel, and J.S. Aitchison, Opt. Lett. **30**, 1027 (2005)
96. J. Meier, G.I. Stegeman, D.N. Christodoulides, R. Morandotti, G. Salamo, H. Yang, M. Sorel, Y. Silberberg, and J.S. Aitchison, Opt. Lett. **30**, 3174 (2005)
97. E. Smirnov, C.E. Rüter, M. Stepić, V. Shandarov, and D. Kip, Opt. Express **14**, 11248 (2006)
98. I.E. Papacharalampous, P.G. Kevrekidis, B.A. Malomed, and D.J. Frantzeskakis, Phys. Rev. E **68**, 046604 (2003)
99. J. Cuevas and J.C. Eilbeck, Phys. Lett. A **358**, 15 (2006)
100. O.F. Oxtoby, D.E. Pelinovsky, and I.V. Barashenkov, Nonlinearity **19**, 217 (2006)
101. A. Maluckov, L. Hadžievski, and M. Stepić, Eur. J. Phys. B **53**, 333 (2006)
102. J. Meier, G.I. Stegeman, Y. Silberberg, R. Morandotti, and J.S. Aitchison, Phys. Rev. Lett. **93**, 093903 (2004)

103. M. Stepić, E. Smirnov, C.E. Rüter, L. Prönneke, D. Kip, and V. Shandarov, Phys. Rev. E **74**, 046614 (2006)
104. E. Smirnov, M. Stepić, C.E. Rüter, V. Shandarov, and D. Kip, Opt. Lett. **32**, 512 (2007)
105. S. Koke, D. Träger, P. Jander, M. Chen, D.N. Neshev, W. Królikowski, Y.S. Kivshar, and C. Denz, Opt. Express **15**, 6279 (2007)
106. M. Belić, D. Jović, S. Prvanović, D. Arsenović, and M. Petrović, Opt. Express **14**, 794 (2006)
107. M. Matsumoto, S. Katayama, and A. Hasegawa, Opt. Lett. **20**, 1758 (1995)
108. A.A. Sukhorukov and Y.S. Kivshar, Opt. Lett. **27**, 2112 (2002)
109. Y.V. Kartashov, V.A. Vysloukh, and L. Torner, J. Opt. Soc. Am. B **22**, 1356 (2005)
110. S. Longhi, M. Marangoni, M. Lobino, R. Ramponi, P. Laporta, E. Cianci, and V. Foglietti, Phys. Rev. Lett. **96**, 243901 (2006)
111. R. Pezer, H. Buljan, G. Bartal, M. Segev, and J.W. Fleischer, Phys. Rev. E **73**, 056608 (2006)
112. H. Buljan, O. Cohen, J.W. Fleischer, T. Schwartz, M. Segev, Z.H. Musslimani, N.K. Efremidis, and D.N. Christodoulides, Phys. Rev. Lett. **92**, 223901 (2004)
113. K. Motzek, A.A. Sukhorukov, and Y.S. Kivshar, Opt. Express **14**, 9873 (2006)
114. K.G. Makris, S. Suntsov, D.N. Christodoulides, G.I. Stegeman, and A. Haché, Opt. Lett. **30**, 2466 (2005)

2

Nonlinear Optical Waves in Liquid Crystalline Lattices

Gaetano Assanto[1] and Andrea Fratalocchi[2]

[1] Nonlinear Optics and OptoElectronics Lab (NooEL), CNISM and INFN, University "Roma Tre", Via della Vasca Navale 84, 00146 Rome, Italy
assanto@uniroma3.it
[2] Research Centers CRS SOFT INFM-CNR and Museo Storico per la Fisica "Enrico Fermi", University "Sapienza", P.le Aldo Moro 2, 00185 Rome, Italy
andrea.fratalocchi@roma1.infn.it

2.1 Introduction

Liquid crystals (LC) are molecular dielectrics encompassing several properties of both liquids and solids; in particular, they are often characterized by an order parameter which can be employed to distinguish among possible LC phases. In the *nematic* phase, liquid crystals show a significant degree of orientational order, their elongated organic molecules being aligned in a mean direction in space, as described by a vectorial field $\boldsymbol{n}$ called *director*. Since most nematics are derivative of benzene, they feature "cigar-like" molecules; hence, the *macroscopic* system can be regarded as an optically uniaxial *crystalline fluid*. The dielectric tensor $\overleftrightarrow{\epsilon}(\boldsymbol{r})$, describing the optical polarization of the medium, can be expressed as $\overleftrightarrow{\epsilon} = \overleftrightarrow{R}^\dagger \cdot \overleftrightarrow{\epsilon}_{\text{NLC}} \cdot \overleftrightarrow{R}$, with $\overleftrightarrow{\epsilon}_{\text{NLC}} = [\epsilon_\perp, \epsilon_\perp, \epsilon_\parallel] \cdot \boldsymbol{I}$, $I_{ij} = \delta_{ij}$ (δ_{ij} is the Kronecker delta) and $\overleftrightarrow{R}(\boldsymbol{n})$ a rotation tensor. The steady-state director configuration is obtained as an extremal point of the action integral $\mathcal{I} = \int \mathcal{L}\, dx\, dy\, dz$, whose density $\mathcal{L}$ defines the energy spent by the molecular system to hold a specific director configuration (Frank free-energy formulation) [1]. The energy density $\mathcal{L}$ can be further expanded into elastic $\mathcal{L}_{\text{el}}$ and electromagnetic $\mathcal{L}_{\text{em}}$ terms: $\mathcal{L} = \mathcal{L}_{\text{el}} + \mathcal{L}_{\text{em}}$. The contribution $\mathcal{L}_{\text{el}}$ can be evaluated in the framework of the elastic continuum theory and, in the *single constant approximation* [2], reads:

$$\mathcal{L}_{\text{el}} = \frac{1}{2} K \left[(\nabla \cdot \boldsymbol{n})^2 + (\boldsymbol{n} \cdot \nabla \times \boldsymbol{n})^2 \right], \tag{2.1}$$

with K accounting for elastic deformations ($[K] = N$). The electromagnetic contribution can be calculated by considering that the electric field induces dipoles on the nematic liquid crystal (NLC) molecules; the latter are then subjected to a torque and change their angular orientation towards a minimum energy configuration (e.g., parallel to the applied field). The contribution describing such *reorientation* process is [2]:

$$\mathcal{L}_{\text{em}} = -\frac{\Delta\epsilon}{2}\langle \boldsymbol{n} \cdot \boldsymbol{E}\rangle , \qquad (2.2)$$

being $\Delta\epsilon = \epsilon_\| - \epsilon_\perp$ the NLC birefringence and $\langle \ldots \rangle$ denoting a square time average. The balance between field-induced reorientation and elastic interactions gives rise to the steady state distribution $\boldsymbol{n}$, found as an extremal of the action integral $\delta\mathcal{L} = 0$.

2.2 Photonic Lattices in Nematic Liquid Crystals

An optical lattice can be realized in NLC by exploiting the electro-optic reorientational response, e.g., by embedding a layer of the material into an electromagnetic lattice (Fig. 2.1). This is obtained by confining a thin film of NLC between two glass plates, properly treated to provide *planar* anchoring in the direction z of light propagation. The mean angular orientation of the NLC molecules (i.e., the molecular director) is conveniently described by the angle θ with the axis z in the plane (y,z), as sketched in Fig. 2.1b. If $n_a^2 = n_\|^2 - n_\perp^2$ is the NLC optical birefringence (with $n_\|$ and $n_\perp$ along or orthogonal to the *director*, respectively), an electric field (static or low frequency) applied across x, constant in z and periodic along y, can reorient the director and determine a one-dimensional optical lattice with index modulation along y for x-polarized waves. To model such a medium, we begin by calculating the steady state director distribution, assuming a director field $\boldsymbol{n} = [\sin\theta, 0, \cos\theta]^T$. The application of the variational derivative to $\mathcal{L}$ gives [3]:

$$K\nabla_{xy}^2\theta + \frac{\Delta\epsilon_{\text{RF}}|E_x|^2}{2}\sin(2\theta) = 0 , \qquad (2.3)$$

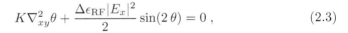

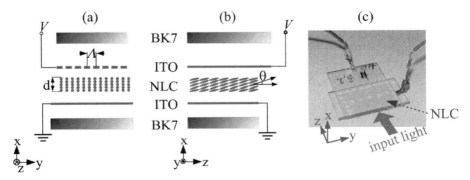

Fig. 2.1. Example of an optical lattice in a cell of thickness d filled with nematic liquid crystals. (**a**) Front view, (**b**) side view, (**c**) an actual NLC sample in the laboratory, the wires allow to bias the cell through the electrodes

2 Nonlinear Optical Waves in Liquid Crystalline Lattices   23

being $\Delta\epsilon_{\rm RF} = \epsilon_\| - \epsilon_\perp$ and E_x the x-component of the static electric field $\boldsymbol{E}$. The static electric field distribution $\boldsymbol{E}$ due to the applied voltage V can be calculated from Maxwell's divergence equation $\nabla \cdot \boldsymbol{D} = 0$ [3]:

$$\frac{\partial}{\partial x}\left[(n_\perp^2 + n_{\rm a}^2 \sin^2\theta)\frac{\partial V}{\partial x}\right] + n_\perp^2 \frac{\partial V}{\partial y} = 0, \qquad \boldsymbol{E} = -\nabla V. \qquad (2.4)$$

The presence of an optical-frequency field can be accounted for by adding the term $\epsilon_0 n_{\rm a}^2 |A_x|^2 \sin(2\theta)/4$ to Eq. (2.3), being A_x the x-component of the optical envelope $\boldsymbol{A}$. Finally, the refractive index *seen* by an x-polarized optical wave is obtained from the xx-component of the dielectric tensor $\overleftrightarrow{\epsilon}_{xx} = n_\perp^2 + n_{\rm a}^2 \sin^2\theta$. Equations (2.3)–(2.4), together with Maxwell's equations, model nonlinear wave propagation in NLC lattices. They can often be reduced to a form of the discrete nonlinear Schrödinger equation, the most general model of discrete optical lattices [4, 5].

The NLC reorientation depends on the strength of the field E_x; hence, the lattice index modulation can be tuned by acting on the input bias V. This property has important consequences on the periodic system, as revealed by its band-gap spectrum. The latter is obtained from the self-adjoint eigenvalue problem [6]:

$$L\Psi_{nk_y} = k_z^2 \Psi_{nk_y}, \qquad L = \frac{\partial^2}{\partial^2 x} + \left(\frac{\partial}{\partial y} - ik_y\right)^2 + \frac{\omega^2 n^2}{c^2}. \qquad (2.5)$$

Equations (2.5) yield the band-gap spectrum (k_z as a function of k_y) and the corresponding Floquet-Bloch (FB) modes Ψ_{nk_y}. We numerically solved Eqs. (2.5) and (2.3)–(2.4) with state-of-the-art numerical methods. Figure 2.2 shows our results for a cell with $\Lambda = d = 6\,\mu$m and bias $0.7\,{\rm V} \leq V \leq 2.0\,{\rm V}$,

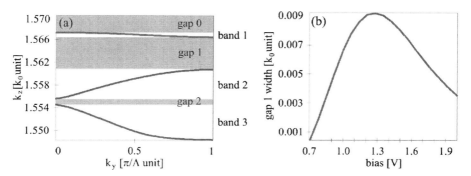

Fig. 2.2. Floquet-Bloch analysis of a dielectric photonic array in PCB: (a) band-gap diagram ($V = 0.9\,{\rm V}$), (b) width of gap 1 versus bias in NLC. The cell had a thickness $d = 6\,\mu$m and a transverse period $\Lambda = 6\,\mu$m. (Adapted from Ref. [7] with permission)

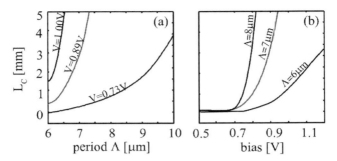

Fig. 2.3. Tuning curves of the voltage-controlled NLC array of thickness $d = 6\,\mu\text{m}$. Coupling distance L_C versus (**a**) period Λ and (**b**) bias V. (Adapted from Ref. [18] with permission)

filled with a standard NLC known as PCB [7]. A characteristic band-gap spectrum is displayed in Fig. 2.2a. It consists of bands separated by gaps (gray areas) in which FB eigen-modes are forbidden. As the bias V increases, the reorientation of the NLC molecules becomes stronger and the resulting index modulation grows. Thereby, in this regime, the width of each band-gap increases (Fig. 2.2b for $V \le 1.3\,\text{V}$). This behavior is counteracted by long-range molecular interactions (or spatial nonlocality [7–9]) which tend to reduce the index contrast between neighboring channels. As a result, above a certain bias the gap shrinks again (Fig. 2.2b for $V > 1.3\,\text{V}$). Such a *gap tunability* has important consequences on the propagation of lattice solitons (see Sec. 2.4.1).

Another important quantity to describe the lattice tunability is the coupling distance L_C, defined as the distance upon which a complete exchange of energy occurs between neighboring channels. We evaluate L_C for variable bias V and lattice period Λ (Fig. 2.3). Similar to the case of 1D coupled slab-waveguides [10], L_C evolves exponentially with guide spacing and bias (Fig. 2.3a–b, respectively). It is apparent that the NLC array exhibits significant electro-optic tunability, inasmuch as the coupling distance (hence discrete difraction) can be adjusted by acting on the external voltage, as visible in Fig. 2.3b. For $\Lambda = 8\,\mu\text{m}$, $L_C < 1\,\text{mm}$ for an external voltage $0.7\,\text{V} \le V \le 0.8\,\text{V}$.

2.3 Discrete Dynamics

2.3.1 Discrete Solitons in NLC

Discrete solitary waves in NLC lattices can be studied either via asymptotic expansions and Lagrangian analysis [11] or by numerical methods, as we will discuss here. Such analysis is performed by coupling Eqs. (2.3)–(2.4) to the paraxial equation describing light propagation [3]. Figures 2.4a–d show

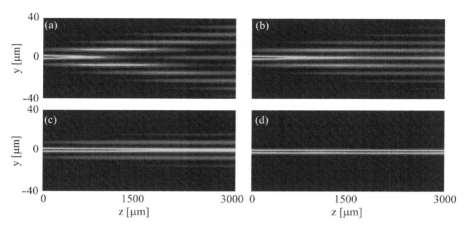

Fig. 2.4. Nonlinear light propagation in an NLC lattice for various input powers. (a) $P = 0.2\,\text{mW}$, (b) $P = 0.4\,\text{mW}$, (c) $P = 0.8\,\text{mW}$, (d) $P = 1\,\text{mW}$. The PCB lattice parameters are $\Lambda = 8\,\mu\text{m}$ and $d = 6\,\mu\text{m}$

some representative results of our numerics for an NLC lattice defined by $V = 0.74\,\text{V}$ in a PCB-filled cell with $\Lambda = 8\,\mu\text{m}$ and $d = 6\,\mu\text{m}$. In the linear regime (Fig. 2.4a for $P = 0.2\,\text{mW}$), power discretely diffracts along the waveguide array, as predicted by coupled mode theory [12]. As the optical excitation grows, light starts reorienting the liquid crystals in the input waveguide(s) thereby *detuning* them from synchronous coupling and quenching discrete diffraction (Fig. 2.4b–c). The nonlinearity, in fact, alters the propagation constants β_n of the guiding channels and breaks the *resonant* condition in the lattice (originated from the *phase-matching* between each site), thereby decreasing the coupling across neighboring waveguides. When the power carried by the beam is large enough, the single input waveguide is completely detuned (i.e. electromagnetically isolated) from the rest of the array and supports a *discrete soliton* confined at one site (Fig. 2.4d).

The experiments are carried out with a near infrared ($\lambda = 1.064\,\mu\text{m}$) Nd:YAG laser source, imaging the light scattered out of the yz-plane with a high resolution CCD camera. Figure 2.5 shows results for an NLC lattice with $\Lambda = 8\,\mu\text{m}$, thickness $d = 6\,\mu\text{m}$ and bias $V = 0.74\,\text{V}$. In the linear regime (input power $P = 1\,\text{mW}$ before the cell), light initially coupled to the fundamental mode of a single waveguide propagates forward and diffracts in a discrete fashion (Fig. 2.5a). By acting on the applied voltage V, the system can be driven from continuous (1D bulk) diffraction (Fig. 2.5b leftmost portion) to discrete diffraction (Fig. 2.5b above 0.7 V), the latter being completely tunable down to nearly no coupling (Fig. 2.5b rightmost portion). As the input power increases and for a constant bias ($V = 0.74\,\text{V}$), conversely, the all-optical shift in propagation constant mismatches the excited waveguide, decreasing

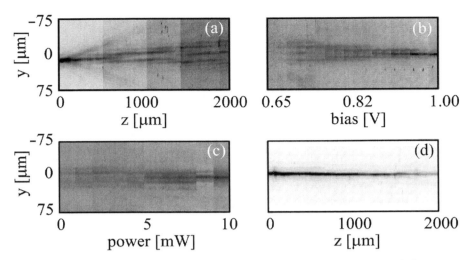

Fig. 2.5. Experimental response of the PCB lattice with $\Lambda = 8\,\mu\text{m}$ and $d = 6\,\mu\text{m}$. (a) Discrete diffraction in (y,z) for $P = 1\,\text{mW}$, transverse light spreading at the segment $1.40\,\text{mm} \leq z \leq 1.55\,\text{mm}$, (b) versus bias for $P = 1\,\text{mW}$ and (c) versus input power for $V = 0.74\,\text{V}$, (d) discrete soliton excited by $P = 10\,\text{mW}$ for $V = 0.74\,\text{V}$ and propagating along the input channel. (Adapted from Ref. [6] with permission)

the (evanescent) coupling between neighbors and, therefore, the amount of discrete spreading (Fig. 2.5c). When the optical excitation is large enough (input power $P = 10\,\text{mW}$), light gets trapped in the launch channel and a discrete soliton is generated in the array, in perfect agreement with theory (Fig. 2.5d) [3,12,13].

2.3.2 Nonlinear Steering in NLC Lattices

The introduction of discreteness is often accompanied by breaking of one or more symmetries, resulting in novel effects. In the case of optical lattices, discreteness breaks the translational symmetry of the medium along y. From a physical perspective, this implies that the lattice can sustain different types of nonlinear waves (each of them with a characteristic energy and a specific spatial configuration) as the site n is shifted by non-integer values. When a one-site discrete soliton tries to propagate obliquely in the lattice, it would have to *jump* from site to site. Assuming that the system Hamiltonian H is bound from below [14, 15], there should be (at least) one stable solitary wave with energy minimizing H, while all the others (stable, unstable) would necessarily require a higher energy to be formed. As a result, any discrete soliton *jumping* to a different site needs to overcome the energy barrier towards the new configuration. Such a barrier plays the role of an *effective nonlinear* potential, known in literature as *Peierls-Nabarro* [16]. As the soliton energy increases, the Peierls-Nabarro barrier increases as well and, beyond a threshold, it forbids soliton motion across n, blocking the self-localized wave around its most stable configuration [6,17,18]. In summary, for solitons of small enough size (size decreases as power increases) a lattice – lacking translational symmetry – cannot sustain discrete solitons propagating with transverse momentum (i.e., a phasefront tilt).

Nonlinear light propagation in the array can be investigated by BPM numerical simulations. Figure 2.6 displays some results for a PCB cell with $\Lambda = 8\,\mu\text{m}$, $d = 6\,\mu\text{m}$ and a bias $V = 0.77\,\text{V}$. We launch a tilted gaussian beam with a 2 μm-waist across x and a 10 μm-waist across y. At low power $P_\text{in} = P_\text{L} = 100\,\mu\text{W}$, for a tilt angle of $\lambda/4\Lambda = 1.90$ degrees (which ensures minimum discrete diffraction in the plane), the beam travels across the array coupling from waveguide to waveguide (Fig. 2.6a). When the beam carries enough power, however, it reorients the NLC molecules, thereby detuning the input channel from the others; hence, the output beam shifts sideways at the output. Figure 2.6b shows the output intensity distribution versus y for increasing input power (vertical axis). Beam diffraction and tilt reduce until light self-confines in the launch channel for a 2 mW power, forming a discrete soliton with zero transverse velocity (Fig. 2.6c): light is guided by a

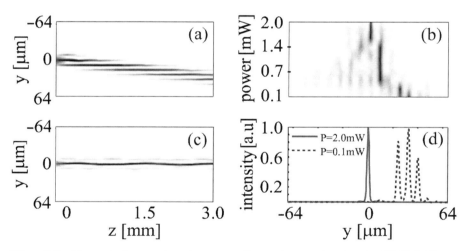

Fig. 2.6. Numerical experiments on nonlinear beam steering in NLC lattices. (a) Linear ($P_\text{L} = 0.1\,\text{mW}$) discrete diffraction of a gaussian beam ($w_{0x} = 2\,\mu\text{m}$, $w_{0y} = 10\,\mu\text{m}$) initially tilted by 1.90 degrees in the yz-plane, (b) transverse intensity distribution in the output segment 1.9 mm $\leq z \leq$ 2.0 mm versus y for various input powers P_in (vertical axis), (c) discrete soliton propagation for $P_\text{in} = P_\text{H} = 2\,\text{mW}$, (d) output intensity profiles in $z = 2\,\text{mm}$ for $P_\text{in} = P_\text{L}$ and $P_\text{in} = P_\text{H}$, respectively. (Adapted from Ref. [18] with permission)

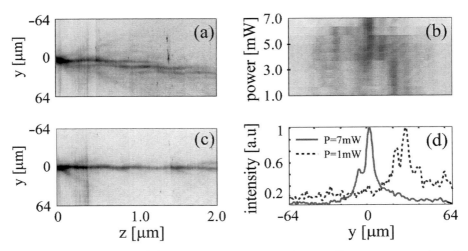

Fig. 2.7. Experiments. (a) Collated photographs showing discrete propagation of a tilted (1.90 degrees) gaussian beam launched at $P_{in} = P_L = 1\,\text{mW}$, (b) light distribution in $1.9\,\text{mm} \leq z \leq 2.0\,\text{mm}$ versus y for various excitations (vertical axis), (c) collated photographs displaying nonlinear propagation for $P_{in} = P_H = 7\,\text{mW}$, (d) transverse intensity profiles in $z = 2\,\text{mm}$ for $P_{in} = P_L$ and $P_{in} = P_H$, respectively. (Adapted from Ref. [18] with permission)

single channel despite the initial tilt and propagates (straight) along it. Such steering is visble in Fig. 2.6d as a transverse displacement (with reshaping).

Figure 2.7 summarizes our results for an input gaussian beam (10 μm-waist along y) launched with a tilt of $\Psi = 1.90$ degrees in an NLC lattice defined by $V = 0.77\,\text{V}$ and realized in PCB. In agreement with our simulations (Fig. 2.6), light diffracts in a discrete fashion at low power ($P_{in} = P_L = 1\,\text{mW}$) and obliquely travels through the array. As the power is raised, a nonlinear beam shift is observed across the plane (y,z). Figure 2.7b displays the beam intensity versus y in a propagation segment between $z_1 = 1.9\,\text{mm}$ and $z_2 = 2\,\text{mm}$ (to be compared with Fig. 2.6b). For $P_{in} = P_H = 7\,\text{mW}$, the all-optical detuning is large enough to trap the injected light into a single channel, giving rise to a discrete soliton with zero transverse momentum (Fig. 2.7c). Such light-driven steering after 2 mm (Fig. 2.7d) is in excellent agreement with numerical predictions (Fig. 2.6d). It should be underlined that the quoted powers are measured in front of the cell, i.e. they are not purged of Fresnel, scattering and coupling losses.

This power-dependent discrete beam steering opens the possibility to realize all-optical switching and routing devices for optical signals. We evaluate transmission T and crosstalk X for applications in switching or multiport routing. Using a finite aperture of the size of a single channel, we estimate $T_L = 0.86$ and $T_H = 0.99$, $X_L = -8.1\,\text{dB}$ and $X_H = -8.2\,\text{dB}$ for low (L) and high (H) excitations, respectively. Such figures show the great potential afforded by this novel design for all-optical information processing.

2.4 Lattice Dynamics

2.4.1 Multi-gap Lattice Solitons

A single gap in the dispersion diagram of a periodic medium is known to support gap solitons. The latter are solitary waves which originates from the nonlinear superposition of Floquet-Bloch modes belonging to distinct bands and close to band-edges [19–23]. Hereby we study multi-gap lattice solitons, i.e. gap solitons originating from two (or more) different gaps and with propagation constants close to the corresponding gap-edges [7]. The resulting localized wave, encompassing distinct propagation constants β_i, has the oscillatory character of a *multigap breather*. For the sake of simplicity, we focus on two-gap solitons.

Using (2+1)D simulations, we investigate breather propagation in the NLC array by exciting superimposed FB modes in distinct bands (Fig. 2.8). Modes of band 1 possess amplitude maxima in the waveguide-core regions (Fig. 2.8a, top) and can be excited by a wide (waist $\omega_y = 8\,\mu\text{m}$ across y) gaussian beam launched on-axis. In this case, light propagating for $P_1 = 0.2\,\text{mW}$ (Fig. 2.8b, top) exhibits the typical discrete diffraction pattern. Modes of band 2, conversely, have maxima between waveguide cores (Fig. 2.8a, bottom) and can

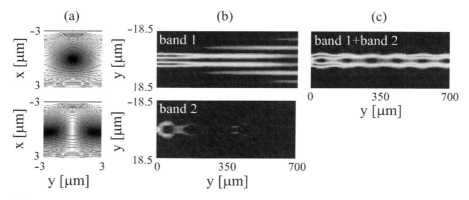

Fig. 2.8. Numerical experiments on multigap breathers. (**a**) FB profiles (density plots) with corresponding index distributions (contour lines) for $V = 1\,\text{V}$ (contour lines) and $k_y = \pm\pi/\Lambda$ in band 1 (*top*) and band 2 (*bottom*), (**b**) light propagation for modes belonging to band 1 (*top*) or band 2 (*bottom*) for $P_1 = P_2 = 0.2\,\text{mW}$, (**c**) multi-gap breather generated by superimposing the previous excitations in band 1 and band 2 with a total $P = P_1 + P_2$

be excited by a narrow gaussian input ($P_2 = 0.2\,\text{mW}$) which is centered between two neighboring channels (Fig. 2.8b, bottom). When co-launched with a total power $P = P_1 + P_2 = 0.4\,\text{mW}$, the FB modes couple via cross phase modulation (XPM) and generate a symmetric breather which oscillates along z in a periodic fashion (Fig. 2.8c).

We perform experiments on an array with $\Lambda = d = 6\,\mu\text{m}$. To ensure an adequate spatial overlap of modes in bands 1 and 2, we employ a single gaussian beam of waist $w_y = 5\,\mu\text{m}$ centered between neighboring waveguides. In this configuration, the overlap between the input and band 2 modes is larger, because the intensity peaks between sites. Figure 2.9 shows our results. At low power ($P = 0.2\,\text{mW}$) we observe diffraction across the array (Fig. 2.9a) as in the simulations (Fig. 2.8b, bottom panel). As the excitation grows, the higher refractive index in the waveguide core improves the spatial superposition of input beam and band 1 modes. When the power coupled to band 1 is large enough, a multigap breather is formed via XPM and propagates in the array (Fig. 2.9b). The oscillation period of the multigap breather, depending on the small difference between the propagation constants of the sourcing FB modes, can be precisely controlled by tuning the width of gap 1 via the input bias V. Figure 2.9c shows the calculated width of gap 1 and the measured period versus applied voltage V. Since the gap-width has a maximum in $V = 1.3\,\text{V}$ (solid line), for that bias the breather period exhibits a minimum (dotted line). For $V > 1.3\,\text{V}$ the NLC nonlocal response introduces molecular reorientation even between neighboring waveguides, thereby reducing the lattice modulation and shrinking the gap once again (see Fig. 2.2b) with a corresponding increase in oscillation period (Fig. 2.9c (dots)).

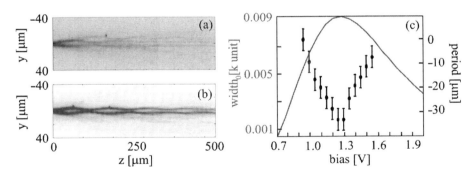

Fig. 2.9. (a) Low power $P = 0.2\,\text{mW}$ and (b) high power $P = 7\,\text{mW}$ light propagation after launching a gaussian beam of waist $w_y = 5\,\mu\text{m}$ between waveguides, (c) measured breather period (dots with error bars) and calculated width of gap 1 (solid line) versus bias V. (Adapted from Ref. [6] with permission)

2.4.2 Light-driven Landau-Zener Tunneling

In a famous paper dated 1932, Clarence Zener disclosed tunneling between energy levels in linear quantum Hamiltonian systems [24]. This phenomenon was originally discussed in the case of a two-level biological system subject to an *external* acceleration. Zener demonstrated that, if the imposed acceleration was *non adiabatic*, different eigenfunctions of the system could *connect*, in spite of their distinct properties and features, allowing the *crossing* of energy levels.

In photonic lattices, such tunneling can occur between FB bands, provided a non-adiabatic *acceleration* (e.g. a refractive index gradient is impressed). In this regime, optical energy initially coupled to a specific band can be tranferred to another, altering properties such as the position of intensity maxima [25] and/or the direction of propagation [26, 27].

We investigate Landau-Zener light tunneling by impressing an all-optical acceleration on a one-dimensional NLC lattice. Figure 2.10 sketches the concept: an intense gaussian beam (the *pump*) is launched straight along the guides (Fig. 2.10 right) and produces a refractive index decrease via reorientation, defining two transition regions (Fig. 2.10 left). These regions can provide a non-adiabatic acceleration to a second beam (the *probe*) (Fig. 2.10 right) which, having initially coupled light to a specific FB band, crosses FB levels and transfers energy to a lower band. To tunnel light from an upper to a lower band, a negative index change is required (Fig. 2.10 left), i.e., a self-defocusing nonlinearity.

We refer to a sample as in Fig. 2.1, with $d = 6\,\mu\text{m}$ and $\Lambda = 4\,\mu\text{m}$. After some lengthy algebra [24], the set of equations describing the NLC

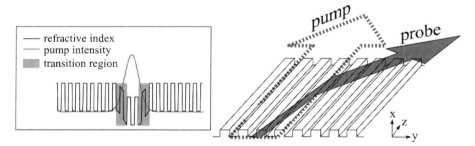

Fig. 2.10. Optically induced Landau-Zener tunneling in an array. An intense pump beam is launched straight into the nonlinear lattice (*right*), giving rise to an index change which forms two transition regions with non-adiabatic acceleration (*left*). A probe, initially coupled to an upper FB band, can undergo acceleration in one of the transition regions and tunnel to a lower band. (Adapted from Ref. [26] with permission)

thermo-optic response can be reduced to a dimensionless "accelerated" Schrödinger like equation:

$$i\frac{\partial \psi}{\partial Z} + \left[\frac{1}{2}\left(\frac{\partial}{\partial Y} + iZ\alpha\right)^2 + V(Y)\right]\psi = 0, \quad (2.6)$$

with periodic *potential* $V(Y) = V(Y + \Lambda_Y)$. We then Fourier-expand the periodic term $V(Y)$ in a series $V = \sum_n v_n \cos(2nY/\Lambda_Y)$, retaining only the first term $v_1 \cos(2Y/\Lambda_Y)$. By writing the field ψ as a sum of counterpropagating plane-waves $\psi = a_1(Z)\exp(iY/\Lambda_Y) + a_2(Z)\exp(-iY/\Lambda_Y)$ and projecting them on Eq. (2.6), we finally obtain the original Zener model [24]:

$$\frac{\partial b_1}{\partial Z} = \frac{iv_1}{2}e^{(-2i\int\Theta(Z)dz)}b_2, \qquad \frac{\partial b_2}{\partial Z} = \frac{iv_1}{2}e^{(2i\int\Theta(Z)dz)}b_1, \quad (2.7)$$

where

$$a_1 = b_1 \exp\left[i\left(\frac{2v_0-1}{2}Z - \frac{\delta^2}{6}Z^3 + \int \Theta(Z)dz\right)\right], \quad (2.8)$$

$$a_2 = b_2 \exp\left[i\left(\frac{2v_0-1}{2}Z - \frac{\delta^2}{6}Z^3 - \int \Theta(Z)dz\right)\right] \quad (2.9)$$

and $\Theta(Z) = -\delta Z$. Equation (2.7) predicts tunneling between bands at the exponential rate $\exp(-\pi v_1^2/4\delta)$.

To verify that tunneling has occurred, the standard practice requires to check the intensity distribution (position of the maxima) of the excited FB modes. In each band, in fact, FB modes possess peaks in characteristic spatial positions; hence, a close inspection of light distribution can reveal tunneling [25]. Such approach, however, cannot be effectively pursued when period Λ and wavelength λ are comparable, as in our case. To overcome this problem, we exploit the dispersion of the periodic lattice and monitor the transverse velocity of the signal before and after its interaction with the accelerated region. Since each Floquet-Bloch band exhibits a maximum in propagation angle (the normal in the band-gap spectrum, see Fig. 2.11a, black arrows), such maximum increasing with band-number (-order), after tunneling to a higher-order band light can travel at a larger angle (with respect to z in the plane (y,z)) than imposed by excitation. To elucidate this idea we numerically solve Eq. (2.6) for $V(Y) = \sin^2(Y)$, $V_0 = 1$ and $\delta = 0.5$. Figure 2.11b displays the propagation of a beam (a linear superposition of FB modes) belonging to band 1 at the maximum transverse velocity. As it reaches the accelerated region, Landau-Zener tunneling to band 2 occurs and the angle of propagation increases (Fig. 2.11c). This *observable* deviation unambiguously demonstrates band-to-band tunneling.

We used samples with $\Lambda = 4\,\mu\text{m}$, $d = 6\,\mu\text{m}$ filled with PCB, injecting mutually incoherent pump and probe beams at $\lambda = 1.064\,\mu\text{m}$. We obtain a self-defocusing response by exploiting the PCB thermal nonlinearity near its

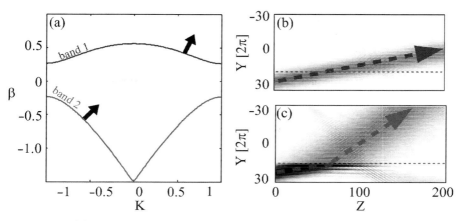

Fig. 2.11. (a) Bandgap diagram of Eq. (2.6) for $V_p = 0$, $V_0 = 1$ and $\delta = 0.5$, the black arrows indicate the maximum transverse velocities, (b) mode in band 1 propagating at the maximum transverse velocity, (c) Zener tunneling from band 1 to band 2. The dashed line refers to the location of the pump. (Adapted from Ref. [26] with permission)

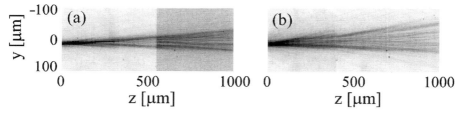

Fig. 2.12. Beam propagating in the NLC lattice for $V = 1.19$ V. Discrete diffraction in the (a) linear regime for a launched power $P = 1.0$ mW and (b) self-defocusing regime for $P = 6.0$ mW. (Adapted from Ref. [26] with permission)

phase-transition temperature [2,28]. To avoid saturation of the camera due to the intense pump, the latter was modulated on/off and the CCD synchronized in order to acquire images of the (cw) probe only when the pump is blocked. In a preliminary series of experiments we characterize the lattice response by injecting light in a single channel. Figure 2.12 displays beam propagation at low ($P = 1$ mW) and high ($P = 6$ mW) power. In the second case thermal self-defocusing reduces the index modulation and the beam widens in propagation (Fig. 2.13b). Finally, the photo sequence displayed in Fig. 2.13a–b shows all-optical Landau-Zener tunneling using a gaussian pump of y-waist $w_y = 15\,\mu$m and Rayleigh distance of about 900 μm (much longer than the observation window). When $P_{\text{pump}} = 0$ mW, probe light initially coupled at the maximum transverse velocity in band 1 (with $P_{\text{probe}} = 1$ mW) discretely diffracts and propagates obliquely in the array plane (y,z). As the pump power increases

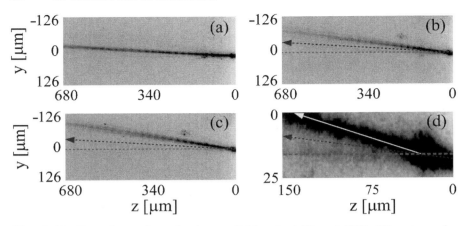

Fig. 2.13. Experimental results in a cell biased at $V = 1.19$ V. Discrete probe propagation for $P_{probe} = 1$ mW and (**a**) $P_{pump} = 0$ mW, (**b**) $P_{pump} = 25$ mW, (**c**), (**d**) $P_{pump} = 30$ mW. Zener tunneling is apparent in (c) and (d) as an increased propagation angle with respect to the maximum allowed in band 1 (black arrows). Figure (**d**) is the enlarged detail of (c). (Adapted from Ref. [27] with permission)

above $P_{pump} = 25$ mW an all-optical acceleration is impressed on the lattice and tunneling of the probe takes place, forcing it to propagate at a larger angle (Fig. 2.13b–d). Since the light remaining in band 1 is hardly distinguishable from the background, the power transfer can be considered highly efficient.

2.5 Summary

We have reported recent progress, theoretical and experimental, on nonlinear light propagation and localization in one dimensional voltage-controlled optical lattices realized in nematic liquid crystals. These structures, taking advantage of both electro-optic and all-optical responses of a reorientational molecular material, exhibit large tunability and enable us to study the transition from bulk to discrete diffraction and to discrete solitons [29]. Using such a versatile system, we have achieved discrete solitons and nonlinear discrete beam steering, the latter enabling broadband all-optical steering and efficient multiport routing at mW inputs. We have been able to generate symmetric multi-gap lattice solitons and to tune their period over several oscillations. Finally, we have investigated all-optically induced Zener tunneling, proving how it can be revealed by observing the angle of propagation (transverse velocity) of a probe across the array; this is a novel approach to all-optical switching in a periodic lattice.

Acknowledgements

The authors thank M. A. Karpierz for the high quality samples.

References

1. P.G. de Gennes and J. Prost, *The Physics of Liquid Crystals*, Oxford Science Publications, Oxford (1995)
2. I.C. Khoo, *Liquid Crystals: Physical Properties and Nonlinear Optical Phenomena*, Wiley, New York (1995)
3. A. Fratalocchi, G. Assanto, K.A. Brzdakiewicz, and M.A. Karpierz, Opt. Lett. **29**, 1530 (2004)
4. A. Fratalocchi and G. Assanto, Phys. Rev. A **76**, 042108 (2007)
5. F. Lederer, G.I. Stegeman, D.N. Christodoulides, G. Assanto, M. Segev, and Y. Silberberg, Phys. Rep. **463**, 1 (2008)
6. A. Fratalocchi, G. Assanto, K.A. Brzdakiewicz, and M.A. Karpierz, Opt. Express **13**, 1808 (2005)
7. A. Fratalocchi, G. Assanto, K.A. Brzdakiewicz, and M.A. Karpierz, Opt. Lett. **30**, 174 (2005)
8. M. Peccianti, K.A. Brzdakiewicz, and G. Assanto, Opt. Lett. **27**, 1460 (2002)
9. M. Peccianti, C. Conti, G. Assanto, A. De Luca, and C. Umeton, J. Nonl. Opt. Phys. Mat. **12**, 525 (2003)
10. A. Yariv, *Optical Electronics in Modern Communications*, Oxford University Press, New York (1997)
11. A. Fratalocchi and G. Assanto, Phys. Rev. E **72**, 066608 (2005)
12. F. Lederer and Y. Silberberg, Opt. Photon. News **2**, 48 (2002)
13. G. Assanto, A. Fratalocchi, and M. Peccianti, Opt. Express **15**, 5428 (2007)
14. S.K. Turitsyn, Theor. Math. Phys. **64**, 226 (1986)
15. E.W. Laedke, K.H. Spatschek, and S.K. Turitsyn, Phys. Rev. Lett. **73**, 1055 (1994)
16. S.V. Dmitiev, P.G. Kevrekidis, and N. Yoshikawa, J. Phys. A **38**, 7617 (2005)
17. A.B. Aceves, C. De Angelis, T. Peschel, R. Muschall, F. Lederer, S. Trillo, and S. Wabnitz, Phys. Rev. E **53**, 1172 (1996)
18. A. Fratalocchi, G. Assanto, K.A. Brzdakiewicz, and M.A. Karpierz, Appl. Phys. Lett. **86**, 051112 (2005)
19. W. Chen and D.L. Mills, Phys. Rev. Lett. **58**, 160 (1987)
20. Y.S. Kivshar and G.P. Agrawal, *Optical Solitons: from Fibers to Photonic Crystals*, Academic Press, San Diego (2003)
21. A.B. Aceves and S. Wabnitz, Phys. Lett. A **141**, 37 (1989)
22. C. Conti, S. Trillo, and G. Assanto, Phys. Rev. Lett. **78**, 2341 (1997)
23. A.V. Buryak, P.D. Trapani, D.V. Skryabin, and S. Trillo, Phys. Rep. **370**, 63 (2002)
24. C. Zener, Proc. R. Soc. A **137**, 696 (1932)
25. R. Khomeriki and S. Ruffo, Phys. Rev. Lett. **94**, 113904 (2005)
26. A. Fratalocchi and G. Assanto, Opt. Express **14**, 2021 (2006)
27. A. Fratalocchi, G. Assanto, K.A. Brzdakiewicz, and M.A. Karpierz, Opt. Lett. **31**, 790 (2006)
28. F. Simoni, *Nonlinear Optical Properties of Liquid Crystals*, World Scientific, Singapore (1997)
29. G. Assanto, A. Fratalocchi, and M. Peccianti, Opt. Express **15**, 5248 (2007)

3

Nonlinear Optics and Solitons in Photonic Crystal Fibres

Dmitry V. Skryabin and William J. Wadsworth

Centre for Photonics and Photonic Materials, Department of Physics, University of Bath, Bath BA2 7AY, UK
d.v.skryabin@bath.ac.uk, w.j.wadsworth@bath.ac.uk

3.1 Introduction

The fibre optics revolution in communication technologies followed the 1950's demonstration of the glass fibres with dielectric cladding [1]. Transmission applications of fibre optics have become a dominant modern day technology not least because nonlinearities present in – or introduced into – glass and enhanced by the tight focusing of the fibre modes allow for numerous light processing techniques, such as amplification, frequency conversion, pulse shaping, and many others. For these reasons, and because of the rich fundamental physics behind it, nonlinear fibre optics has become a blossoming discipline in its own right [1]. The 1990's witnessed another important development in fibre optics. Once again it came from a new approach to the fibre cladding, comprising a periodic pattern of air holes separated by glass membranes forming a photonic crystal structure [2, 3]. This prompted the name Photonic Crystal Fibres (PCFs). The fascinating story behind the invention of PCF and research into various fibre designs can be found, e.g., in [4]. Our aim here is to review the role played by PCFs in nonlinear and quantum optics, which is becoming the mainstream of the PCF related research and applications. Our focus will be on the areas where PCFs have brought to life effects and applications which were previously difficult, impossible to observe or simply not thought about.

There are currently two main PCF types. One has a solid silica core with a periodic air-glass or glass-glass structure around it, so called solid-core PCFs (SC-PCFs) [2], see inset in Fig. 3.5. The usual, but not exclusive, guidance mechanism in SC-PCFs is the classical total internal reflection. Second type of PCF is hollow core PCF (HC-PCF), where light is guided in an air core by the photonic band gap mechanism [3,5], see inset in Fig. 3.6. In SC-PCFs most of the light propagates in the silica glass, as in conventional fibres. Therefore the nonlinearity mechanisms are the instantaneous electronic response and non-instantaneous Raman response associated with molecular motion. The wavelength at which interesting effects occur is determined by the PCF's group velocity dispersion (GVD), which can be very different from GVD in conventional fibres [6]. The strength of the nonlinear effects is inversely proportional to the effective area of the fundamental fibre mode, which in a telecom fibre is $\approx 100\,\mu m^2$, while in the PCFs it can be reduced down to $\approx 2\,\mu m^2$ if the core diameter is close to $1\,\mu m$ [7]. Thus the same nonlinear effects as everyone is used to in telecom fibres become important at different wavelengths and for the powers 10 to 100 times less when a SC-PCF is used instead. The nonlinearity may also be enhanced in non-silica PCFs [8,9] with consequent changes to the dispersion. The combination of strong nonlinearity with flexible dispersion control via cladding and core designs has been the main reason for the continuing stream of new experiments, modeling and theories relying on opportunities offered by PCFs. Production of HC-PCFs guiding via a photonic-band gap with losses order of few to tens of dB/km [5,10] has matured relatively recently. Therefore their exploitation for nonlinear applications is still in its initial stage. HC-PCFs can be used for the guided delivery of high power pulses [11,12], offering such obvious advantages as tight focusing, choice of dispersion, long interaction length and low loss to out-perform capillary waveguides in frequency conversion [13], pulse shaping [14] and quantum optics applications [15].

In the second section we briefly review the vast area of supercontinuum generation and frequency conversion in PCFs, which has been previously described in excellent reviews, see, e.g., [16–19].

In the third and most detailed section we focus on properties of optical solitons in SC-PCFs and their prominent role in supercontinuum generation. There we summarize our very recent and few year old results and give a fairly detailed account of the work done by other groups. In the next two sections we review results on pulse compression and nonlinear and quantum optics with HC-PCFs.

3.2 Supercontinuum Generation and Frequency Conversion: Techniques and Applications

3.2.1 Femtosecond Supercontinua

An area of great recent interest has been supercontinuum generation in SC-PCF and in conventional optical fibres, see Fig. 3.1a. The initial impetus came from the work of Ranka *et al.* [6] demonstrating a supercontinuum spanning more than two octaves from 400 nm to 1600 nm, see Fig. 3.1b. Broad continuum spectra had been observed before, but only with high energy ($\sim$ tens of μJ or above) pulses in bulk solids or liquids [20–22]. What was new was the low energy pulses used, only a few nJ, in a small-core optical fibre [6,23]. The other important factor for PCFs is that the chromatic dispersion required can be achieved in the visible and near IR [24], and particularly at the 700–900 nm range of the ubiquitous Ti:sapphire modelocked laser. In this section

3 Nonlinear Optics and Solitons in Photonic Crystal Fibres   39

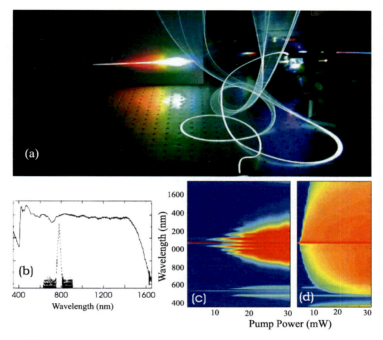

Fig. 3.1. (a) Photograph of a supercontinuum. 100 fs, 800 nm pulses enter a 2 μm core PCF from the bottom. (b) Femtosecond supercontinuum generated under conditions similar to (a). Dashed line is the pump spectrum. Reprinted from [6]. (c) and (d) Output spectra as a function of power for a 5 μm core PCF pumped with 0.6 ns, 1064 nm pulses with 1 m and 20 m length respectively. The MI peaks are clearly visible in (c), converting into a broad flat continuum over a longer length (d). Reprinted from [25]

we shall review the state of the art for femtosecond supercontinua, similar to this initial result, and also for continuum and frequency conversion using longer, picosecond to nanosecond pulses and even cw. This division arises from the underlying physics of the initiation of the continuum: with short, femtosecond, pulses the rapidly changing intensity, dI/dt, at the rising and falling edges of the pulse gives rise to strong self-phase modulation (SPM). For picosecond and longer pulses the pulse edges are less important, we can consider the pulses as quasi-cw undergoing classical four-wave mixing (FWM) under a third-order nonlinear susceptibility, $\chi^{(3)}$.

There are two pressing reasons that femtosecond supercontinua may be necessary. The first is in modelocked laser carrier-envelope phase stabilisation and frequency metrology; the second is in time resolved studies, where the sliced continuum can provide 200 fs pulses at selected wavelengths. Femtosecond supercontinua are also often used for optical coherence tomography (OCT) [26], where excellent spatial resolution can be obtained. In this case there is less need for a femtosecond source. An OCT system requires bandwidth (for axial resolution) and high repetition rate for fast scanning with at least one pulse per pixel.

We shall consider first supercontinuum for femtosecond laser pulse stabilisation and frequency metrology [27, 28]. In this application there is less emphasis placed on the spectral envelope of the supercontinuum; as long as there is sufficient light at the required f and $2f$ frequencies to achieve a phase-locked loop, that is enough. What is vitally important though is that the coherence is maintained. Dudley and Coen calculated numerically [29] how the first order coherence in the supercontinuum spectrum varies with propagation distance and with input pulse duration. These simulations clearly show that although the spectrum may be very similar for input pulses of different duration, only pulses of ≤ 50 fs duration will yield good coherence over the full spectral bandwidth. The predictions are borne out by experiment [30] and by the evidence of the many successful metrology or stabilization [27, 28] schemes using short pulses, and the failure of attempts to use 75 fs laser pulses to the same ends, even by the same research teams [28]. The simulations also point to a direction where good coherence is possible, even with long pulses, namely with very short fibre lengths, where the spectrum appears as a SPM spectrum. As well as being useful for frequency metrology, continuum which maintains coherence can be compressed into shorter pulses [31].

For some applications the spectral flatness is of as great importance as the spectral bandwidth and coherence. Long-pulse continua tend to give flatter spectra [6, 23], but where short pulses are required smooth spectra can be obtained by remaining in the regime of SPM, without soliton formation and breakup [32]. This has been demonstrated by using a PCF with a small normal dispersion [26], or by using very short lengths of PCFs with anomalous dispersion [8, 29]. Multiphoton microscopy [33], fluorescence lifetime imaging [34] and coherent anti-stokes Raman scattering (CARS) spectroscopy [35] have all be demonstrated using femtosecond continua generated in PCF.

3.2.2 Long-pulse Supercontinua

For pulses longer than a few picoseconds modulation instability (MI) overtakes SPM as the dominant spectral broadening process at low power [36], see Fig. 3.1c. In contrast to the fs pulses considered above, longer pulses have comparatively small dI/dt, and pulse walk-off between different spectral components is not significant. The dynamics observed can be considered using a cw approximation [36]. MI of cw or quasi-cw beams is a well known phenomenon in optical fibre in this regime [1]. In the frequency domain MI corresponds to FWM in which two degenerate pump photons create signal and idler photons equally spaced in frequency about the pump, see Fig. 3.1c. The $\chi^{(3)}$ nonlinearity provides coupling of the phasematched fields, and for a pump wavelength in the anomalous GVD regime broad MI bands appear close to the pump wavelength. Then generated frequencies serve as the pump fields and the MI process develops in the cascade manner, see Figs. 3.1c, 3.1d.

The first ps/ns continua, based simply on placing the fibre zero GVD close to the pump wavelength, yielded spectra spanning only to 500 nm from a 1064 nm pump [25,37], see Fig. 3.1c, 3.1d. Extension into the visible has been achieved by dual wavelength pumping [38], by pumping nano-core fibres at 532 nm [39], by pump wavelength conversion by FWM followed by supercontinuum generation [40], by fibres with changing zero GVD wavelength, either piecewise using multiple fibres [41] or continuously using a tapered fibre [42] and most recently and simply by using uniform fibres designed to match the group velocity of the long wavelength parts of the continuum to the shortest possible wavelengths [43–45]. Development of MI along the fiber length results in generation of a train of quasi-solitonic pulses, which implies that the cw generated continuum can ultimately acquire some features of the femtosecond one. Applications of ps-ns continua include confocal microscopy [46] and fibre analysis [12], with sufficient impact for commercial product offerings.

It is often considered that phasematching for MI can only occur in the anomalous dispersion regime, however this is only so if one simply considers dispersion terms up to $\beta^{(2)}$ [1]. With sufficiently large higher even-order dispersion, $\beta^{(4)}$, $\beta^{(6)}$, there can be phasematching for pump wavelengths in the normal dispersion regime [25, 36, 47, 48]. Typical PCFs with zero dispersion wavelengths in the visible/near-IR range do have large and positive $\beta^{(4)}$, which allows phasematching for pump wavelengths in the normal dispersion regime. In contrast to the broad MI gain bands close to the pump wavelength in the anomalous regime, the phasematched wavelengths for the normal regime are far from the pump, and much narrower band [25,36]. For this reason it is often referred to by the frequency domain nomenclature FWM. In the time domain the period of modulation of the MI is only a few cycles of the carrier pump wave (e.g. in [25] can be as low as 2 cycles).

In the long-pulse regime there has been a revisiting of the possibility of generating correlated and entangled photon pairs in fibre. This was attempted using the MI available in conventional fibre, but the signal was swamped by spontaneous Raman generation in the idler band. The FWM/MI signal grows quadratically with pump power, whereas the spontaneous Raman only grows linearly. At high power this favours FWM/MI, but at the very low powers required for generating single photon pairs Raman generation can dominate. A solution is offered with the widely spaced FWM phasematching available in the normal dispersion regime of PCF, yielding high count rates and low background [49] compared with previous results for the anomalous regime in PCF or conventional fibres [50,51]. Further rich phasematching is attainable in birefringent fibres through vector modulation instability, with pump, signal and idler fields not necessarily on the same axis of the fibre [48], which has been seen in conventional fibres [52] and in PCF [53]. This has been suggested as a means of reducing raman noise in pair-photon generation [54]. PCF is a particularly interesting fibre for the study of these effects because of the ease with which birefringence may be introduced through asymmetry in the core or cladding.

As well as the $\chi^{(3)}$ process discussed above, the acousto-optic interaction has also been found to be significantly affected by the PCF cladding structure [55]. In particular, the nano-sized core of PCFs increases five fold the threshold of the stimulated Brillouin and significantly alters the Brillouin spectrum.

3.3 Solitons in Solid-core PCFs and Their Role in Supercontinuum Generation

3.3.1 Soliton Fission and Intrapulse Raman Scattering

The nonlinearity enhancement in small solid core PCFs described in the introduction means that for the same dispersion, pulse duration and energy the single soliton pulse in a telecom fibre becomes a multi-soliton one (~10's order soliton) in a small core PCF. Evolution of the multi-soliton pulse over a length shorter than the GVD length is dominated by the self-phase modulation producing symmetric spectral lobes and accompanied by the compression in time domain [56]. After the point of maximal compression the Raman effect and higher order dispersions break the pulse up into single solitons and residual radiation. This process is often referred to as soliton fission [32]. Emerging solitons are still imperfect due to perturbations and continue to shake off radiation. They travel through the fibre embedded into a sea of dispersive waves, which leads to generation of yet new frequencies and further spectral broadening. The above scenario has been modeled and seen before the invention of PCFs, see, e.g., [57], but most of the how and why questions have remained without the answers until recently. PCFs arrived at a time when experimental techniques allowing simultaneous temporal and spectral studies of complex optical signals have become widely known [58] and the theoretical tools of soliton science have matured to a level which allows us to approach complex problems confidently. The combination of these factors has led to the significant advances in our knowledge about physics of solitons in optical fibres, which is reviewed here.

Solid-core PCFs can, uniquely, provide anomalous dispersion in the region 700–1100 nm, covering modelocked Ti:sapphire, Nd and Yb lasers, and solitons were observed in the earliest PCF experiments [59]. In the most typical experiments when small core PCFs are pumped not far from the zero GVD wavelength with 100 fs pulses having peak powers of few to 10s kW the duration of the most intense of the emerging solitons is order of few 10s of fs. Such short pulses are readily influenced by the intrapulse Raman scattering, which downshifts the soliton frequency [1]. The red shift of the carrier soliton frequency is by no means a perturbation in PCFs and fibre parameters such as GVD and effective area felt by the soliton can vary substantially over even a short fibre length, profoundly affecting soliton evolution [60–62] and collisions [63, 64]. The soliton self-frequency shift in PCFs, controlled by the choice of the appropriate fibre length and power of the input pulse, has been used in imaging systems based on CARS and fluorescence microscopy [65].

3.3.2 Resonant Radiation from Solitons

Solitons do not disperse, which implies that the dependence of the propagation constants of the plane waves constituting the soliton on frequency makes a straight line whose angle determines the soliton group velocity. The straight line representing the soliton dispersion is tangent (with a small offset) to the dispersion of linear waves [32, 66]. The curvature of the dispersion of linear waves changes sign at the zero GVD frequency which makes the intersection of the soliton dispersion and of the dispersion of the linear waves topologically unavoidable and placed in the range of the normal GVD, see Fig. 3.2a. The intersection point determines the frequency of the dispersive wave emitted by the solitons (so called resonant or Cherenkov radiation) [67]. It has been demonstrated that the resonant radiation from solitons contributes to supercontinuum generation [32, 68–71]. Emission of resonant radiation is not a noise amplifying parametric process. It is rather analogous to the signal of an oscillator driven by a resonant external force. This amongst other things implies that the radiation together with the soliton form a coherent part of the supercontinuum. Resonant radiation makes a dominant contribution into spectral broadening for sufficiently short pump pulses, when modulational instability and other noise amplifying processes are suppressed [29]. When the frequency detuning between the soliton and the resonance increases the amplitude of the resonant radiation exponentially decreases and vice versa [60].

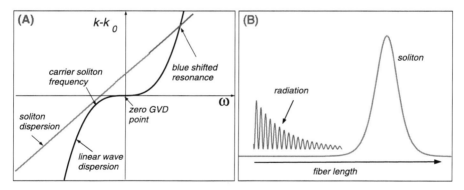

Fig. 3.2. Emission of the resonance radiation by a soliton in fibre with positive third order dispersion. (**A**) Resonance conditions. Changing colors of the straight lines schematically show detuning between the carrier soliton frequency and the resonance. k_0 is the propagation constant at the zero GVD frequency. (**B**) Soliton and emitted radiation. Radiation amplitude falls down with propagation

In the case where GVD increases with wavelength, i.e. the 3rd order dispersion is positive, see Fig. 3.2a (which is the most typical case for supercontinuum generation with pump around 800 nm) the self-frequency shift pulls the solitons away from the zero GVD point so that the resonant frequency gets detuned progressively further from the solitons towards the blue side of the spectrum. Hence the efficiency of the emission of radiation falls down quickly as soliton propagates along the fibre, which is schematically illustrated in Fig. 3.2b. Therefore one needs to look for other reasons explaining formation of the continuously blue shifting with propagation short-wavelength edge of the continuum spectra generated with femtosecond pump [43–45, 72, 73].

3.3.3 Mixing of Solitons with Dispersive Radiation, Radiation Trapping and Short-wavelength Edge of Supercontinuum

Mechanisms of the continuous blue shift of the shortwavelength edge of the supercontinuum spectra obtained with femtosecond pump sources have been understood only recently [44, 45]. First we should recall once again that the Raman effect continuously pulls solitons towards redder frequencies, which for anomalous GVD implies an increase of the group index felt by the soliton. Hence solitons slow down with a constant acceleration. At the same time the wave packet of the blue radiation emitted behind the soliton propagates with constant group velocity, see Fig. 3.2b. Thus the slowing soliton will collide with the blue radiation. The soliton locally increases the refractive index felt by the blue radiation. It is well known that the refractive index increase acts as an attractive potential for diffracting optical beams. According to the principle of spatio-temporal analogy in paraxial optics, diffraction of beams is analogous to the anomalous GVD of pulses. However, GVD of the blue radiation is normal. Therefore the latter is reflected from (not attracted towards) the effective potential barrier created by the accelerating soliton. Acceleration of the potential is equivalent to presence of an effective constant force or in our context to the linear increase of the refractive index, superimposed on the index hump created by the soliton, see Fig. 3.3a, and [44, 45]. Thus the radiation emitted by the soliton gets reflected from the linear potential back towards the soliton and reflects from the soliton again. This bouncing continues with propagation implying trapping of the radiation. Every collision of the radiation wave packet with the soliton is accompanied by the blue shift of the scattered radiation [43]. This frequency transformation is in its nature a FWM process, which is disabled for solitons interacting with dispersive waves in the ideal NLS equation, but becomes activated by the strong higher order dispersions typical for PCFs [43, 66]. The cascade of distinguishable FWM events converges to a continuous intrapulse FWM process [43]. The emerging waveform is a two-color soliton-radiation bound state, with a red shifting soliton part and a blue shifting trapped radiation component [44, 45, 74, 75]. Dispersive spreading of the radiation and its nonlinear phase modulation by the strong red shifting soliton (cross-phase modulation) are both suppressed

3 Nonlinear Optics and Solitons in Photonic Crystal Fibres   45

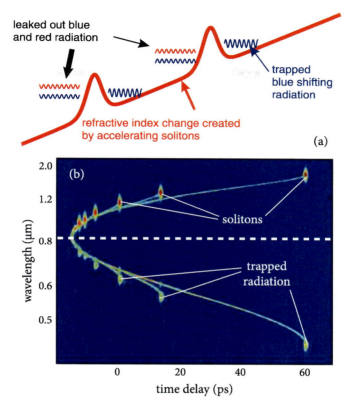

Fig. 3.3. Trapping of the blue radiation and supercontinuum formation in PCFs with positive 3rd order dispersion. (**a**) Refractive index profile created by the accelerating solitons for the radiation. (**b**) Computed spectrogram for a typical supercontinuum experiment showing formation of the bound soliton-radiation states. Doted line marks the zero GVD wavelength. Data used for (**b**) are from [44]

by the trapping mechanism [44, 45]. Obviously the radiation trapped by the soliton is delayed together with the latter. However, for normal GVD the group velocity decreases towards shorter wavelengths, which is consistent with the blue shift of the radiation.

It is important to realize that solitons can robustly trap radiation at short wavelengths, provided that both are group-velocity matched. As the group delay at short wavelengths is mainly determined by the material dispersion of silica and is not changed by the fibre structure, a practical fibre design counter intuitively needs to engineer the group delay of the longest wavelengths in the continuum in order to broaden the spectrum on the short wavelength side. The trapping of blue shifting radiation by red shifting solitons can be clearly seen on the time-frequency spectrograms, see Fig. 3.3b. Since the soliton creates

46   D.V. Skryabin and W.J. Wadsworth

only finite potential barrier a small part of the blue radiation leaks through it forming dispersive wave joining all the trapped radiation bunches in Fig. 3.3b. Raman shifted solitons are themselves constantly shaking off some radiation even without presence of the higher order dispersions [76]. This radiation is not reflected by the solitons (because its GVD is anomalous) and it fills the space between them, see Fig. 3.3a, 3.3b.

3.3.4 Red Shifted Radiation and Soliton Self-frequency Shift Cancelation

In the opposite case, where the GVD is decreasing with wavelength (negative third order dispersion), the resonant radiation is red detuned from the soliton. The radiation escapes from the soliton on the side where the linear potential due to acceleration decreases, i.e. emitted wave is faster than the soliton. Hence the radiation can not get trapped by the slowing soliton, see Fig. 3.4. However, the radiation is amplified with propagation in this case, because the Raman effect pulls the soliton closer and closer to the zero GVD point [60]. Eventually the radiation gets so strong that the soliton recoils against it. As a result the soliton acceleration and the frequency shift drop significantly and nearly disappear [60, 77, 78], see Fig. 3.4. The powerful

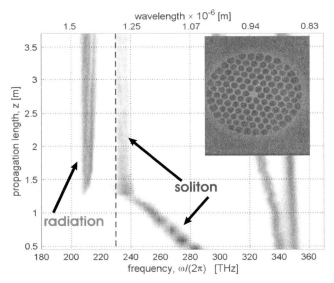

Fig. 3.4. Experimentally measured spectral evolution along the fibre length showing emission and amplification of the red shifted resonant radiation, which recoil on the soliton compensates the soliton self-frequency shift. Inset shows the fibre used in the experiments. Dashed line marks the zero GVD wavelength. Reprinted from [60]

3 Nonlinear Optics and Solitons in Photonic Crystal Fibres   47

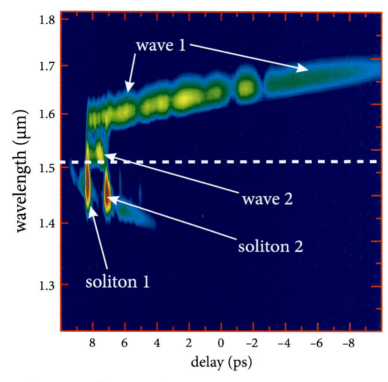

Fig. 3.5. Experimentally measured spectrogram showing emission of the strong red radiation and radiation trapping effect in PCFs with negative 3rd order dispersion. The wave 1 of the resonant radiation is emitted by the soliton 1. Soliton 2 reflects part of this wave, traps it on its tail and simultaneously transforms frequency of the reflected radiation creating the wave 2. Doted line marks the zero GVD wavelength. Reprinted from [77]

radiation wave reflects from the second soliton, see Fig. 3.5. The frequency of the reflected wave blue shifts and the wave itself faces the rising linear potential and gets trapped, see Fig. 3.5. The frequency of the trapped radiation increases with propagation, while the soliton frequency decreases. Trapping and frequency conversion mechanisms are exactly as described above. The difference here is, however, that initially the radiation has smaller frequency than the soliton. It implies that the frequencies of the soliton and trapped radiation are converging with propagation along the fibre. Therefore the generated spectrum fills the middle part of the continuum not its edges. This scenario is shown on the spectrogram in Fig. 3.5.

3.3.5 Other Soliton Effects in Solid-core PCFs

Such rich zoology of the interactions of solitons with dispersive waves appearing in practically important experimental settings creates fertile ground for further fundamental and applied studies in this area. Moreover, the solid core PCFs have been and are planned to be used for realization of variety of other ideas involving optical solitons. Low loss (0.3 dB/km) solid-core PCFs have been used for soliton propagation over hundreds of kilometers in dispersion managed transmission lines [79]. Amongst the soliton related research which is likely to be within experimental reach soon are the spatio-temporal nonlinear effects in multi-core fibres. Fibres potentially suitable for this purpose have been fabricated, see, e.g. [80, 81], but no experiments with spatially discrete solitons similar to the ones accomplished in planar waveguide arrays and photonic lattices in photorefractives [82] have been reported so far. A strong drive towards this goal exists because the fibres will allow us to observe interplay of temporal and spatial degrees of freedom, see, e.g., recent experiments with two-core PCFs [83]. An interesting alternative to using multiple silica cores could be filling the holes of photonic crystal fibres with nonlinear polymers or liquids having large refractive index, so that the liquid channels become waveguides in the silica surrounding [84–86].

3.4 Pulse Compression in PCFs

Once a soliton is formed in a fibre, it is very robust. Gradual changes in power (from fibre attenuation), wavelength (from the soliton self-frequency shift), dispersion and nonlinearity (from deliberate nonuniformity in the fibre) will not cause the soliton pulse to break up. Rather the available energy will be redistributed in time and frequency to yield a fundamental soliton under the local conditions as the soliton propagates along the fibre. This robustness has been used for adiabatic soliton compression in both SC- [87] and HC-PCF [88]. High order solitons exhibit pulse compression in the first stages of propagation [1], which has been successfully demonstrated in both solid-core [56] and HC-PCF [89]. Non-solitonic compression of pre-chirped pulses by factors 20 to 80 in HC-PCFs has been demonstrated in [14]. In these applications PCF provides two unique properties, firstly it can be designed with the necessary anomalous dispersion at the pump wavelength and secondly HC-PCF has a sufficiently low nonlinearity for high energy fundamental solitons, up to $0.5\,\mu$J. Pulse compression has also been achieved by spectral broadening by SPM [90] or supercontinuum generation [31] with subsequent compensation for the chirp or spectral phase variations using dispersive fibres [90] or adaptive phase control [91]. Pulses as short as 5 fs have been generated in these ways from 12 to 30 fs input pulses [56, 91], and temporal compression ratios more than 10 [87]. In all of these compression schemes the ultimate limit on the short pulse duration is higher order dispersion and the Raman shift.

3.5 Nonlinear and Quantum Optics in Hollow-core PCFs

One of the initial stimuli driving research into HC-PCFs has been development of waveguides where the guided light has only small overlap with solid materials and therefore nonlinearity and loss are minimized. However, currently opportunities offered by HC-PCFs for research into new nonlinear and quantum optical effects are becoming a dominant theme. Indeed instead of using gas cells or capillaries one can fill a HC-PCF with a required gas and conduct experiments on nonlinear and quantum optics in an environment where a constant intensity of the diffraction free beam is sustained along the desirable length. The fibre also can be bent so that the entire device can be made compact. Using an enhancement factor $\lambda l/a^2$ [92], where l is the constant intensity interaction length and a is the beam radius, it has been demonstrated that the low loss HC-PCFs can beat capillary guides by the factor $\sim 10^6$ [13]. This remarkable efficiency boost is due to the fact that in hollow capillary waveguides l is limited by loss and scales as $l \sim a^3/\lambda^2$ [93], while in high quality PCFs l practically equals the fibre length. This advantage of PCFs has been used to dramatically reduce the threshold of stimulated Raman scattering [13, 94, 95]. First, vibrational Raman lines in hydrogen have been generated in a relatively high loss (but wide bandwidth) HC-PCF, beating the previous best pump threshold by the factor of 100 [94] and later a low loss PCF with a narrow bandwidth has been used to demonstrate one million times threshold reduction in generation of rotational Raman lines [13]. Driving the Raman transition away from the resonance makes the effective Kerr nonlinearity dominate strongly over the nonlinear gain and loss. This effective nonlinearity is about two orders of magnitude stronger than the intrinsic Kerr nonlinearity of the HC-PCFs [96] and, in combination with the fibre dispersion, can be used for various nonlinear applications [97] and observation of new types of solitons [96].

HC-PCFs represent potentially useful structures linking photonics and telecom applications with quantum and atom optics [95, 98–101]. Acetylene and rubidium filled HC-PCFs have been used to demonstrate electromagnetically induced transparency [98, 99] (see Fig. 3.6), slow light propagation [98] and increased resolution of the saturated absorption signal [100]. Acetylene has a very weak oscillator strength of the molecular transition and strong enhancement of light-matter interaction in PCFs promotes practical applications of acetylene spectroscopy. It has been recently demonstrated that the SC-PCF having a ≈ 200 nm air hole running along its center supports guided modes with pronounced intensity peak in the air core [101]. This geometry promises further enhancement of light interaction with gases. The idea of light confinement in air by periodic dielectric structures has been recently applied for planar optical chips [102], where the micron-sized waveguides with periodic cladding and cores filled with rubidium have been created and used to demonstrate on-chip atomic spectroscopy.

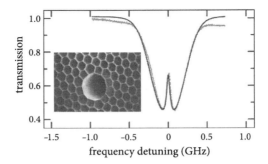

Fig. 3.6. Electromagnetically induced transparency signal in acytelene filled hollow-core PCF (see inset). Reprinted from [98]

The soliton energies in SC-PCFs are rather low, corresponding to around 1mW from a typical 100fs oscillator. This is not acceptable for many practical applications requiring high energy delivery. However, higher energy solitons can be observed in HC-PCF. Femtosecond solitons with megawatt peak powers propagating in air-core PCFs and suitable for medical and micro-machining applications have been demonstrated in Refs. [11, 12, 103–105]. The dispersion in HC-PCFs is anomalous at the red side of the transmission band, whatever the wavelength of the guidance band, because of the presence of the bandgap edge [5, 103] and focusing nonlinearity comes from combined contribution of the glass cladding and gas in the core. Solitons in HC-PCFs carry energy of one to few hundreds nJ, corresponding to the output from low energy or high repetition rate amplified laser pulses, or the pulse energies for amplified fibre lasers. There is some simple tunability of the soliton energy attainable through fibre design (overlap of the guided mode with the glass) [10, 104, 106], or through evacuating the hollow core to remove the nonlinearity of the air, or by increasing the core pressure to add nonlinearity. The soliton frequency also can be tuned by the intrapulse Raman scattering, which can be a combination of glass and gas contributions [11].

3.6 Summary

Nonlinear and quantum optical effects in the waveguides enhancing light-matter interaction are going to be the driving force behind developments in photonics for years to come. Classical applications of optical nonlinearities for light manipulation and emerging applications for processing and transmission of quantum information will continue to benefit from and feedback onto research into micro- and nano-structured waveguides, where photonic-crystal fibers is an established and still very promising player.

References

1. G.P. Agrawal, *Nonlinear Fiber Optics*, Academic Press (2001)
2. J.C. Knight, T.A. Birks, P.S. Russell, and D.M. Atkin, Opt. Lett. **21**, 1547 (1996)
3. R.F. Cregan, B.J. Mangan, J.C. Knight, T.A. Birks, P.S. Russell, P.J. Roberts, and D.C. Allan, Science **285**, 1537 (1999)
4. P. Russell, Science **299**, 358 (2003)
5. C.M. Smith, N. Venkataraman, M.T. Gallagher, D. Muller, J.A. West, N.F. Borrelli, D.C. Allan, and W.W. Koch, Nature **424**, 657 (2003)
6. J.K. Ranka, R.S. Windeler, and A.J. Stentz, Opt. Lett. **25**, 25 (2000)
7. V. Finazzi, T.M. Monro, and D.J. Richardson, J. Opt. Soc. Am. B **20**, 1427 (2003)
8. F.G. Omenetto, N.A. Wolchover, M.R. Wehner, M. Ross, A. Efimov, A.J. Taylor, V.V.R.K. Kumar, A.K. George, J.C. Knight, N.Y. Joly, and P.S.J. Russell, Opt. Express **14**, 4928 (2006)
9. P. Petropoulos, H. Ebendorff-Heidepriem, V. Finazzi, R.C. Moore, K. Frampton, D.J. Richardson, and T.M. Monro, Opt. Express **11**, 3568 (2003)
10. P.J. Roberts, D.P. Williams, B.J. Mangan, H. Sabert, F. Couny, W.J. Wadsworth, T.A. Birks, J.C. Knight, and P.S.J. Russell, Opt. Express **13**, 8277 (2005)
11. D.G. Ouzounov, F.R. Ahmad, D. Muller, N. Venkataraman, M.T. Gallagher, M.G. Thomas, J. Silcox, K.W. Koch, and A.L. Gaeta, Science **301**, 1702 (2003)
12. G. Humbert, J.C. Knight, G. Bouwmans, P.S. Russell, D.P. Williams, P.J. Roberts, and D.J. Mangan, Opt. Express **12**, 1477 (2004)
13. F. Benabid, G. Bouwmans, J.C. Knight, P. St Russell, and F. Couny, Phys. Rev. Lett. **93**, 123903 (2004)
14. C.J.S. de Matos, S.V. Popov, A.B. Rulkov, J.R. Taylor, J. Broeng, T.P. Hansen, and V.P. Gapontsev, Phys. Rev. Lett. **93**, 103901 (2004)
15. S. Ghosh, A.R. Bhagwat, C.K. Renshaw, S. Goh, A.L. Gaeta, and B.J. Kirby, Phys. Rev. Lett. **97**, 023603 (2006)
16. S. Smirnov, J.D. Ania-Castanon, T.J. Ellingham, S.M. Kobtsev, S. Kukarin, and S.K. Turitsyn, Opt. Fiber Techn. **12**, 122 (2006)
17. J.M. Dudley, G. Genty, and S. Coen, Rev. Mod. Phys. **78**, 1135 (2006)
18. A. Zheltikov, J. Opt. A **8**, S47 (2006)
19. P. Russell, J. Light-Wave Tech. **24**, 4729 (2006)
20. R.R. Alfano and S.L. Shapiro, Phys. Rev. Lett. **24**, 584 (1970)
21. W.L. Smith, P. Liu, and N. Bloembergen, Phys. Rev. A **15**, 2396 (1977)
22. R.L. Fork, C.V. Shank, C. Hirlimann, R. Yen, and W.J. Tomlinson, Opt. Lett. **8**, 1 (1983)
23. W.J. Wadsworth, A. Ortigosa-Blanch, J.C. Knight, T.A. Birks, T.P.M. Man, and P.S. Russell, J. Opt. Soc. Am. B **19**, 2148 (2002)
24. J.C. Knight, J. Arriaga, T.A. Birks, A. Ortigosa-Blanch, W.J. Wadsworth, and P.S. Russell, IEEE Photon. Technol. Lett. **12**, 807 (2000)
25. W.J. Wadsworth, N. Joly, J.C. Knight, T.A. Birks, F. Biancalana, and P.S.J. Russell, Opt. Express **12**, 299 (2004)
26. G. Humbert, W.J. Wadsworth, S.G. Leon-Saval, J.C. Knight, T.A. Birks, P.S.J. Russell, M.J. Lederer, D. Kopf, K. Wiesauer, E.I. Breuer, and D. Stifter, Opt. Express **14**, 1596 (2006)

27. D.J. Jones, S.A. Diddams, J.K. Ranka, A. Stentz, R.S. Windeler, J.L. Hall, and S.T. Cundiff, Science **288**, 635 (2000)
28. R. Holzwarth, T. Udem, T.W. Hansch, J.C. Knight, W.J. Wadsworth, and P.S.J. Russell, Phys. Rev. Lett. **85**, 2264 (2000)
29. J.M. Dudley and S. Coen, Opt. Lett. **27**, 1180 (2002)
30. X. Gu, M. Kimmel, A.P. Shreenath, R. Trebino, J.M. Dudley, S. Coen, and R.S. Windeler, Opt. Express **11**, 2697 (2003)
31. G.Q. Chang, T.B. Norris, and H.G. Winful, Opt. Lett. **28**, 546 (2003)
32. J. Herrmann, U. Griebner, N. Zhavoronkov, A. Husakou, D. Nickel, J.C. Knight, W.J. Wadsworth, P.S.J. Russell, and G. Korn, Phys. Rev. Lett. **88**, 173901 (2002)
33. B. von Vacano, W. Wohlleben, and M. Motzkus, Opt. Lett. **31**, 413 (2006)
34. C. Dunsby, P.M.P. Lanigan, J. McGinty, D.S. Elson, J. Requejo-Isidro, I. Munro, N. Galletly, F. McCann, B. Treanor, B. Onfelt, D.M. Davis, M.A.A. Neil, and P.M.W. French, J. Phys. D **37**, 3296 (2004)
35. H.N. Paulsen, K.M. Hilligsoe, J. Thogersen, S.R. Keiding, and J.J. Larsen, Opt. Lett. **28**, 1123 (2003)
36. S. Coen, A.H.L. Chan, R. Leonhardt, J.D. Harvey, J.C. Knight, W.J. Wadsworth, and P.S.J. Russell, Opt. Lett. **26**, 1356 (2001)
37. M. Seefeldt, A. Heuer, and R. Menzel, Opt. Commun. **216**, 199 (2003)
38. P.A. Champert, V. Couderc, P. Leproux, S. Fevrier, V. Tombelaine, L. Labonte, P. Roy, C. Froehly, and P. Nerin, Opt. Express **12**, 4366 (2004)
39. S.G. Leon-Saval, T.A. Birks, W.J. Wadsworth, P.S.J. Russell, and M.W. Mason, Opt. Express **12**, 2864 (2004)
40. C. Xiong, A. Witkowska, S.G. Leon-Saval, T.A. Birks, and W.J. Wadsworth , Opt. Express **14**, 6188 (2006)
41. J.C. Travers, S.V. Popov, and J.R. Taylor, Opt. Lett. **30**, 3132 (2005)
42. A. Kudlinski, A.K. George, J.C. Knight, J.C. Travers, A.B. Rulkov, S.V. Popov, and J.R. Taylor, Opt. Express **14**, 5715 (2006)
43. A.V. Gorbach, D.V. Skryabin, J.M. Stone, and J.C. Knight, Opt. Express **14**, 9854 (2006)
44. A.V. Gorbach and D.V. Skryabin, Nature Photonics **1**, 653 (2007)
45. A.V. Gorbach and D.V. Skryabin, Phys. Rev. A **76**, 053803 (2007)
46. R. Borlinghaus, H. Gugel, P. Albertano, and V. Seyfried, Three-Dimensional and Multidimensional Microscopy: Image Acquisition and Processing XIII, 6090:T900 (2006)
47. W.H. Reeves, D.V. Skryabin, F. Biancalana, J.C. Knight, P.S. Russell, F.G. Omenetto, A. Efimov, and A.J. Taylor, Nature **424**, 511 (2003)
48. F. Biancalana and D.V. Skryabin, J. Opt. A **6**, 301 (2004)
49. J.G. Rarity, J. Fulconis, J. Duligall, W.J. Wadsworth, and P.S. Russell, Opt. Express **13**, 534 (2005)
50. M. Fiorentino, P.L. Voss, J.E. Sharping, and P. Kumar, IEEE Phot. Techn. Lett. **14**, 983 (2002)
51. J. Fan and A. Migdall, Opt. Express **13**, 5777 (2005)
52. S.G. Murdoch, R. Leonhardt, and J.D. Harvey, Opt. Lett. **20**, 866 (1995)
53. J.S.Y. Chen, G.K.L. Wong, S.G. Murdoch, R.J. Kruhlak, R. Leonhardt, J.D. Harvey, N.Y. Joly, and J.C. Knight, Opt. Lett. **31**, 873 (2006)
54. Q. Lin, F. Yaman, and G.P. Agrawal, Phys. Rev. A **75**, 023803 (2007)
55. P. Dainese, P.S.J. Russell, N. Joly, J.C. Knight, G.S. Wiederhecker, H.L. Fragnito, V. Laude, and A. Khelif, Nature Physics **2**, 388 (2006)

56. M. Foster, A. Gaeta, Q. Cao, and R. Trebino, Opt. Express **13**, 6848 (2005)
57. P. Beaud, W. Hodel, B. Zysset, and H.P. Weber, IEEE J. Quant. Electr. **23**, 1938 (1987)
58. R. Trebino, *Frequency-Resolved Optical Gating: The measurement of Ultrashort Laser Pulses*, Kluwert (2000)
59. W.J. Wadsworth, J.C. Knight, A. Ortigosa-Blanch, J. Arriaga, E. Silvestre, and P.S.J. Russell, Elect. Lett. **36**, 53 (2000)
60. D.V. Skryabin, F. Luan, J.C. Knight, and P.S. Russell, Science **301**, 1705 (2003)
61. E.E. Serebryannikov, A. M. Zheltikov, S. Köhler, N. Ishii, C.Y. Teisset, T. Fuji, F. Krausz, and A. Baltuska, Phys. Rev. E **73**, 066617 (2006)
62. F. Luan, A.V. Yulin, J.C. Knight, and D.V. Skryabin, Opt. Express **14**, 6550 (2006)
63. F. Luan, D.V. Skryabin, A. Yulin, and J.C. Knight, Opt. Express **14**, 9844 (2006)
64. A. Podlipensky, P. Szarniak, N.Y. Joly, C.G. Poulton, and P.S.J. Russell, Opt. Express **15**, 1653 (2007)
65. E.R. Andresen, V. Birkedal, J. Thogersen, and S.R. Keiding, Opt. Lett. **31**, 1328 (2006)
66. D.V. Skryabin and A.V. Yulin, Phys. Rev. E **72**, 016619 (2005)
67. P.K.A. Wai, H.H. Chen, and Y.C. Lee, Phys. Rev. A **41**, 426 (1990)
68. J.M. Dudley, X. Gu, L. Xu, M. Kimmel, E. Zeek, P. O'Shea, R. Trebino, S. Coen, and R.S. Windeler, Opt. Express **10**, 1215 (2002)
69. I. Cristiani, R. Tediosi, L. Tartara, and V. Degiorgio, Opt. Express **12**, 124 (2004)
70. G. Genty, M. Lehtonen, H. Ludvigsen, and M. Kaivola, Opt. Express **12**, 3471 (2004)
71. D.R. Austin, C.M. de Sterke, B.J. Eggleton, and T.G. Brown, Opt. Express **14**, 11997 (2006)
72. G. Genty, M. Lehtonen, and H. Ludvigsen, Opt. Express **12**, 4614 (2004)
73. M.H. Frosz, P. Falk, and O. Bang, Opt. Express **13**, 6181 (2005)
74. T. Hori, N. Nishizawa, T. Goto, and M. Yoshida, J. Opt. Soc. Am. B **21**, 1969 (2004)
75. N. Nishizawa and T. Goto, Opt. Express **8**, 328 (2001)
76. N. Akhmediev, W. Krolikowski, and A.J. Lowery, Opt. Commun. **131**, 260 (1996)
77. A. Efimov, A.J. Taylor, F.G. Omenetto, A.V. Yulin, N.Y. Joly, F. Biancalana, D.V. Skryabin, J.C. Knight, and P.S. Russell, Opt. Express **12**, 6498 (2004)
78. F. Biancalana, D.V. Skryabin, and A.V. Yulin, Phys. Rev. E **70**, 016615 (2004)
79. K. Kurokawa, K. Tajima, K. Tsujikawa, K. Nakajima, T. Matsui, I. Sankawa, and T. Haibara, J. Lightwave Technol. **24**, 32 (2006)
80. L. Li, A. Schülzgen, S. Chen, V.L. Temyanko, J.V. Moloney, and N. Peyghambarian, Opt. Lett. **31**, 2577 (2006)
81. U. Röpke, H. Bartelt, S. Unger, K. Schuster, and J. Kobelke, Opt. Express **15**, 6894 (2007)
82. D.N. Christodoulides, F. Lederer, and Y. Silberberg, Nature **424**, 817 (2003)
83. A. Betlej, S. Suntsov, K.G. Makris, L. Jankovic, D.N. Christodoulides, G.I. Stegeman, J. Fini, R.T. Bise, and D.J. DiGiovanni, Opt. Lett. **31**, 1480 (2006)

84. C. Kerbage, P. Steinvurzel, P. Reyes, P.S. Westbrook, R.S. Windeler, A. Hale, and B.J. Eggleton, Opt. Lett. **27**, 842 (2002)
85. S. Lebrun, P. Delaye, R. Frey, and G. Roosen, Opt. Lett. **32**, 337 (2007)
86. P. Lesiak, T.R. Wolinski, K. Brzdakiewicz, K. Nowecka, S. Ertman, M. Karpierz, A.W. Domanski, and R. Dabrowski, Opto-Electronics Review **15**, 27 (2007)
87. J.C. Travers, B.A. Cumberland, A.B. Rulkov, S.V. Popov, J.R. Taylor, J.M. Stone, A.K. George, and J.C. Knight, Conference On Lasers & Electro-Optics (CLEO), paper CFK1 (2007)
88. F. Gerome, K. Cook, A.K. George, W.J. Wadsworth, and J.C. Knight, Opt. Express **15**, 7126 (2007)
89. D.G. Ouzounov, C.J. Hensley, A.L. Gaeta, N. Venkateraman, M.T. Gallagher, and K.W. Koch, Opt. Express **13**, 6153 (2005)
90. R.E. Kennedy, S.V. Popov, and J.R. Taylor, Electron. Lett. **41**, 234 (2005)
91. B. Schenkel, R. Paschotta, and U. Keller, J. Opt. Soc. Am. B **22**, 687 (2005)
92. R.B. Miles, G. Laufer, and G.C. Bjorklun, Appl. Phys. Lett. **30**, 417 (1977)
93. S.O. Konorov, A.B. Fedotov, and A.M. Zheltikov, Opt. Lett. **28**, 1448 (2003)
94. F. Benabid, J.C. Knight, G. Antonopoulos, and P.S.J. Russell, Science **298**, 399 (2002)
95. F. Benabid, F. Couny, J.C. Knight, T.A. Birks, and P.S. Russell, Nature **434**, 488 (2005)
96. D.V. Skryabin, F. Biancalana, D.M. Bird, and F. Benabid, Phys. Rev. Lett. **93**, 143907 (2004)
97. I.V. Fedotov, A.B. Fedotov, and A.M. Zheltikov, Opt. Lett. **31**, 2604 (2006)
98. S. Ghosh, J.E. Sharping, D.G. Ouzounov, and A.L. Gaeta, Phys. Rev. Lett. **94**, 093902 (2005)
99. F. Couny, P.S. Light, F. Benabid, and P.S. Russell, Opt. Commun. **263**, 28 (2006)
100. J. Hald, J.C. Petersen, and J. Henningsen, Phys. Rev. Lett. **98**, 213902 (2007)
101. G.S. Wiederhecker, C.M.B. Cordeiro, F. Couny, F. Benabid, S.A. Maier, J.C. Knight, C.H.B. Cruz, and H.L. Fragnito, Nature Photonics **1**, 115 (2007)
102. W. Yang, D.B. Conkey, B. Wu, D. Yin, A.R. Hawkins, and H. Schmidt, Nature Photonics **1**, 331 (2007)
103. G. Bouwmans, F. Luan, J.C. Knight, P.S.J. Russell, L. Farr, B.J. Mangan, and H. Sabert, Opt. Express **11**, 1613 (2003)
104. F. Luan, J.C. Knight, P.S. Russell, S. Campbell, D. Xiao, D.T. Reid, B.J. Mangan, D.P. Williams, and P.J. Roberts, Opt. Express **12**, 835 (2004)
105. J.D. Shephard, J.D.C. Jones, D.P. Hand, G. Bouwmans, J.C. Knight, P.S. Russell, and B.J. Mangan, Opt. Express **12**, 717 (2004)
106. D.V. Skryabin, Opt. Express **12**, 4841 (2004)

4

Spatial Switching of Slow Light in Periodic Photonic Structures

Andrey A. Sukhorukov

Nonlinear Physics Centre and Centre for Ultrahigh-bandwidth Devices for Optical Systems (CUDOS), Research School of Physics and Engineering, Australian National University, Canberra, ACT 0200, Australia
ans124@rsphysse.anu.edu.au

4.1 Introduction

The speed of light sets the maximum possible rate for transmission of information, in excess of 10^8 meters per second. Light pulses in optical fibers carry bits of data around the world in sub-second time frame, enabling interactive global communications. There is a constant demand for increasing the network performance in view of steadily growing information flows. It is envisioned that the presently required multiple conversions between optical pulses and electronic signals at network hubs may be eliminated in future when routing and switching of data flows is performed all-optically. This vision can be realized if the speed of light is dynamically controlled, allowing for synchronization and multiplexing of signals. Furthermore, by temporarily making the light slower it becomes possible to compress optical signals and perform their manipulation in compact photonic chips. Additionally, in the regime of slow light the photon-matter interactions are dramatically enhanced, enabling the active control of light and nonlinear transformations of signals.

Slowing down the light is a challenging physical problem. In conventional dielectrics, the speed of light can only be reduced by a factor less than four which is limited by the optical refractive index of available materials. The most dramatic slowing down of light to a complete stop was reported in the regime of electromagnetically induced transparency [1]. This phenomenon is based on a resonant interaction of light with an atomic system and accordingly the speed of light is very sensitive to the frequency detuning. This restricts the effect to narrow frequency ranges limiting its applicability to communication networks with demands for data rates in excess of 100 Gb per second. In contrast, dielectric photonic structures with a periodic modulation of the optical refractive index at a sub-micrometer scale can be engineered to operate at any frequency range. Periodic modulation results in resonant light scattering, and reduction of pulse speed by more than 100 times was registered experimentally in photonic crystals [2–4]. Ultra-slow light propagation can also be realized based on the coupling between high-Q optical cavities [5–10]. However, in all static structures the maximum usable delay decreases for shorter pulses which have larger bandwidth and exhibit stronger distortions due to the effect of frequency dispersion [11]. It was suggested that the restriction on the delay-bandwidth product, the key parameter characterizing the device capacity to store optical signals [12], can be overcome in dynamically tunable structures, with the exciting possibility to completely stop and then release light pulses all-optically [13, 14].

Slow light was mostly studied in configurations where the propagation direction is fixed by the waveguide geometry. In contrast, novel ways to direct the flow of light can be realized in extended periodic photonic structures where the familiar rules of refraction and reflection may be manipulated, including such unconventional effects as negative refraction [15] in the direction opposite to normal and self-collimation [16] of beams. The potential for slow light propagation in two spatial dimensions was demonstrated experimentally [17] with plane-wave excitations, yet the effect of diffraction needs to be considered for tightly focused beams. On the other hand, in nonlinear periodic structures a light wave can alter the refractive index and thereby adjust its own velocity, enabling all-optical switching and beam steering [18]. Most importantly, the effect of nonlinearity is enhanced in the slow-light regime [19–21]. These considerations have motivated our investigations on all-optical manipulation of both the magnitude and the direction of the speed of light in nonlinear periodic structures.

In this Chapter, we present our recent theoretical results [22–24] demonstrating the potential for dynamically tunable slowing down and spatial switching of optical pulses in specially designed nonlinear photonic structures. In Sec. 4.2, we overview the methods for the reduction of the speed of light based on dispersion control in periodic structures. In Sec. 4.3, we introduce the photonic structures in the form of nonlinear Bragg-grating and photonic-crystal couplers that can be used to simultaneously slow down the pulses and redirect them between the output ports. Finally, in Sec. 4.4 we discuss pulse routing in defect-free extended periodic structures in the form of nonlinear Bragg-grating waveguide arrays, where the propagation direction is selected by the internal phase structure of the optical pulse.

4.2 Dispersion and Tuning of the Speed of Light in Nonlinear Periodic Structures

Dielectric structures with micro-scale modulations of the refractive index can behave as metamaterials with unconventional characteristics. These structures offer new possibilities for control over the fundamental properties of electromagnetic waves, including the tuning of the group velocity defining the speed of optical pulses. One of the mechanisms that can slow down optical waves is Bragg scattering from periodic inhomogeneities of the refractive

4 Spatial Switching of Slow Light in Periodic Photonic Structures 57

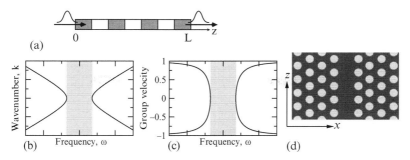

Fig. 4.1. (a) Schematic of a Bragg-grating structure, and the corresponding (b) dispersion and (c) group velocity dependencies. Shading in plots (b) and (c) marks the frequency band-gap. (d) Sketch of a photonic-crystal waveguide, where circles indicate the regions with reduced values of the optical refractive index

index [25], such as a Bragg grating shown schematically in Fig. 4.1a. In periodic structures, the wave spectrum contains band-gaps at certain frequency ranges, similar to energy gaps experienced by electrons in crystalline potentials (Fig. 4.1b). In periodic structures where the propagation direction is fixed by the waveguide geometry, the dispersion relation between the optical frequency (ω) and the Bloch wavenumber (k) in the vicinity of band-edges is commonly expressed as $\omega \simeq \omega_0 + D_2(k-k_0)^2$, where the ω_0 and k_0 denote the values at the band-edge, and D_2 is the second-order dispersion coefficient. The group velocity is found as $V_g = d\omega/dk \simeq \pm 2[D_2(\omega-\omega_0)]^{1/2}$, and it gradually reduces to zero as the frequency is tuned towards the edge of a transmission band (Fig. 4.1c). Recent experiments [2–4] reported the reduction of the group velocity by factors exceeding 100 in photonic crystal waveguides, such as the W1 waveguide schematically shown in Fig. 4.1d. In optical fibers with Bragg gratings the experimentally demonstrated slow-down factor [26,27] is smaller (up to 3), yet the overall pulse delay can be significant due to a much longer propagation distance.

An optical pulse occupies a spectral region that is inversely proportional to its duration, $\delta\omega \sim \tau^{-1}$. Propagating pulses tend to broaden due to dispersion which appears because the speed of light depends on frequency. The sensitivity to frequency detuning is inversely proportional to the group velocity, $dV_g/d\omega \simeq 2D_2/V_g$. Therefore, the pulse distortion is especially strong for slow light in the vicinity of band edges [28]. It was suggested that dispersion can be suppressed in specially designed structures, supporting propagation of broadband slow light inside the transmission band away from the gap edges, where the dispersion curve contains a point with $D_2 = 0$. In this case, the pulse distortion is defined by higher-order dispersion effects, which can also be minimized [29–36].

An important aspect of slow-light systems is their tunability. Let us consider the effect of the overall modification of the refractive index of the dielectric material or special inclusions [37], which can be induced through the electro-optic, thermal, photorefractive, or other mechanisms. Such a change can be understood, in the first-order approximation, as an effective shift of band-gaps and transmission bands by $\delta\omega$. The sensitivity to such tuning of the phase velocity ($V_{\rm p} = \omega/k$), which defines the characteristics of phase-sensitive devices such as Mach-Zehnder interferometer, is enhanced by a factor inversely proportional to the group velocity, allowing one to construct very compact optical switches [2, 20]. For communication networks, the capability to dynamically adjust the group velocity is desirable for the realization of tunable pulse delays. Such tunability may be realized when the group velocity is sensitive to the effective shift of band-gaps, i.e. $dV_{\rm g}/d\omega \neq 0$. However, in this case the pulses inevitably experience dispersion-induced broadening, and there appears a fundamental limit on the delay-bandwidth product [11].

The delay-bandwidth product can be made infinitely large if the refractive index is modified in local regions at the same time as the pulse is propagating through the photonic structure [38]. In a recent experiment, the pulse trapping and subsequent release was controlled by an external pump [14], and the signal distortion was minimized. Such scheme requires precise synchronization between the signal and pump waves. On the other hand, the need for a pump wave may be avoided in nonlinear media, where the signal pulse can itself induce the refractive index change.

It was shown that the simultaneous tuning of the propagation velocity and suppression of dispersion-induced pulse broadening can be realized in media with fast nonlinear response, where pulse may change its own propagation. The nonlinear pulse self-action in Bragg gratings can result in the formation of gap solitons [39], which envelope profiles remain undistorted as they propagate through the photonic structure (see sketches in Fig. 4.2). On the other hand, the propagation velocity of gap solitons can be theoretically reduced down to zero, depending on the excitation conditions. Most recent experimental observations have demonstrated that the gap-soliton velocity may be tuned by varying the optical power [27]. Additionally, the efficiency of light coupling into the photonic structure can be improved in the nonlinear regime.

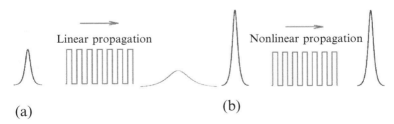

Fig. 4.2. Schematic illustration of pulse propagation in the slow-light regime though a nonlinear periodic structure, such as Bragg-grating waveguide: (**a**) Broadening of pulse with small peak intensity due to the linear group-velocity dispersion, and (**b**) Nonlinear dispersion compensation and formation of a gap soliton when the pulse energy is increased

4.3 Slow-light Switching in Waveguide Couplers

Directional waveguide coupler has attracted a great deal of attention as a major candidate for creation of ultra-fast all-optical switches [40–42]. This device utilizes light tunneling between two optical waveguides placed in close proximity to each other. In the linear regime, light is switched from one to another waveguide at the distance called coupling length. At high input powers, intensity-dependent change of the refractive index through optical non-linearity creates detuning between the waveguides which can suppress power transfer between coupler arms, such that light remains in the input waveguide. In the following, we describe the configuration of waveguide couplers enabling the switching of slow-light pulses between the output ports. In Sec. 4.3.1, we describe the potential of nonlinear Bragg-grating couplers for all-optical switching of slow-light pulses, combined with delay tuning and dispersion compensation. Then, we show in Sec. 4.3.2 that the routing of slow-light pulses is also possible in specially designed photonic-crystal couplers, where the switching distance can be reduced by several orders of magnitude compared to Bragg-grating structures.

4.3.1 All-optical Switching in Bragg-grating Couplers

We consider the pulse propagation along two parallel waveguides, where each waveguide contains a Bragg grating. Such structures were previously suggested for the applications in mode conversion and add-drop filtering [43–47], and their dynamical tuning through the nonlinearly-induced shift of Bragg resonance was demonstrated [48]. It was also shown that stationary gap solitons can exist in the nonlinear regime [49–51]. In the following, we reveal the potential of such structures for all-optical switching and manipulation of slow-light pulses [23].

The concept of the conventional coupler [40–42] is based on the effect of complete tunneling of light between the waveguides in the linear regime. It is therefore essential to achieve the same kind of tunneling in the slow-light regime, when the optical frequency is tuned in the vicinity of the band-edge associated with the resonant Bragg-reflection from the periodic grating. The pulse dynamics under such conditions can be modeled by a set of coupled-mode equations [52] for the normalized slowly varying envelopes of the forward (u_n) and backward (w_n) propagating modes in each of waveguides $n = 1, 2$,

$$\mathrm{i}\frac{\partial u_n}{\partial t} + \mathrm{i}\frac{\partial u_n}{\partial z} + C u_{3-n} + \rho_n w_n + \gamma(|u_n|^2 + 2|w_n|^2)u_n = 0 \,, \quad (4.1)$$

$$\mathrm{i}\frac{\partial w_n}{\partial t} - \mathrm{i}\frac{\partial w_n}{\partial z} + C w_{3-n} + \rho_n^* u_n + \gamma(|w_n|^2 + 2|u_n|^2)w_n = 0 \,, \quad (4.2)$$

where t and z are the dimensionless time and propagation distance normalized to t_s and z_s, respectively, C is the coupling coefficient for the modes of the neighboring waveguides, ρ_n characterizes the amplitude and phase of the Bragg gratings, γ is the nonlinear coefficient, and the group velocity far from the Bragg resonance is normalized to unity. The scaling coefficients are $t_s = \lambda_0^2 |\rho_1|/(\pi c \Delta\lambda_0)$ and $z_s = t_s c/n_0$, where c is the speed of light in vacuum, λ_0 is the wavelength in vacuum, $\Delta\lambda_0$ is the width of Bragg resonance for an individual waveguide, n_0 is the effective refractive index in the absence of a grating. To be specific, in numerical examples we set $\gamma = 10^{-2}$, $\lambda_0 = 1550.63$ nm, $\Delta\lambda_0 = 0.1$ nm, $t_s \simeq 12.8$ ps, $z_s \simeq 1.8$ mm corresponding to characteristic parameters of fiber Bragg gratings [27, 46] with refractive index contrast $\Delta n \simeq 1.3 \times 10^{-4}$.

We consider the case of identical waveguides and analyze the effect of a *phase shift* (φ) between the otherwise equivalent waveguide gratings with $\rho_1 = \rho$ and $\rho_2 = \rho \exp(i\varphi)$ (with no loss of generality, we take ρ to be real and positive), see schematic illustrations in Figs. 4.3a, 4.3b. It was shown that the grating shift can strongly modify the reflectivity of modes with different symmetries [44, 46, 47], and we investigate how this structural parameter affects the properties of slow-light modes.

In the linear regime, wave propagation is fully defined through the Floquet-Bloch eigenmode solutions of the form, $u_n = U_n \exp(ikz - i\omega t)$, $w_n = W_n \exp(ikz - i\omega t)$. After substituting these expressions into the linearized coupler equations (4.1) and (4.2) (with $\gamma = 0$), we obtain the dispersion relation between the frequency ω and the corresponding wavenumber k, $\omega^2(k) = k^2 + C^2 + |\rho|^2 \pm 2\, C[k^2 + |\rho|^2 \cos^2(\varphi/2)]^{1/2}$.

Slow-light propagation can be observed due to the reduction of the normalized group velocity ($V_g = d\omega/dk$) when the pulse frequency is tuned close to

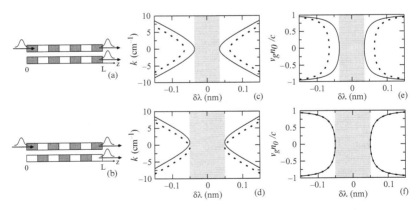

Fig. 4.3. (a), (b) Schematic of directional couplers with (a) in-phase ($\rho_1 = \rho_2 = 0.5$) or (b) out-of-phase ($\rho_1 = -\rho_2 = 0.5$) Bragg gratings. (c), (d) Characteristic dispersion, and (e), (f) normalized group velocity dependence on wavelength detuning for the case of in-phase ((c) and (e)) and out-of-phase ((d) and (f)) gratings. For all the plots $C \simeq 0.144$. Solid and dotted lines mark different branches, note their overlap in (f)

the bandgap edge. We find that different regimes of slow light can be realized depending on the structural parameters.

(i) If $|\rho\cos(\varphi/2)/C| > 1$, the bandgap appears for $\omega^2 < \omega_g^2 = C^2 + |\rho|^2 - 2C|\rho\cos(\varphi/2)|$, and only a single forward propagating mode (with $V_g > 0$) exists for the frequencies near the gap edges. This situation can be realized for in-phase gratings with $\varphi = 0$, see Figs. 4.3c, 4.3e.

(ii) If $|\rho\cos(\varphi/2)/C| < 1$, the bandgap appears for $|\omega| < \omega_g = |\rho\sin(\varphi/2)|$, and two types of the forward propagating modes (with $V_g > 0$) exist simultaneously (in the regions with $k > 0$ and $k < 0$) for the frequencies arbitrarily close to the gap edges. This situation is always realized for out-of-phase gratings with $\varphi = \pi$, see Figs. 4.3d, 4.3f.

We now analyze linear propagation of pulses in a semi-infinite Bragg grating coupler. When the optical frequency is detuned from the bandgap, light periodically tunnels between the waveguides with the characteristic period $L_c \simeq \pi/(2C)$ defined for a conventional coupler without the Bragg grating, see examples in Figs. 4.4a and 4.4b. The periodic tunneling appears due to the beating of even and odd modes, which correspond to different branches of the dispersion curves. When the pulse frequency is tuned closer to the gap edge and (i) only one slow mode is supported, then periodic beating disappears and light is equally distributed between the waveguides irrespective of the input excitation, see Figs. 4.4c and 4.4e. Note that the light intensity at the boundary of the second waveguide is non-zero due to the strong reflection of forward-propagating even mode. The periodic coupling can only be sustained in the slow-light regime when (ii) two modes co-exist at the gap edge, see Figs. 4.4d and 4.4f. Therefore, the configuration with out-of-phase shifted

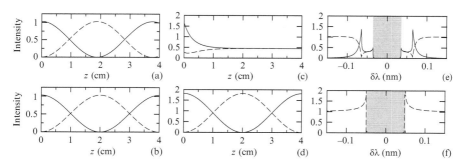

Fig. 4.4. Linear transmission of incident wave coupled to the first waveguide of a semi-infinite ($z \geq 0$) coupler with in-phase (*top*) and out-of-phase (*bottom*) Bragg gratings: (a)–(d) Intensity distribution (averaged over grating period) shown in the first (solid line) and second (dashed line) waveguides for (**a**), (**b**) large frequency detuning from the resonance and (**c**), (**d**) frequency tuned close to the band edge with slow group velocity $V_g = 0.1$. (**e**), (**f**) Intensities at $z = 2$ cm vs. wavelength detuning. Parameters correspond to Fig. 4.3, and the intensities are normalized to the input intensity

Bragg gratings is the most preferential for switching of slow-light pulses, since for $\varphi = \pi$ the dispersion of the type (ii) is always realized for any values of the grating strength and the waveguide coupling, and simultaneously the bandgap attains the maximum bandwidth.

At higher optical powers, nonlinear effects become important, and we perform numerical simulations of equations (4.1) and (4.2) to model pulse propagation. We find that the optimal regime of spatio-temporal control can be realized, in particular, when the structure size is equal to three coupling lengths, $L = 3\,L_c$. In the linear regime, the pulse tunnels three times between the waveguides and switches accordingly to the other waveguide at the output, however, it significantly broadens due to the group-velocity dispersion, see Fig. 4.5a. As the input pulse energy is increased, nonlinearity may support

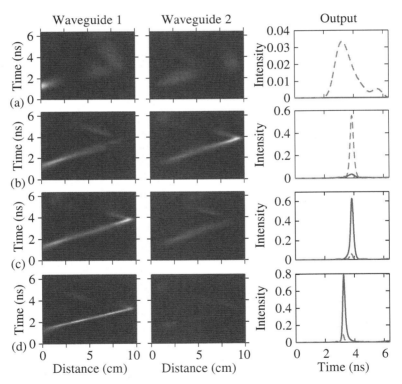

Fig. 4.5. (a)–(d) Pulse dynamics inside the nonlinear coupler for different values of the normalized peak input intensities $I_0 = 10^{-4}, 3.33, 3.37, 4$. Shown are the density plots of intensity in the first (*left column*) and second (*middle column*) waveguides. Output intensity profiles normalized to I_0 at the first (solid line) and second (dashed line) waveguides are shown in the right column. Input Gaussian pulse has full width at half-maximum of intensity of 577 ps, and its central wavelength is tuned to the gap edge at $\lambda_0 - \Delta\lambda_0/2$

dispersionless slow-light pulses in the form of gap solitons, studied previously in single [27] and coupled waveguides with in-phase gratings [49–51]. Most remarkably, we find that the presence of two types of slow-light modes in the structure with out-of-phase gratings gives rise to gap solitons which periodically tunnel between the waveguides while preserving a constant width, see Figs. 4.5b–4.5d. This effect is analogous to tunneling of fast-light solitons in twin-core rocking filters [53]. In agreement with the properties of conventional nonlinear couplers [40–42], the coupling length is gradually extended as the optical power is increased, resulting in the pulse switching between the output waveguides. As the input power is further increased, we observe a *sharp switching* when the output is highly sensitive to small changes of the input intensity (less than 1%), cf. Figs. 4.5b and 4.5c. At the same time, the pulse delay is also varied with optical power. The power tunability of the pulse delay and switching dynamics can be adjusted by selecting parameters such as waveguide coupling, and choosing the frequency detuning from the gap edge. Such switching can be realized in compact planar devices created with highly nonlinear materials such as AlGaAs [54] or chalcogenide glass [55].

4.3.2 Tunneling of Slow Light in Photonic-crystal Couplers

Waveguides created in planar photonic crystals offer many unique opportunities for manipulating optical pulses. We present a general approach to the design of directional couplers in photonic crystals where dispersionless routing of slow light may be realized [24]. The pulses are fully switched between parallel waveguides at the fixed coupling length, even as the group velocity is varied by orders of magnitude. The additional advantage of suggested structure is the short coupling length, which is equal to just several unit cells. Such remarkable performance is enabled by the co-existence of forward and backward modes which band-edge dispersion is exactly matched, realizing a fundamentally different physical regime compared to the previously considered [30, 35, 56–59] photonic-crystal couplers.

To illustrate the general concept, we consider two-dimensional photonic crystals created by a hexagonal array of holes in a Si membrane with the hole radius of $0.3d$, where d is the lattice constant. The W1 waveguide is created when a single row of holes is absent, see Fig. 4.1d. Due to the hexagonal lattice geometry, the coupler symmetry critically depends on the number of rows (N) between the two W1 waveguides. When N is odd, then the coupler is symmetric with respect to reflection about a central line between the waveguides ($x \rightarrow -x$). When N is even, then the coupler becomes anti-symmetric, as it maps onto itself only when reflection is performed simultaneously along the two axes ($x \rightarrow -x$ and $z \rightarrow -z$), see example for $N = 2$ in Fig. 4.6a.

The tunneling of light between the coupled waveguides is possible through the beating of two modes which are (i) co-propagating and (ii) have different symmetries. Therefore, to realize the routing of slow-light pulses with largely

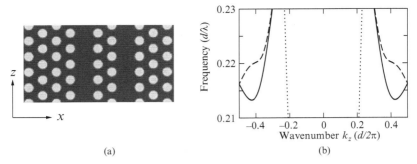

Fig. 4.6. Example of one realization of antisymmetric photonic-crystal couplers supporting dispersionless tunneling of slow-light pulses between the waveguides. (a) The coupler geometry, (b) dispersion of the fundamental modes (additionally dotted curves mark the light lines)

varying group velocities, it is necessary to have two distinct branches of dispersion curves with the same slope (i.e. the same sign of group velocities) arbitrarily close to the edge of the photonic band-gap. We note that in dielectric waveguide couplers, the following relation is always fulfilled, $\omega(k) \equiv \omega(-k)$. Then, the condition (i) can be satisfied when the band-edge is reached at a point with nonzero wavenumber inside the Brillouin zone, i.e. $d\omega/dk|_{k=k_0} = 0$ at $0 < |k_0| < k_b$, where $k_b = \pi/d$ is the Bloch wavenumber. We show in Fig. 4.6b that such situation is indeed realized in the anti-symmetric coupler at $d/\lambda \simeq 0.214$. We check that the condition (ii) is also satisfied only in the anti-symmetric coupler by calculating the mode profiles close to the band-edge, see Fig. 4.7a, 4.7b. Whereas the intensity patterns practically coincide, the phase structures have opposite symmetries. As a result, the beating of these modes realizes light switching between the waveguides, see Fig. 4.7c. Such tunneling is facilitated solely by the coupler symmetry, without the need for any special structure optimization. We note that the out-of-phase Bragg-grating coupler described in Sec. 4.3.1 belongs to the class of antisymmetric structures, which explains its optimal performance for slow-light switching.

The operation of the anti-symmetric directional coupler is based on the beating of forward ($k \simeq +k_0$) and backward ($k \simeq -k_0$) waves. At the band-edge, the dispersion around these points can be expanded as $\omega \simeq \omega_0 + D_2(|k| - k_0)^2 + D_3(|k| - k_0)^3$, where D_2 and D_3 are the second- and third-order dispersion coefficients. By inverting these expressions, we obtain the asymptotic dependence of wave-numbers on frequency as $k_{\omega \to \omega_0} \simeq sk_0 + \sigma[(\omega - \omega_0)/D_2]^{1/2} - s(D_3/2)(\omega - \omega_0)(D_2)^{-2}$, where the values $s = \pm 1$ and $\sigma = \pm 1$ correspond to four different modes. The corresponding group velocities are $V_g = d\omega/dk \simeq 2\sigma D_2[(\omega - \omega_0)/D_2]^{1/2} + 2s(\omega - \omega_0)D_3/D_2$. We see that the group velocities of branches with positive ($s = +1$) and negative ($s = -1$) wave-numbers asymptotically coincide in the slow-light regime when $\omega \to \omega_0$, as confirmed by numerical calculations presented in Fig. 4.8a.

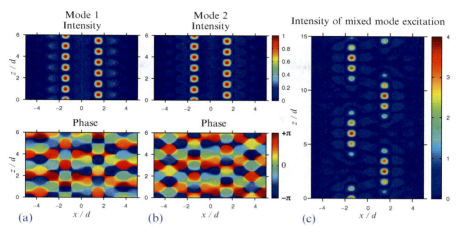

Fig. 4.7. (a), (b) Intensity (*top*) and phase (*bottom*) of the transverse magnetic field distributions for the band-edge modes at $d/\lambda \simeq 0.214$ with (a) positive ($k = 0.88\,\pi/d$) and (b) negative ($k = -0.82\,\pi/d$) wavenumbers, respectively. (c) Intensity of the simultaneously excited modes

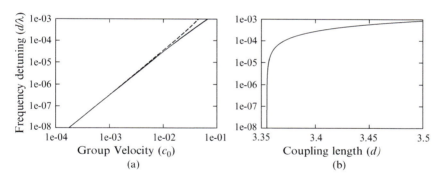

Fig. 4.8. (a) Dependence of the group velocities on the frequency detuning from the band-edge shown in logarithmic scale and (b) the corresponding dependence of the coupling length

The coupling length is defined as the distance where the phase between the co-propagating modes changes by π, and close to the band-edge we obtain $L_c \simeq \pi d \, |\arg \exp\{i[k_{s=+1} - k_{s=-1}]d\}|^{-1} \simeq \pi d \, |\arg \exp\{i[2\,k_0 - D_3(D_2)^{-2}(\omega - \omega_0)]d\}|^{-1}$. According to this expression, the coupling length approaches a constant value in the slow-light regime, as shown in Fig. 4.8b. It is this remarkable feature which enables dispersionless tunneling of slow light, where the same dynamics as shown in Fig. 4.7c is preserved even under the variation of the speed of light by several orders of magnitude.

4.4 Slow Optical Bullets

In this section, we discuss the possibility for the spatio-temporal control of slow-light pulses in an array of coupled Bragg-grating waveguides. In such structures, it becomes possible to simultaneously slow down light pulses and perform their spatial steering. Additionally, we show how to engineer independently the strength of diffraction and dispersion in the slow-light regime, providing the optimal conditions for the nonlinear control of spatio-temporal pulse dynamics. In particular, we predict and demonstrate numerically the formation of strongly localized slow-light optical bullets in such structures [22], suggesting a way to overcome the issue of pulse broadening in extended linear systems.

Similar to the case of two Bragg-grating waveguides discussed in Sec. 4.3.1, we find that the optimal conditions for the control of slow pulses are realized by introducing a phase shift between the otherwise equivalent waveguide gratings, as illustrated in Fig. 4.9a. We calculate the linear dispersion of Floquet-Bloch waves in such periodic structure, $\omega(K,k) = \pm\{\rho^2 + [k - 2C\cos(K)]^2\}^{1/2}$, where K is the transverse wave-number defining the phase difference between the neighboring waveguides and k is the propagation constant along the waveguides. Most importantly, for any propagation angle defined by the Bloch wavevector component K, the width and position of the one-dimensional frequency gap remains the same, $|\omega| < \rho$. This unusual property leads to remarkable spectral features. First, the 2D (quasi-)gap is always present in the spectrum irrespectively to the grating strength (ρ) and coupling between the waveguides (C). Second, the shape of isofrequency contours does not depend on frequency in the transmission band, see Fig. 4.9b. This means that the beam refraction and diffraction remain the same even for slow light when the band edge is approached. In particular, when the frequency is detuned from the Bragg resonance and the effect of the grating is negligible, it was shown that the beam diffraction is reduced to zero [60] at the incident angle

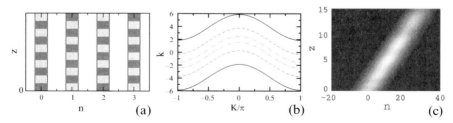

Fig. 4.9. (a) Schematic of a waveguide array with out-of-phase neighboring Bragg gratings, (b) corresponding isofrequency contours for different detunings from the gap edge: $\omega = 4$ (solid), $\omega = 2$ (dashed), $\omega = 1.1$ (dashed-dotted). (c) Beam self-collimation for the incident angle corresponding to $K = \pi/2$. For all the plots $\rho = 1$ and $C = 1$

4 Spatial Switching of Slow Light in Periodic Photonic Structures   67

Fig. 4.10. Snapshots of field intensities for an optical pulse propagating in a waveguide array structure shown in Fig. 4.9: (**a**)–(**d**) Linear broadening due to spatial diffraction and temporal dispersion, (**e**)–(**h**) Nonlinear self-trapping in space and time and formation of an optical bullet. Input pulse has Gaussian profile with FWHM $\Delta t \simeq 13$

corresponding to $K = \pi/2$. Most remarkably, for the waveguide array with phase-shifted gratings, such self-collimation behavior is preserved in the slow light regime, see Fig. 4.9c.

The unique features of linear spectrum in arrays with phase-shifted gratings suggest that these structures provide optimal conditions for a nonlinear control of the pulse dynamics. In particular, since the 2D gap appears for any values of the grating strength and waveguide coupling, it is possible to choose these parameters independently in order to balance the rates of dispersion and diffraction. This allows for *simultaneous compensation* of the pulse broadening in space and time and formation of light bullets [61–63] through the nonlinear self-trapping effect. Indeed, numerical simulations confirm the possibility to perform spatial steering of pulses across the array. In the linear regime, the pulse broadens in both transverse and longitudinal directions (Figs. 4.10a–4.10d). Nonlinear self-action results in the pulse self-trapping in both space and time. In Figs. 4.10e–4.10h, the velocity of the generated light bullet is 30% of the speed of light in the absence of the Bragg grating, and smaller velocities can be accessed as well by controlling the central frequency and bandwidth of the input pulse. Taking the characteristic experimental values for a single Bragg-grating waveguide fabricated in AlGaAs [54] $\Delta\lambda_0 = 0.2\,\text{nm}$ and $\lambda_0 = 1550\,\text{nm}$, we have $t_s \simeq 12.8\,\text{ps}$ and $z_s \simeq 1.8\,\text{mm}$. Accordingly, simulations in Fig. 4.10 correspond to experimentally feasible conditions of the input pulse duration 170 ps and device length 27 mm. Control of shorter pulses may be realized in deeper gratings with larger bandwidth.

4.5 Summary

In conclusion, we have shown that both the spatial and temporal dynamics of slow-light pulses can be controlled in specially designed nonlinear periodic photonic structures with optimized band-edge dispersion characteristics. We have presented the approaches to pulse switching and steering combined with delay tuning and dispersion compensation in arrays of nonlinear Bragg-grating waveguides, and shown that slow light pulses can be efficiently manipulated in photonic-crystal couplers. We anticipate that even more flexibility in pulse control can be achieved by combining the nonlinear pulse self-action and external tunability of photonic structures, and expect rapid progress in theoretical studies and experimental demonstrations of such concepts.

Acknowledgements

A.S. thanks Sangwoo Ha, Yuri Kivshar, Kokou Dossou, Lindsay C. Botten, Dmitry Chigrin, and Andrei Lavrinenko for their substantial and valuable contributions to the results summarized in this chapter. A.S. also acknowledges useful discussions with Martijn de Sterke, Mike Steel, and Benjamin Eggleton.

References

1. C. Liu, Z. Dutton, C.H. Behroozi, and L.V. Hau, Nature **409**, 490 (2001)
2. Y.A. Vlasov, M. O'Boyle, H.F. Hamann, and S.J. McNab, Nature **438**, 65 (2005)
3. H. Gersen, T.J. Karle, R.J.P. Engelen, W. Bogaerts, J.P. Korterik, N.F. van Hulst, T.F. Krauss, and L. Kuipers, Phys. Rev. Lett. **94**, 073903 (2005)
4. R.S. Jacobsen, A.V. Lavrinenko, L.H. Frandsen, C. Peucheret, B. Zsigri, G. Moulin, J. Fage-Pedersen, and P.I. Borel, Opt. Express **13**, 7861 (2005)
5. A. Yariv, Y. Xu, R.K. Lee, and A. Scherer, Opt. Lett. **24**, 711 (1999)
6. S. Mookherjea and A. Yariv, IEEE J. Sel. Top. Quantum Electron. **8**, 448 (2002)
7. J.K.S. Poon, L. Zhu, G.A. DeRose, and A. Yariv, Opt. Lett. **31**, 456 (2006)
8. Y. Akahane, T. Asano, B.S. Song, and S. Noda, Nature **425**, 944 (2003)
9. T. Asano, W. Kunishi, B.S. Song, and S. Noda, Appl. Phys. Lett. **88**, 151102 (2006)
10. T. Tanabe, M. Notomi, E. Kuramochi, A. Shinya, and H. Taniyama, Nature Photonics **1**, 49 (2007)
11. J.B. Khurgin, Opt. Lett. **30**, 643 (2005)
12. R.S. Tucker, P.C. Ku, and C.J. Chang-Hasnain, J. Lightwave Technol. **23**, 4046 (2005)
13. M.F. Yanik and S. Fan, Phys. Rev. Lett. **92**, 083901 (2004)
14. Q.F. Xu, P. Dong, and M. Lipson, Nature Physics **3**, 406 (2007)
15. S. Foteinopoulou and C.M. Soukoulis, Phys. Rev. B **67**, 235107 (2003)

16. P.T. Rakich, M.S. Dahlem, S. Tandon, M. Ibanescu, M. Soljacic, G.S. Petrich, J.D. Joannopoulos, L.A. Kolodziejski, and E.P. Ippen, Nature Materials **5**, 93 (2006)
17. H. Altug and J. Vuckovic, Appl. Phys. Lett. **86**, 111102 (2005)
18. R.E. Slusher and B.J. Eggleton (eds.), *Nonlinear Photonic Crystals*, Springer Series in Photonics, vol. 10, Springer-Verlag, Berlin (2003)
19. M. Soljacic and J.D. Joannopoulos, Nature Materials **3**, 211 (2004)
20. R.S. Jacobsen, K.N. Andersen, P.I. Borel, J. Fage-Pedersen, L.H. Frandsen, O. Hansen, M. Kristensen, A.V. Lavrinenko, G. Moulin, H. Ou, C. Peucheret, B. Zsigri, and A. Bjarklev, Nature **441**, 199 (2006)
21. M. Roussey, F.I. Baida, and M.P. Bernal, J. Opt. Soc. Am. B **24**, 1416 (2007)
22. A.A. Sukhorukov and Y.S. Kivshar, Phys. Rev. Lett. **97**, 233901 (2006)
23. S. Ha, A.A. Sukhorukov, and Y.S. Kivshar, Opt. Lett. **32**, 1429 (2007)
24. S. Ha, A.A. Sukhorukov, K.B. Dossou, L.C. Botten, A.V. Lavrinenko, D.N. Chigrin, and Y.S. Kivshar, Opt. Express **16**, 1104 (2008)
25. J.D. Joannopoulos, R.D. Meade, and J.N. Winn, *Photonic Crystals: Molding the Flow of Light*, Princeton University Press, Princeton (1995)
26. B.J. Eggleton, C.M. de Sterke, and R.E. Slusher, J. Opt. Soc. Am. B **16**, 587 (1999)
27. J.T. Mok, C.M. de Sterke, I.C.M. Littler, and B.J. Eggleton, Nature Physics **2**, 775 (2006)
28. R.J.P. Engelen, Y. Sugimoto, Y. Watanabe, J.P. Korterik, N. Ikeda, N.F. van Hulst, K. Asakawa, and L. Kuipers, Opt. Express **14**, 1658 (2006)
29. A.Y. Petrov and M. Eich, Appl. Phys. Lett. **85**, 4866 (2004)
30. D. Mori and T. Baba, Opt. Express **13**, 9398 (2005)
31. J.B. Khurgin, Opt. Lett. **30**, 513 (2005)
32. A. Figotin and I. Vitebskiy, Waves in Random and Complex Media **16**, 293 (2006)
33. L.H. Frandsen, A.V. Lavrinenko, J. Fage-Pedersen, and P.I. Borel, Opt. Express **14**, 9444 (2006)
34. M.D. Settle, R.J.P. Engelen, M. Salib, A. Michaeli, L. Kuipers, and T.F. Krauss, Opt. Express **15**, 219 (2007)
35. S.C. Huang, M. Kato, E. Kuramochi, C.P. Lee, and M. Notomi, Opt. Express **15**, 3543 (2007)
36. T. Baba and D. Mori, J. Phys. D **40**, 2659 (2007)
37. K. Busch and S. John, Phys. Rev. Lett. **83**, 967 (1999)
38. M.F. Yanik and S.H. Fan, Stud. Appl. Math. **115**, 233 (2005)
39. C.M. de Sterke and J.E. Sipe, in *Progress in Optics*, vol. XXXIII, ed. by E. Wolf, pp. 203–260, North-Holland, Amsterdam (1994)
40. S.M. Jensen, IEEE Trans. Microw. Theory Tech. **MTT-30**, 1568 (1982)
41. A.A. Maier, Kvantov. Elektron. **9**, 2296 (1982), in Russian, English translation: Quantum Electron. **12**, 1490 (1982)
42. S.R. Friberg, Y. Silberberg, M.K. Oliver, M.J. Andrejco, M.A. Saifi, and P.W. Smith, Appl. Phys. Lett. **51**, 1135 (1987)
43. S.S. Orlov, A. Yariv, and S. VanEssen, Opt. Lett. **22**, 688 (1997)
44. G. Perrone, M. Laurenzano, and I. Montrosset, J. Lightwave Technol. **19**, 1943 (2001)
45. S. Tomljenovic-Hanic, J.D. Love, J. Opt. Soc. Am. A **22**, 1615 (2005)
46. M. Aslund, J. Canning, L. Poladian, C.M. de Sterke, and A. Judge, Appl. Optics **42**, 6578 (2003)

47. J.M. Castro, D.F. Geraghty, S. Honkanen, C.M. Greiner, D. Iazikov, and T.W. Mossberg, Appl. Optics **45**, 1236 (2006)
48. M. Imai and S. Sato, in *Photonics Based on Wavelength Integration and Manipulation*, IPAP Books, vol. 2, ed. by K. Tada, T. Suhara, K. Kikuchi, Y. Kokubun, K. Utaka, M. Asada, F. Koyama, and T. Arakawa, pp. 293–302 (2005)
49. W.C.K. Mak, P.L. Chu, and B.A. Malomed, J. Opt. Soc. Am. B **15**, 1685 (1998)
50. W.C.K. Mak, B.A. Malomed, and P.L. Chu, Phys. Rev. E **69**, 066610 (2004)
51. A. Gubeskys and B.A. Malomed, Eur. Phys. J. D **28**, 283 (2004)
52. G.P. Agrawal, *Nonlinear Fiber Optics*, Academic Press, New York (1988)
53. D.C. Psaila and C.M. de Sterke, Opt. Lett. **18**, 1905 (1993)
54. P. Millar, R.M. De la Rue, T.F. Krauss, J.S. Aitchison, N.G.R. Broderick and D.J. Richardson, Opt. Lett. **24**, 685 (1999)
55. M. Shokooh-Saremi, V.G. Ta'eed, N.J. Baker, I.C.M. Littler, D.J. Moss, B.J. Eggleton, Y.L. Ruan, and B. Luther-Davies, J. Opt. Soc. Am. B **23**, 1323 (2006)
56. Y. Sugimoto, Y. Tanaka, N. Ikeda, T. Yang, H. Nakamura, K. Asakawa, K. Inoue, T. Maruyama, K. Miyashita, K. Ishida, and Y. Watanabe, Appl. Phys. Lett. **83**, 3236 (2003)
57. D. Mori and T. Baba, Appl. Phys. Lett. **85**, 1101 (2004)
58. N. Yamamoto, T. Ogawa, and K. Komori, Opt. Express **14**, 1223 (2006)
59. Y.J. Quan, P.D. Han, X.D. Lu, Z.C. Ye, and L. Wu, Opt. Communications **270**, 203 (2007)
60. D.N. Christodoulides, F. Lederer, and Y. Silberberg, Nature **424**, 817 (2003)
61. Y. Silberberg, Opt. Lett. **15**, 1282 (1990)
62. H.S. Eisenberg, R. Morandotti, Y. Silberberg, S. Bar-Ad, D. Ross, and J.S. Aitchison, Phys. Rev. Lett. **87**, 043902 (2001)
63. B.A. Malomed, D. Mihalache, F. Wise, and L. Torner, J. Opt. B **7**, R53 (2005)

Part II

Nonlinear Effects in Multidimensional Lattices: Solitons and Light Localization

5

Introduction to Solitons in Photonic Lattices

Nikolaos K. Efremidis[1], Jason W. Fleischer[2], Guy Bartal[3], Oren Cohen[3], Hrvoje Buljan[4], Demetrios N. Christodoulides[5], and Mordechai Segev[3]

[1] Department of Applied Mathematics, University of Crete, 71409 Heraclion, Crete, Greece
nefrem@tem.uoc.gr
[2] Electrical Engineering Department, Princeton University, Princeton, NJ 08544, USA
jasonf@princeton.edu
[3] Physics Department and Solid State Institute, Technion-Israel Institute of Technology, Haifa 32600, Israel
msegev@techunix.technion.ac.il
[4] Department of Physics, University of Zagreb, PP 332, Zagreb, Croatia
hbuljan@phy.hr
[5] College of Optics and Photonics, University of Central Florida, Orlando, Florida 32813, USA
demetri@creol.ucf.edu

5.1 Introduction to Optical Periodic Systems

In this chapter, we present a review of optical systems that have a periodic variation in their index transverse to the direction of propagation. Such photonic systems include photonic crystal fibers, which have a large index variation that controls frequency dispersion, and coupled waveguide arrays, which have a relative small index variation that controls spatial diffraction. Here, we will focus on the latter case and consider 1+1D and 2+1D dynamics. A photonic lattice has the advantage that the refractive index contrast requirements are low, and thus, for example, a 2D bandgap can be established for index modulations of the order 10^{-3}. Light excitation is quite simple because the optical wave is launched in a direction that is almost perpendicular to the direction of the index modulations. Also, nonlinearity can quite easily manifest in such systems simply because of low refractive index modulations. As a result, nonlinear self-localized structures or solitons are possible for nonlinear index modulations of the order of 10^{-4}.

An optical periodic system with periodicity along the transverse direction was first theoretically studied in 1965 [1] in the linear regime. In that work the diffraction pattern of an array of identical fibers has been found using coupled mode theory in terms of Bessel functions. Experimentally this behavior was reported in 1973 in an array of optical waveguides [2].

In 1988, for the first time a nonlinear waveguide array was considered [3]. It was shown that nonlinearity can counteract waveguide coupling, leading to suppressed diffraction and the formation of optical discrete solitons. Discrete solitons were experimentally observed ten years later [4] in AlGaAs arrays.

The transition from one-dimensional to two-dimensional lattices came a few years later. Using an optical induction technique [5, 6] allowed for the creation of two-dimensional periodic topologies in photosensitive crystals. This let to the first experimental observation of two-dimensional discrete/lattice solitons in any physical system in nature [7].

During the last years the research interest for studying nonlinear optical periodic systems has grown rapidly (see the recent review articles [8,9]). There are several factors for this rapid growth in this area. From the physical point of view, the behavior of nonlinearity and periodicity arises in a wide variety of fields, ranging from nonlinear optics and photonic crystals to biology, solid-state physics, and Bose-Einstein condensates in lattice potentials. The common ground in these problems is the co-existence of a band structure along with nonlinearity that in turn allows for inter and intra band interactions.

Considering applications, waveguide lattices are potential candidates for optical switching applications [10–13]. By engineering regions with different periodicities, which have different band structure properties, it is possible to control of the flow of light. In addition, nonlinearity is a necessary ingredient for performing routing and logic operations.

It has only been recently that the experimental techniques have grown to the point where experiments can be successively performed in two spatial dimensions for a variety of different settings. These methods include, the aforementioned optical induction technique [5,6] (see section 5.2 for details), the use of arrays of optical fibers [14], and writing optical waveguides in bulk glasses using femtosecond laser beams [15].

5.2 Optically Induced Lattices

Until 2002, waveguide arrays were only fabricated by etching the top of a substrate, creating a series of ridged structures [4]. However, this procedure limits the allowed topologies to only one transverse spatial dimension. Considering applications, higher dimensionality provides the possibility to route information in an optical network from a point of origin A to a final destination Z, something that is not possible using only one spatial dimension [10–12]. In [5] a new method was suggested to induce a two-dimensional optical lattice in photosensitive crystals. Experimentally the method was first verified in one dimension [6] and subsequently in two dimensions [7], leading to the first observation of two-dimensional lattice solitons.

Optically induced lattices, are periodic index configurations which are established in biased photorefractive crystals. Let us assume that the crystal is biased with voltage V along the extraordinary crystalline c-axis (which

5 Introduction to Solitons in Photonic Lattices

is along the x direction). a is the ordinary axis (y direction) and the optical field propagates along the z direction. Lattice stability requires that the photo-induced crystal is highly anisotropic, being essentially linear and highly nonlinear along the two polarizations. Crystals with such properties are available in photorefractives such as the Strontium Barium Niobate (SBN), which is highly nonlinear along the c-axis and essentially linear along the a-axis.

Following [16] the normalized propagation of two incoherent beams in a biased photorefractive crystal in one dimension is given by

$$iu_z + \frac{1}{2}u_{xx} - \beta_u \left[\frac{1}{1+I} - \delta\frac{\partial I/\partial x}{1+I}\right]u = 0, \tag{5.1}$$

$$iv_z + \frac{1}{2}v_{xx} - \beta_v \left[\frac{1}{1+I} - \delta\frac{\partial I/\partial x}{1+I}\right]v = 0, \tag{5.2}$$

where u, v are the extraordinary and ordinary (respectively) polarized waves, and $\beta_u/\beta_v = r_{33}/r_{13}$ (r_{13} and r_{33} are the relevant electro-optic coefficients), $I = |u|^2 + |v|^2$ is the total intensity, and the nonlinearity is of the self-focusing or self-defocusing type if $\beta_u, \beta_v > 0$ or < 0, respectively. Below, we will establish a simplified model that accurately describes the behavior of the system of Eqs. (5.1)–(5.2).

Notice that the term proportional to δ represents small diffusion effects that can be, in general, ignored. On the other hand, the requirement for large electro-optic anisotropy $r_{33} \gg r_{13}$ results in a parameter region, where the nonlinear term of Eq. 5.2 can be ignored, leading to the linear diffraction equation for the v field

$$iv_z + \frac{1}{2}v_{xx} = 0. \tag{5.3}$$

Thus, propagation along the ordinary axis is essentially linear. Equation (5.3) admits periodic exact solutions of the form

$$v = \sum_j A_j \exp(-i\lambda_j^2 z/2 + i\lambda_j x + i\phi_j), \tag{5.4}$$

where A_j, λ_j, and ϕ_j are real constants. Experimentally such patterns can be achieved by superimposing plane waves. The only coherent interference pattern in 1D that does not travel in the x direction and remains invariant along z is obtained by interfering two plane waves with $\lambda_1 = -\lambda_2 = \pi/L$:

$$v = A\cos(\pi x/L)\exp(-i\pi^2 z/(2L^2)) \tag{5.5}$$

Thus, under the above assumptions, Eqs. (5.1)–(5.2) describing the evolution of two orthogonally polarized waves can be approximated by

$$iu_z + \frac{1}{2}u_{xx} - \frac{\beta u}{1 + V(x) + |u|^2} = 0, \tag{5.6}$$

where
$$V(x) = A^2 \cos^2(\pi x/L). \tag{5.7}$$

The field u, which is the probe beam, sees a highly nonlinear environment due to the large electro-optic coefficient. Another generic and experimentally feasible model is that of Kerr nonlinear media. The corresponding model equation is given by
$$iu_z + \frac{1}{2}u_{xx} + V(x)u + \sigma|u|^2 u = 0, \tag{5.8}$$

where $V(x)$ is proportional to the index modulations and $\sigma = \pm 1$ accounts for self-focusing (+) and self-defocusing (−) nonlinearities. Equation (5.8) describes a variety of optical systems, such as periodic waveguide arrays [4], fiber arrays [14], and Bose-Einstein condensates [17]. In the case of photorefractives, Eq. (5.8) can be obtained from Eq. (5.6) in the low intensity limit (assuming $V(x), |u|^2 \ll 1$, $-u/(1 + V(x) + |u|^2) \approx -u + V(x)u + |u|^2 u$). Equation (5.8) is associated with two conserved quantities, namely the total power
$$P = \int_{-\infty}^{\infty} |u|^2 \, dx \tag{5.9}$$

and the Hamiltonian
$$H = \frac{1}{2}\int_{-\infty}^{\infty} \left(|u_x|^2 - |u|^4 - 2V(x)|u|^2\right) dx. \tag{5.10}$$

Thus, Eq. (5.8) can be written as
$$i\frac{\partial u}{\partial z} = \frac{\delta H}{\delta u^*}. \tag{5.11}$$

In two dimensional settings, by employing similar arguments the original model can be simplified to take the form
$$iu_z + \frac{1}{2}\nabla^2 u - \frac{\beta u}{1 + V(x,y) + |u|^2} = 0, \tag{5.12}$$

or for Kerr nonlinear media
$$iu_z + \frac{1}{2}\nabla^2 u + V(x,y)u + \sigma|u|^2 u = 0. \tag{5.13}$$

In Fig. 5.1 the experimental configuration of an optically induced lattice is schematically shown. In two spatial dimensions the freedom to select the optically induced lattice is much higher. A generic integral form of the field generated by the interfering beams is given by
$$v = \iint_{-\infty}^{\infty} A(k_x, k_y) e^{i\phi(k_x, k_y)} e^{-i(k_x^2 + k_y^2)z/2} e^{ik_x x + ik_y y} \, dk_x \, dk_y. \tag{5.14}$$

5 Introduction to Solitons in Photonic Lattices   77

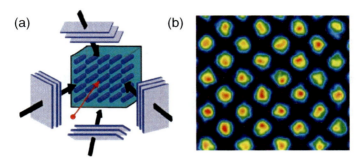

Fig. 5.1. Configuration of an optically induced lattice. Two pairs of plain waves interfere along the (essentially linear) ordinary polarization, establishing a periodic intensity pattern. On the other hand, the probe beam (red color) polarized along the extraordinary axis propagates in a highly nonlinear environment with a periodic index lattice which is proportional to the intensity pattern of the interfering beams

Notice that the condition for an invariant, along z, lattice intensity is satisfied when the "Bloch momentum" vectors k_x, k_y lie on a circle, i.e., $k_x^2 + k_y^2 = \rho^2$. In cylindrical coordinates $k_x = \rho\cos\theta$, $k_y = \rho\sin\theta$, $A(k_x, k_y) = \delta(\rho - R)f(\theta)/\rho$, and thus

$$v = e^{iR^2 z/2} \int_0^{2\pi} f(\theta) \exp(i\phi(\theta)) \exp[iR(\cos\theta x + \sin\theta y)] \, d\theta. \tag{5.15}$$

All possible coherent non-diffracting beams can be obtained from Eq. (5.15) by specifying the amplitude and the phase of the interfering plane waves $(f(\theta), \phi(\theta))$. Another degree of freedom can be introduced when the lattice is established from mutually incoherent plane waves all of which are polarized along the ordinary axis. In that latter case, the potential is given by

$$V(x) = \sum_j |v_j|^2, \qquad j = 1, \ldots, n. \tag{5.16}$$

where v_j are the mutually incoherent fields, each one of them being determined by an equation of the form (5.15). Different types of one and two-dimensional diffraction free lattices have been studied in the literature. These include coherent ($f(\theta) = \sum_{j=0}^{3} \delta(\theta - j\pi/2)$, $\phi(\theta) = 0$) and incoherent ($f_1(\theta) = \sum_{j=0}^{1} \delta(\theta - j\pi)$, $f_2(\theta) = \sum_{j=0}^{1} \delta(\theta - j\pi + \pi/2)$, along with $\phi(\theta) = 0$) square lattices [5,7], hexagonal lattices [18,19] for which $f(\theta) = \sum_{j=0}^{5} \delta(\theta - j\pi/3)$, $\phi(\theta) = 0$, and Bessel lattices [20,21] with weight functions $f(\theta) = c$ and $\phi(\theta) = m\theta$, where m is the vorticity number. Quasicrystals which are structures with long-range order but no periodicity can be formed by the interference of five different plane waves separated by angles $2\pi/5$ [22], i.e., ($f(\theta) = \sum_{j=0}^{4} \delta(\theta - j2\pi/5)$ and $\phi(\theta) = 0$). In Fig. 5.1b the experimental picture of the square lattice intensity pattern established by the coherent interaction of two pairs of orthogonal plane waves is shown.

5.3 Coupled-mode Theory

Coupled-mode theory (CMT), provides an approximate model for the description of equations with periodic potentials such as (5.8) and (5.13). Although CMT is not exact, it is quite accurate, providing both qualitative and quantitative results, as long as the original assumption from which it was derived are satisfied. Coupled-mode theory originates in the theory of solids, where it is known as tight binding approximation [23]. In optics, it has been extensively used in a variety of linear and nonlinear problems [1, 3, 24]. For simplicity here we consider a one-dimensional model although the same formal analysis can be repeated for higher dimensionalities. CMT is based on the following expansion of the optical field

$$u = \sum_n c_n(z)\psi_n(x). \tag{5.17}$$

In (5.17) $\psi_n(x) = \Psi(x - nx_0)$, where $\Psi(x)$ is the linear lowest order localized mode of a single waveguide in isolation having potential $V_0(x)$, i.e.,

$$q\Psi + \frac{1}{2}\Psi_{xx} + V_0(x)\Psi = 0. \tag{5.18}$$

Notice that Ψ has zero nodes and is even when the index function is even. In Eq. (5.17) $c_n(z)$ represents the amplitude of the local mode ψ_n at distance z. Substituting Eq. (5.17) to (5.8) an evolution Equation for the amplitudes is derived

$$i\frac{dc_n}{dz} + Ec_n + \kappa(c_{n+1} + c_{n-1}) + \gamma|c_n|^2 c_n = 0, \tag{5.19}$$

where $\kappa = (1/2)\langle\phi_n|\phi_{n\pm 1}\rangle + \langle\phi_n|V(x)\phi_{n\pm 1}\rangle$, $E = \langle\phi_0|(V(x) - V_0(x))\phi_0\rangle$, $\gamma = \sigma\langle\phi_n|\phi_n^3\rangle$, assuming that $\langle\phi_n|\phi_n\rangle = 1$. In a normalized form Eq. (5.19) reads

$$i\frac{dc_n}{dz} + (c_{n+1} + c_{n-1}) + \sigma|c_n|^2 c_n = 0, \tag{5.20}$$

where σ is the sign of the nonlinearity. Equation (5.20) is known as the Discrete Nonlinear Schrödinger (DNLS) equation. The regimes of validity of Eq. (5.20) as an approximate model for periodic systems is discussed in [25]. In particular, DNLS-type models are single band approximations of the lattice NLS equation. Thus, self-defocusing lattice solitons predicted in DNLS models exist in a finite instead of a semi-infinite gap. Notice that coupled-mode theory approximations are accurate, as long as the nonlinear index change is much smaller than the linear index modulations. Besides optics, DNLS type models appear in diverse settings such as Biology [26], in molecular crystals [27], in atomic chains [28], and in Bose-Einstein condensates [29] (see also the review papers [30–32]).

In two transverse dimensions a two-dimensional DNLS equation is derived

$$i\frac{dc_{m,n}}{dz} + (c_{m,n+1} + c_{m,n-1} + c_{m+1,n} + c_{m-1,n}) + \sigma |c_n|^2 c_n = 0 \quad (5.21)$$

for the case of square lattices. For lattices of different symmetries the coupling coefficients can take different form (see for example the case of a honeycomb lattice in [33]).

5.4 Linear Properties

A fundamental difference between homogeneous and periodic media is that in homogeneous media the dispersion/diffraction curves characterize the regions of continuous/plane wave solutions. On the other hand, media with periodic index modulations do not possess such solutions, but instead support Floquet-Bloch modes. Understanding the linear properties of the system, is not only important per se, but also is fundamental for analyzing the nonlinear properties of the system. In the literature, linear properties have been studied both in the context of the periodic Eqs. (5.8) and (5.13) or using approximate CMT descriptions (5.20) and (5.21).

Let us start by considering the approximate, CMT description in one-dimension as given by Eq. (5.20). The discrete Fourier transform of the field amplitude c_n is defined by

$$\tilde{c}_n(\kappa) = \frac{1}{\sqrt{2\pi}} \sum_n c_n(z) e^{-i\kappa n}, \quad c_n(z) = \frac{1}{\sqrt{2\pi}} \int_0^{2\pi} \tilde{c}_n(\kappa) e^{i\kappa n} \, d\kappa. \quad (5.22)$$

The diffraction relation is directly obtained by assuming low amplitude plane-wave solutions of the form

$$c_n = \exp(-iqz + ikn) \quad (5.23)$$

resulting in

$$q = -2\cos k. \quad (5.24)$$

From Eq. (5.24) the group velocity and the (second order) diffraction are given by $v_g = q' = dq/dk = 2\sin k$ and $g_2 = q'' = 2\cos k$ respectively. Perhaps the most interesting feature of the diffraction relation is the possibility to engineer the second order diffraction term as a function of the incident angle inside the array. For $|k| < \pi/2$ the diffraction of the array is normal ($q'' > 0$), whereas for $\pi/2 < |k| < \pi$ the diffraction becomes anomalous ($q'' < 0$). Furthermore, for $|k| = \pi/2$ the second order diffraction term is identical to zero [34, 35]. Thus a beam propagating with this value of the "Bloch momentum" k is going to experience minimal diffraction, arising only from higher order diffraction terms. The phenomena of negative and zero second order diffraction are not possible in uniform media where the diffraction is always positive (since $q = k^2/2$).

The diffraction of optical beams in waveguide arrays when an initial index ramp is applied to the waveguides can be obtained in closed form in the case of discrete models. For the sake of generality, let us consider the linear part of Eq. (5.20)

$$ic_n + (c_{n+1} + c_{n-1}) + \gamma n f(z) c_n = 0, \tag{5.25}$$

where the linear index ramp across the array γn is multiplied by a function $f(z)$. Equation (5.25) can be solved in the Fourier domain using the method of characteristics [36]. In particular, assuming single waveguide excitation at the input ($c_m(0) = \delta_{m,0}$) the intensity of the optical field along z is given by

$$I_m(z) = |c_m(z)|^2 = J_m^2(w), \tag{5.26}$$

where

$$w = \sqrt{u^2 + v^2}, \quad \begin{pmatrix} u \\ v \end{pmatrix} = \int_0^z \begin{pmatrix} \cos \\ \sin \end{pmatrix} [\gamma \eta(z')] \, dz', \quad \eta = \int_0^z f(z') \, dz'. \tag{5.27}$$

Because Eq. (5.25) is linear, the propagation of more complicated patterns is obtained analytically by superposition.

In optics, the diffraction in an array of linear fibers without an additional index tilt ($\gamma = 0$) has been theoretically studied in [1]. In this case, the evolution of a single waveguide excitation is given by $c_n(z) = J_n(2z) \exp(i\pi n/2)$. It is interesting to notice that the amplitude of the field as it propagates is maximum at the edges and not at the center. Experimentally, this behavior has been observed in [2].

A waveguide array with propagation constants that vary linearly in the transverse direction (i.e., $f(z) = 1$ in Eq. (5.25)) exhibits solutions which are called "Bloch oscillations" [37]. Independently of the form of the initial condition, such waves exhibit periodic revivals along z, and thus remain localized as they propagate. The required linear variation of the propagation constant is usually achieved by a linear index ramp. Solving Eq. (5.25) in the special case $f(z) = 1$ results to

$$c_n(z) = J_n \left[\frac{4}{\gamma} \sin \left(\frac{\gamma z}{2} \right) \right] \exp \left[\frac{in}{2} (\gamma z + \pi) \right]. \tag{5.28}$$

From Eq. (5.28) one can find that the (intensity) period of the oscillations is equal to $2\pi/\gamma$. In optics, the existence of Bloch oscillations in waveguide arrays has been predicted in [38]. The experimental observation came one year later independently from two-different groups: In [39] the linear variation of the index contract is created by applying a temperature gradient, whereas in [40] the thickness of the central layer is varied. In [34], the diffraction properties of a waveguide arrays were exploited, and it was shown that by using a waveguide array in a zig-zag configuration the second and third order diffraction terms can be canceled out. This can be achieved by using different short segments of waveguides with average zero diffraction. The normal and anomalous refraction and diffraction properties of arrays have been experimentally verified in [35]. AC Bloch oscillations were predicted in [36] ($f(z)$ being a sinusoidal function of z) in the case where the ratio of the field amplitude and field spatial frequency is a root of the ordinary Bessel function of order 0. Such periodic modulations in $f(z)$ have not been implemented thus far in optical waveguides. However, it has been shown that an array of optical waveguides whose curvature periodically changes can give rise to AC optical Bloch oscillations [41, 42]. Such periodic oscillations have been observed in [43, 44].

Coupled-mode theory equations are single band approximations of Eqs. (5.8) and (5.13) which exhibit a periodic potential. To consider higher-band behavior, these equations can be analyzed by making use of Floquet-Bloch theory. Let us consider the linear two-dimensional case. Looking for wave solutions, whose amplitude is stationary along z, i.e., $u(x,y,z) = \psi(x,y)\exp(-iqz)$, we obtain

$$q\psi_z + \frac{1}{2}(\psi_{xx} + \psi_{yy}) + V(x)\psi = 0. \quad (5.29)$$

Floquet-Bloch's theorem [23,37,45] states that the eigenfunctions of Eq. (5.29) for a periodic potential are the products of a plane wave $\exp(i\mathbf{k}\cdot\mathbf{r})$ multiplied with a function $\Psi_{\mathbf{k}}(\mathbf{r})$ with the periodicity of the crystal lattice, or

$$\psi_{\mathbf{k}} = \Psi_{\mathbf{k}}(\mathbf{r})e^{i\mathbf{k}\cdot\mathbf{r}}, \quad (5.30)$$

where $\Psi_{\mathbf{k}}(\mathbf{r}+\mathbf{R}) = \Psi_{\mathbf{k}}(\mathbf{r})$, $\mathbf{R} = m\mathbf{R}_1 + n\mathbf{R}_2$, $m,n \in \mathbb{Z}$ and $\mathbf{R}_1$, $\mathbf{R}_2$ are the primitive vectors of the lattice such that $V(\mathbf{r}+\mathbf{R}) = V(\mathbf{r})$. A related concept is that of a Brillouin zone which is defined as the primitive cell in the reciprocal lattice. Several methods have been developed in condensed matter physics [23] to find the band structure of Eq. (5.29). A simple approach, known as the plane-wave method, relies on the Fourier series decomposition of $\Psi_q(\mathbf{r})$ under periodic boundary conditions. The resulting system of algebraic equations can then be solved numerically. Some specific types of potentials admit exact solutions, such as the Kronig-Penney model in one-dimension [46] and the sinusoidal potential which can be solved in one or higher dimensions with the use of Mathieu functions [47].

Let us consider the one-dimensional case with a typical sinusoidal potential of the form

$$V(x) = -V_0 \sin^2(\pi x/2). \quad (5.31)$$

The period of the potential is 2 and its band structure is shown in Fig. 5.2 for $V_0 = 10$ in the reduced zone scheme (the band structure is "folded" inside the first Brillouin zone). The bands of Fig. 5.2 are the regions where periodic

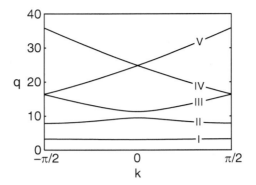

Fig. 5.2. A typical one-dimensional band structure for the lattice potential of Eq. (5.31) for $V_0 = 10$. Curves I–V correspond to the first five bands of the complete band structure

Floquet-Bloch modes exist, i.e. modes that have allowed values of the propagation constant q. The regions outside the bands are called bandgaps. Inside the bandgaps exponentially decaying/growing modes are supported. These modes can be obtained for example using the plane-wave method for complex values of the Bloch momentum. Notice that between two successive bands, a bandgap always exist in the q-domain. In waveguide arrays, the band structure and the corresponding Floquet-Bloch modes were investigated in [48] for one spatial dimension.

The band structure properties become more complicated in more than one dimension. We analyze two different type of lattices [49]. The first index potential is sinusoidal (Fig. 5.3a)

$$V(x,y) = -(V_0/2)\left[\sin^2 \pi x + \sin^2 \pi y\right], \tag{5.32}$$

whereas the second lattice has a backbone index profile

$$V(x,y) = -\frac{V_0}{1 + A^2 \cos^2(\pi x)\cos^2(\pi y)}. \tag{5.33}$$

The lattice of Eq. (5.33) (shown in Fig. 5.3b) is established by the interference of two pairs of plane waves that are coherently superimposed.

A complete bandgap is defined as a finite region in q between two successive (in increasing q order) Floquet-Bloch modes. As can be seen from the three-dimensional band structure shown in Fig. 5.3, different bands can overlap with each other in the q axis restricting, or even eliminating altogether, the number of complete bandgaps. This property of higher dimensional lattices is in contract to the one-dimensional arrays, which are, in general, associated with an infinite number of complete Bragg resonance bandgaps. In Fig. 5.3 we display the band (gray) and bandgap (white) regions for varying potential depth. In Fig. 5.3a we find that no complete bandgap exist for $V_0 \lesssim 13.8$.

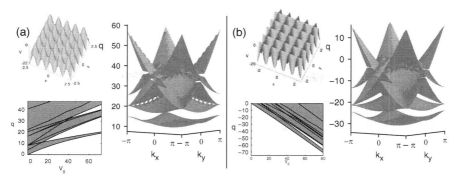

Fig. 5.3. Index potentials of Eqs. (5.32) and (5.33) and their correspond band structure properties are shown in (**a**), (**b**), respectively. The numerically calculated band structures for (a) $V_0 = 21.6$ and (b) $A^2 = 1.21$ and $V_0 = -36.3$ are shown. The regions of bands (gray) and bandgaps (white) are depicted as a function of the potential depth

By increasing the potential depth further, a second bandgap opens up when $V_0 \simeq 40.7$. The backbone lattice of Fig 5.3b exhibits different behavior. When the potential depth is less than 28.4 (and $A^2 = 1.21$) all the bands overlap, i.e. no complete band gap exists. On the other hand, for bigger values of V_0, one gap opens up between the first and the second bands. Unlike the sinusoidal lattice case, we find that no other band gaps emerge for even greater values of V_0.

The presence of a linear index ramp (say $\gamma x + \delta y$ in two spatial dimensions) in addition to the periodic lattice is expected to give rise to Bloch oscillations. However, Zener predicted [50] that Bloch oscillations are not ideal (they persist for finite distances) in periodic lattices due to interband interactions. Furthermore, Bloch oscillations are expected to breakdown when the index difference imposed on a period of a lattice due to the linear potential is of the order of the gap to the next band. Zener tunneling in one-dimensional waveguide arrays has been observed in [51]. In [52] Bloch oscillations and Zener tunneling in two-dimensional periodic systems have been reported. In this latter work, an optical induction technique was applied and, in order to create a transverse refractive index gradient, the crystal was illuminated from the top with incoherent white light.

A method of optical waveguiding, which relies on Bragg diffractions from a 1D grating that gives rise to waveguiding in the direction normal to the grating wave vector was proposed in [53]. The waveguide structure consists of a shallow 1D grating that has a bell- or trough-shaped amplitude in the confinement direction. In [54] non-diffracting beams in two-dimensional periodic systems were identified. Such beams are constructed by superposition of Floquet-Bloch modes.

In dispersion curves, there are specific points known as diabolic points which are singular. In [18] it was demonstrated that such a diabolical point

5.5 One-dimensional Lattice Solitons

In the early works on optical discrete solitons, the theoretical analysis was based on CMT description. This single band model, can provide valuable information about the behavior of optical periodic systems (5.8), (5.13). In one-dimension Eq. (5.20) is not integrable unlike its continuous analogue, the nonlinear Schödinger equation. An integrable version of the DNLS equation, which is known as the Ablowitz-Ladik DNLS exists [55].

We are going to analyze basic families of discrete soliton solutions by utilizing the DNLS equation in one transverse dimension. We assume that Eq. (5.20) admits solutions of the form

$$c_n = a_n \exp(-\mathrm{i}qz), \qquad (5.34)$$

and thus

$$qa_n + (a_{n+1} + a_{n-1}) + \sigma a_n^3 = 0. \qquad (5.35)$$

Asymptotic analysis of soliton solutions can be carried out in the limit of nonlinear modes which are highly localized inside the lattice. We are going to analyze two basic families of discrete solitons of Eq. (5.35) known as "on-site" and "off-site". We isolate these solutions from many other types because, by analytic continuation, they represent single hump soliton solutions in the long wavelength limit. In the strongly nonlinear limit the first family of solutions [3] has the approximate form

$$a_0 = \alpha, \qquad a_{\pm 1} = \frac{1}{\sigma \alpha}, \qquad (5.36)$$

whereas $u_n = 0$ for $n \neq 0, \pm 1$ and $q = -\sigma\alpha^2$. This family of solution is called "on-site" because the maximum of an imaginary envelope is located exactly at the lattice site $n = 0$ (Fig. 5.4). The second family of solutions is given by

$$a_j = \alpha \left(\mathrm{sgn}(\sigma)\right)^j, \qquad j = 0, 1, \qquad (5.37)$$

$$a_j = \frac{\alpha \left(\mathrm{sgn}(\sigma)\right)^j}{1 + |\sigma|\alpha^2}, \qquad j = -1, 2, \qquad (5.38)$$

and $q = -\mathrm{sgn}(\sigma)(1 + |\sigma|\alpha^2)$. This family of solutions is called "intra-site" or "off-site" because an imaginary envelope has its maximum between lattice sites 0 and 1 [56]. For both families of solitons $\mathrm{sgn}(\sigma) = -\mathrm{sgn}(q)$. The "on-site" family of solitons is energetically favorable, and thus stable whereas "intra-site" discrete solitons are unstable [57,58]. The exponential decay (for large $|n|$) of these solutions is given by

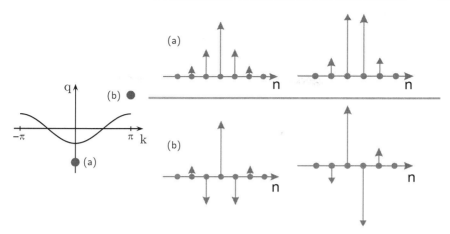

Fig. 5.4. Diffraction relation (*left*) and families of discrete solitons supported by the DNLS equation. The two families shown in (**a**) ("on-site" (*left*) and "off-site" (*right*)) exist for self-focusing nonlinearities and have eigenvalues in the bandgap below the base of the 1st Brillouin zone. The families shown in (**b**) ("on-site" (*left*) and "off-site" (*right*)) exist for self-defocusing nonlinearities and have eigenvalues in the bandgap above the edge of the 1st Brillouin zone

$$a_n = (\text{sgn}(\sigma))^n \exp(-\lambda n), \tag{5.39}$$

where $\lambda = \text{arccosh}(k/2)$.

Notice that these two families of solutions exist for both signs of the nonlinearity [59]. This result is in contrast to the nonlinear Schrödinger equation which admits bright soliton solutions in the self-focusing case only. In Fig. 5.4 these two families of discrete solitons are schematically illustrated for self-focusing (Fig. 5.4a) and self-defocusing (Fig. 5.4b) nonlinearity. We would like to point out that for self-focusing nonlinearity ($\sigma > 0$) the soliton eigenvalue resides in the semi-infinite bandgap below the 1st Brillouin zone ($q < -1$). On the other hand, self-defocusing solitons ($\sigma < 0$) have eigenvalues residing above the Brillouin zone ($q > 1$). In addition, the phase difference between adjacent lattice sites of self-focusing discrete solitons is zero, and thus these solitons reside at the base of the Brillouin zone ($k = 0$). The adjacent lattice site phase difference of self-focusing discrete solitons is π, i.e., these solitons reside at the edge of the Brillouin zone ($k = \pi$). The mathematical proof for the existence of discrete solitons was derived in [60].

The analysis of lattice solitons in periodic lattices beyond the DNLS limit reveals that the behavior of the system is more complex [5,25,61]. Families of lattice solitons (LS) can exist in the semi-infinite (or total internal reflection (TIR)) gap and in the finite (Bragg) bandgaps of the band structure. Thus, an infinity of families of LS in principle exist in one-dimensional periodic lattices.

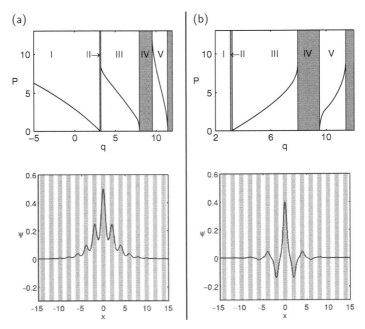

Fig. 5.5. Lattice solitons existence curves (total power P vs. propagation constant q) for self-focusing (**a**) and self-defocusing (**b**) Kerr media. The gray (shaded) regions II and IV depict the first two bands. Regions I, III, IV correspond to the semi-infinite and the first two Bragg (finite) gaps of the system. Typical soliton profiles with an eigenvalue in the semi-infinite gap (**a**) and in the first finite bandgap (**b**) are shown on the bottom. Gray (white) areas represent high (low) refractive index regions

In the case of self-focusing nonlinearity (Fig. 5.5a) families of lattice solitons exist both in the TIR gap and in every Bragg resonance gap. A typical TIR lattice soliton is shown on the bottom of Fig. 5.5a. Notice the absence of nodes (the field does not become zero) in the profile which is characteristic for TIR LS. In Fig. 5.5b the existence curves of self-defocusing lattice solitons are shown [25]. In this case lattice solitons do not exist in the TIR bandgap, but only in the Bragg gaps. The field profile of a typical lattice soliton with eigenvalue inside the first Bragg resonance is shown on the bottom of Fig 5.5b.

Figure 5.6 depicts experimental results showing lattice soliton formation in optically induced lattices [6]. In Fig. 5.6I the transition of the signal beam from discrete diffraction to a discrete soliton for on-axis (corresponding to zero Bloch momentum, or at the base of the 1st Brillouin zone) input as a function of self-focusing nonlinearity is depicted. When the nonlinearity is small, the signal beam experiences diffraction. On the other hand, in the strongly nonlinear regime, a highly localized lattice soliton is formed (Fig. 5.6Ie–f). This family of solutions reside in the semi-infinite bandgap of the band structure.

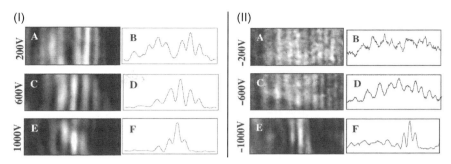

Fig. 5.6. Signal beam output intensity as a function of increasing focusing (**I**) and defocusing (**II**) nonlinearity. (A), (B) show discrete diffraction, (C), (D) show intermediate self-focusing, and (E), (F) depict lattice soliton formation. The period of the waveguide lattice is 8.8 μm and 9.3 μm in (I) and (II), respectively

Such solitons were first observed in optical waveguide arrays in [4]. Figure 5.6II shows lattice soliton formation when the signal beam is incident at an angle very close to the first Bragg resonance (or at the edge of the 1st Brillouin zone) of the system with self-defocusing nonlinearity. When the nonlinearity is small, the beam diffracts. Increasing the applied voltage (thus increasing the nonlinearity) a self-defocusing gap soliton residing in the first bandgap of the band structure is formed. This family of solitons was first observed in [6].

There is a plethora of works based on DNLS type Equations and, during the last years, on lattice NLS equations. Pairs of out-of-phase solitons, which resemble twisted localized modes were observed in [62]. Self-focusing gap solitons were reported in waveguide arrays in [63]. Properties of gap solitons such as Bloch wave interactions [64] and controlled generation and steering [65] have also been studied.

Modulational instability of plane wave solutions has been theoretically analyzed in [3, 66]. Modulational instability was observed in AlGaAs waveguide arrays with self-focusing nonlinearity [67]. Such instabilities occur as long as the spatial Bloch momentum of the initial excitation is within the normal diffraction region of the Brillouin zone. Modulation instability in the anomalous-diffraction regions of a photonic lattice has been observed in [68]. The experiments were carried out in a 1D waveguide array fabricated in a lithium niobate crystal displaying the photovoltaic self-defocusing nonlinearity.

Another basic family of solutions supported by a lattice is that of dark solitons. Dark discrete solitons have the form of a dip in a uniform background of light. Theoretically, their properties have been studied in the discrete domain [69]. Experimentally, they have been observed in self-focusing waveguide arrays, that support dark gap solitons [63, 70]. Notice that in order to excite a dark soliton, a π step needs to be introduced at the center of the interference pattern that results in a dark narrow notch in the shaped input beam.

The propagation of two different waves, which can be optical fields of different colors or polarizations or mutually incoherent beams, is described by a system of coupled nonlinear Schrödinger equations. Solutions of such models are known as vector solitons [71]. The families of vector solitons in discrete lattices have been analyzed and their stability has been studied in [72] for strongly localized modes. Experimentally, discrete vector solitons were reported in [73] in Kerr nonlinear waveguide arrays. The vector elements consisted of two coherently coupled orthogonal polarizations. In spite of four-wave mixing effects, such solitons were found to be stable. In the continuous periodic model [74] or in discrete superlattices [75] the presence of multiple Bragg gap can allow for more complicated localized vector structures, such as multigap vector solitons, which have components with eigenvalues in different gaps of the band structure.

Another two-component vector family is that of quadratic solitons. In such media, the fundamental and the second harmonic interact to form a localized soliton solution. Theoretically, discrete quadratic solitons were analyzed in [76]. Discrete solitons with two frequency components mutually locked by a quadratic nonlinearity have been observed in [77]. Experiments have been performed in waveguide arrays with tunable quadratic nonlinearity.

Spatiotemporal effects have been analyzed theoretically in [78] and experimentally in [79]. Temporal dispersion results to a sharp transition from strong diffraction at low powers to strong localization at high powers. In [80] it was shown that two components consisting of a periodic and a localized wave, such as that of an optical lattice, are exact vector soliton solutions of the system. The analytic form of the solution was derived in [81]. Families of exact solutions were also derived in [82] for the case of a linear-nonlinear structure.

Recently, the study of discrete/lattice solitons at interfaces has attracted considerable attention. Such solutions were first predicted to exist at the edge of an array above a certain power threshold [83]. Surface lattice solitons can also exist in the Bragg gap of an optical lattice [84]. Experimentally, surface solitons were observed in [85] at the interface between a nonlinear self-focusing waveguide lattice and a continuous medium. Surface gap solitons were observed in [86].

Discrete solitons traveling in the transverse plane are known to decelerate in periodic systems due to the presence of the Peierls-Nabarro potential [56, 87]. In [88] it was shown that by using a special prechirped ansatz traveling waves are more robust as compared to regular linear chirp. In [89] it was shown that in discrete systems with saturable nonlinearity traveling modes can exist for specific values of the spectrum. Such solutions can be considered as embedded solitons.

Two-dimensional X-wave nondiffracting solutions are known to exist in linear bidispersive optical systems [90]. This family of optical waves has been excited in waveguide arrays, by using the interplay between discrete diffraction and normal temporal dispersion, in the presence of Kerr nonlinearity [91].

In the case of a periodic potential with a low-index defect, localized defect modes exist as a result of repeated Bragg reflections [92]. Strongly confined defect modes appear when the lattice intensity at the defect site is nonzero rather than zero. Furthermore, it is possible to construct a waveguide lattice that relies on the effect of bandgap (Bragg) guidance, rather than total internal reflection, in the regions between defects [93]. In the nonlinear regime the Kerr effect can counteract diffraction leading to the formation of gap lattice solitons.

Another family of solutions is that of dissipative discrete solitons. Such solitons were studied first in the context of the discrete cubic-quintic Ginzburg-Landau equation [94]. In that work, the basic families of solutions and their stability were analyzed. Dissipative lattice solitons were also studied theoretically in waveguide array configurations that involve periodically patterned semiconductor optical amplifiers and saturable absorbers [95]. Exact solutions for dissipative discrete solitons can be found, when the discretization of the Ginzburg-Landau equation is similar to the Ablowitz-Ladik model [96].

The theoretical and experimental investigation of optical beam interactions was reported on [97]. Discrete solitons in periodic diffraction managed systems were studied in [98]. Lattice solitons have been studied in other settings, such as nematic liquid crystals [99] which are supported due to a nonlocal nonlinearity. For reviews on the properties of one-dimensional discrete solitons see also [8, 100–102].

5.6 Two-dimensional Lattice Solitons

Two-dimensional settings allow better control of the flow of light as compared to a planar geometry. For example, in [10–12] it was shown theoretically that discrete solitons can be navigated in two-dimensional networks of nonlinear waveguide arrays. This can be accomplished via vector interactions between two classes of discrete solitons: signals and blockers. Discrete solitons in such two-dimensional networks can exhibit a rich variety of functional operations, e.g., blocking, routing, logic functions, and time gating.

Following [49], we are going to analyze basic properties of solitons in 2D periodic lattices. In Fig. 5.7 typical existence curves as well as field profiles of two-dimensional LS are shown. The self-focusing soliton shown in Fig. 5.7a exists in the semi-infinite (TIR) bandgap below the first band. On the other hand, the self-defocusing solitons shown in Fig. 5.7b exist in the finite (Bragg) bandgap between the first and the second band. It is important to notice that a complete bandgap is always required for gap lattice solitons to exist, i.e., shallow potentials do not support gap lattice solitons. If the bandgap is only partial (a situation not encountered in 1D), an input beam will radiate due to the interactions with the linear spectrum.

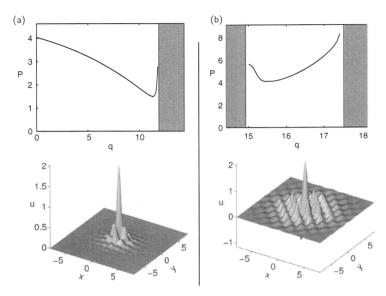

Fig. 5.7. The existence (power vs. eigenvalue) curves (*top*) for self-focusing (**a**) and self-defocusing (**b**) Kerr nonlinearity. Gray areas represent bands. On the bottom typical field profiles of these soliton families are depicted

The power P conveyed by the solitons versus the eigenvalue q is shown on the top of Fig. 5.7. Note that there is a minimum power threshold required in order to observe a lattice soliton (in a finite or infinite bandgap) in two dimensions. In the 1D case, such a threshold does not exist [25, 61] (see Fig. 5.5). In the case of a semi-infinite band gap these results are in agreement with the discrete nonlinear Schrödinger case [103, 104] as rigorously proven in [105].

The existence curve also provides information on the stability of the solitons. For the self-focusing case the stability can be determined by a straightforward application of the Vakhitov-Kolokolov criterion [106]. More specifically, when $\partial P/\partial q < 0$ the solutions are stable, while, close to the band $\partial P/\partial q > 0$ and the lattice solitons become *unstable*. This analysis cannot be applied directly to the defocusing case (the soliton amplitude has nodes).

In a Kerr nonlinear NLS equation in two (critical) or three (supercritical) spatial dimensions an input beam collapses in finite time to a singularity. A characteristic property of DNLS lattices is the absence of collapse irrespectively of the dimensionality of the problem [107]. However, this behavior does not always convey to periodic systems [108]. Specifically, when the soliton is highly confined into the lattice, it becomes unstable in the supercritical case. In the critical case, although the soliton is mathematically stable, its basin of attraction is so small that the lattice soliton is physically unstable. On the other hand, broad solitons in critical and supercritical dimensions can

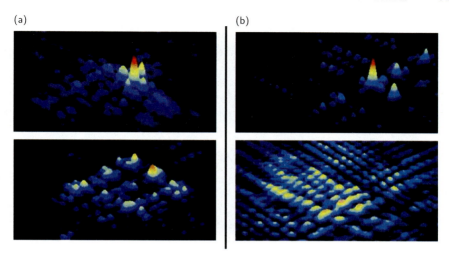

Fig. 5.8. Experimental results depicting the intensity structure of the probe beam at the exit facet of the 6 mm long crystal. (**a**) self-focusing nonlinearity, discrete diffraction at low intensity (*bottom*), and formation of self-focusing TIR lattice solitons at high intensities (*top*), (**b**) self-defocusing nonlinearity, discrete diffraction at low intensity (*bottom*), formation of self-defocusing gap lattice solitons at high intensities (*top*)

be stable if the lattice confinement is high, in agreement with the discrete model.

Following [7] experimental results of 2D lattice solitons are shown in Fig. 5.8. The waveguide array is induced in real time, in a photosensitive material from two pairs of plane waves. A separate "probe" beam is launched into the periodic waveguide array. The experimental results for self-focusing nonlinearity are shown in Fig. 5.8a. The "Bloch momentum" of the probe beam is zero (base of the 1st Brillouin zone). The probe beam is launched into a single waveguide. At low voltages, the beam propagates, essentially, linearly and as a result discrete diffraction concentrates the signal intensity into the outer perimeter of a square (Fig. 5.8a, bottom). For a stronger nonlinearity (at higher voltage), self-focusing dominates and a lattice soliton is formed (Fig. 5.8a, top). An interferogram of this soliton (not shown), obtained by interfering the soliton output beam with a plane wave, shows constructive interference of all the elements; that is, the central peak is in-phase with its neighbors.

The formation of self-defocusing lattice solitons is shown in Fig. 5.8b. In this case the Bloch momentum of the "probe" beam lies in the vicinity of the M point of the first Brillouin zone, so as to excite gap lattice solitons with eigenvalues inside the first bandgap. At low voltages, a diffuse diffraction pattern occurs (Fig. 5.8a, bottom), while a self-defocusing gap lattice soliton is observed (Fig. 5.8a, top) at a higher nonlinearity. Self-focusing solitons in the TIR regime have also been shown in [109]. An interferogram (not shown) confirms the π phase difference between first neighbors: the central peak is lowered while the surrounding lobes increase their intensity, indicating destructive and constructive interference, respectively.

Interaction of a 2D lattice soliton with a lattice was shown in [110]. In this work, other phenomena such as lattice dislocation, and creation of structures akin to optical polarons were demonstrated. Two-dimensional gap lattice solitons can also exist in the presence of self-focusing nolninearity. Such a family of solutions was studied theoretically and experimentally in [111].

Two-dimensional lattice solitons at interfaces have been studied theoretically in [112, 113]. Experimentally, surface lattice solitons were observed at the boundaries of a finite optically induced photonic lattice [114] and at the edge of femtosecond laser-written waveguide arrays in fused silica [115].

Families of two-dimensional dissipative Ginzburg-Landau solitons have been studied in [116]. Discrete solitons and their stability in Honeycomb lattices were examined theoretically in [33]. Two-dimensional TIR and gap solitons in such lattices were observed in [18, 19]. In square lattices, dipole like modes [117] and two-dimensional lattice vector solitons [118] have been studied theoretically and experimentally.

Nondiffractive rotary Bessel lattices can support families of localized waves. In particular, in addition to the lowest order soliton trapped in the center of the lattice, solitons can be trapped at different lattice rings [20, 21].

Quasicrystals are structures with long range order but no periodicity. The lack of periodicity excludes the possibility of describing quasicrystals with analytical tools, such as Bloch's theorem and Brillouin zones. In [22] it was demonstrated that light launched at different quasicrystal sites travels through the lattice in a way equivalent to quantum tunneling of electrons in a quasiperiodic potential. At high intensities lattice solitons are formed.

Anderson localization theory predicts that an electron may become immobile when placed in a disordered lattice. The origin of localization is interference between multiple scatterings of the electron by random defects in the potential altering the eigenmodes from being extended (Floquet-Bloch waves) to exponentially localized. In [119] the experimental observation of Anderson localization in a perturbed periodic potential was reported.

5.7 Vortex Solitons in Lattices

Vortex solitons are self-localized solutions of nonlinear wave equations, which are characterized by a phase singularity at the pivotal point. The phase charge of a simple closed curve surrounding the vortex core is equal to $2\pi m$, where m is the integer vorticity of the solution.

The optical case of discrete vortices was considered in Kerr nonlinear waveguide arrays, where on-site vortices (vortices whose singularity is located

5 Introduction to Solitons in Photonic Lattices   93

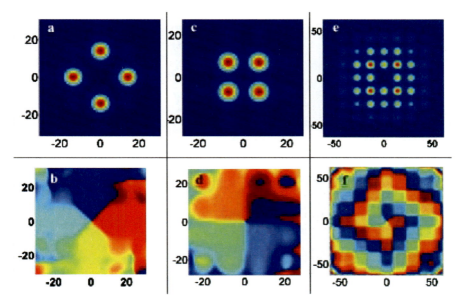

Fig. 5.9. Calculated intensity and phase of (**a**), (**b**) the on-site and (**c**), (**d**) the off-site vortex lattice solitons, along with the output diffraction pattern of (**e**), (**f**) the off-site vortex at $z = 800$

on a lattice site) [120] and off-site vortices (vortices whose singularity is located between sites) [109,121] were studied. Both cases were found to be stable within a certain range of parameters. Experimental results on optical vortex solitons were presented in [122,123]. Here, we present results from [123].

Typical beam propagation results showing on-site and off-site vortex solitons for $z = 800$ are given in Figs. 5.9a–5.9d. The main four "lobes" all have the same peak intensity and, importantly, each lobe is $\pi/2$ out of phase with its neighbors. Note again that the singularity of the on-site vortex is centered on a lattice site (Figs. 5.9a and 5.9b), whereas the singularity of the off-axis vortex is centered between four lattice sites (Figs. 5.9c and 5.9d). The soliton exhibits stationary propagation, and the shapes of the vortices remain unchanged, i.e., these are, indeed, vortex lattice solitons. For comparison, when the nonlinearity is set to zero, the vortices diffract by tunneling between lattice sites, as shown in Figs. 5.9e and 5.9f for the off-site vortex after $z = 800$. Note that the phase of the diffracting beam maintains its spiral structure throughout diffraction (Fig. 5.9f).

Experimental results are shown in Fig. 5.10. The photorefractive screening nonlinearity is controlled by applying voltage against the crystalline c-axis. At a low voltage (~ 100 V), the output diffraction of both the on-site and off-site vortices looks similar. Figures 5.10a and 5.10b show the diffraction pattern of an on-site vortex after 5 mm of propagation, showing that both the hole

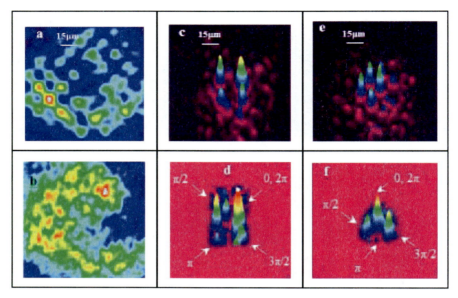

Fig. 5.10. Experimental results at the output face of crystal. (**a**), (**b**) Diffraction at 100 V: (a) intensity and (b) phase information formed by interference of output with a plane wave. (**c**) Typical on-site soliton at 700 V and (**d**) relative phase from an interferogram. (**e**) Typical off-site soliton at 700 V and (**f**) relative phase from an interferogram. Note that phase information is extremely sensitive to background noise, so only the interference patterns from the relevant lattice sites are shown

and the width of the ring expand through the lattice. The ring expands to roughly 3 times its original size (Fig. 5.10a), while an interferogram, created by interfering the output pattern with a plane wave, clearly shows the $0 \to 2\pi$ spiral phase structure of the vortex (Fig. 5.10b). At around 700 V, the input vortices become vortex lattice solitons and maintain their structure (intensity and phase) while propagating. Figures 5.10c and 5.10d show the intensity and phase of an on-site soliton, while Figs. 5.10e and 5.10f show those features for an off-site soliton. Because phase information is very sensitive to noise in the system, only the relative phase of the relevant (soliton) lattice sites is shown for clarity.

Other families of vortices have been subsequently studied. These include higher-band vortex solitons [124, 125], asymmetric vortex solitons [126], and multivortex solitons in triangular photonic lattices [127].

5.8 Random-phase lattice solitons

This chapter has dealt with coherent nonlinear phenomena in discrete lattices. It should be pointed out that many interesting nonlinear phenomena with partially incoherent light in photonic lattices have been discovered in the past few years. Lattice solitons made of partially-incoherent light were theoretically predicted in [128]. The first experimental observation of such solitons [129] has shown that due to the simultaneous excitation of multiple bands, and the interplay of dispersion in a lattice and the nonlinearity, the spatial power spectra of the observed solitons is multi-humped, with humps being located in the normal diffraction regions for self-focusing nonlinearity and vice versa for the self-defocusing nonlinearity. The dynamics of incoherent light in nonlinear lattices, which leads to such intricate structures has lead to a technique for Brillouin zone spectroscopy [130], and studies of other nonlinear phenomena with incoherent light in photonic lattices (e.g., see [131] and Refs. therein).

References

1. A.L. Jones, J. Opt. Soc. Am. **55**, 261 (1965)
2. S. Somekh, E. Garmire, A. Yariv, H.L. Garvin, and R.G. Hunsperger, Appl. Phys. Lett. **22**, 46 (1973)
3. D.N. Christodoulides and R.I. Joseph, Opt. Lett. **13**, 794 (1988)
4. H.S. Eisenberg, Y. Silberberg, R. Morandotti, A.R. Boyd, and J.S. Aitchison, Phys. Rev. Lett. **81**, 3383 (1998)
5. N.K. Efremidis, D.N. Christodoulides, S. Sears, J.W. Fleischer, and M. Segev, Phys. Rev. E **66**, 046602 (2002)
6. J.W. Fleischer, T. Carmon, M. Segev, N.K. Efremidis, and D.N. Christodoulides, Phys. Rev. Lett. **90**, 023902 (2003)
7. J.W. Fleischer, M. Segev, N.K. Efremidis, and D.N. Christodoulides, Nature **422**, 147 (2003)
8. D.N. Christodoulides, F. Lederer, and Y. Silberberg, Nature **424**, 817 (2003)
9. F. Lederer, G.I. Stegeman, D.N. Christodoulides, G. Assanto, M. Segev, and Y. Silberberg, Phys. Rep. **463**, 1 (2008)
10. D.N. Christodouldes and E.D. Eugenieva, Phys. Rev. Lett. **87**, 233901, (2001)
11. D.N. Christodouldes and E.D. Eugenieva, Opt. Lett. **23**, 1876 (2001)
12. E.D. Eugenieva, N.K. Efremidis, and D.N. Christodoulides, Opt. Lett. **26**, 1978 (2001)
13. T. Pertsch, U. Peschel, and F. Lederer, Opt. Lett. **28**, 102 (2003)
14. T. Pertsch, U. Peschel, J. Kobelke, K. Schuster, H. Bartelt, S. Nolte, A. Tünnermann, and F. Lederer, Phys. Rev. Lett. **93**, 053901 (2004)
15. T. Pertsch, U. Peschel, F. Lederer, J. Burghoff, M. Will, S. Nolte, and A. Tünnermann, Opt. Lett. **29**, 468 (2004)
16. S.R. Singh, M.I. Carvalho, and D.N. Christodoulides, Opt. Lett. **20**, 2177 (1995)
17. F. Dalfovo, S. Giorgini, L.P. Pitaevskii, and S. Stringari, Rev. Mod. Phys. **71**, 463 (1999)
18. O. Peleg, G. Bartal, B. Freedman, O. Manela, M. Segev, and D.N. Christodoulides, Phys. Rev. Lett. **98**, 103901 (2007)
19. C.R. Rosberg, D.N. Neshev, A.A. Sukhorukov, W. Krolikowski, and Y.S. Kivshar, Opt. Lett. **32**, 397 (2007)
20. Y.V. Kartashov, V.A. Vysloukh, and L. Torner, Phys. Rev. Lett. **93**, 093904 (2004)
21. X. Wang, Z. Chen, and P.G. Kevrekidis, Phys. Rev. Lett. **96**, 083904 (2006)

22. B. Freedman, G. Bartal, M. Segev, R. Lifshitz, D.N. Christodoulides, and J.W. Fleischer, Nature **440**, 1166 (2006)
23. C. Kittel, *Introduction to Solid State Physics*, Wiley, New York (1986)
24. P. Yeh, *Optical waves in layered media*, John Wiley and Sons, New York (1988)
25. N.K. Efremidis and D.N. Christodoulides, Phys. Rev. A **67**, 063608 (2003)
26. A.S. Davydov, J. Theor. Biol. **38**, 559 (1973)
27. W.P. Su, J.R. Schrerffer, and A.J. Heeger, Phys. Rev. Lett. **42**, 1698 (1979)
28. A.J. Sievers and S. Takeno, Phys. Rev. Lett. **61**, 970 (1988)
29. A. Trombettoni and A. Smerzi, Phys. Rev. Lett. **86**, 2353 (2001)
30. A. Scott, Phys. Rep. **217**, 1 (1992)
31. S. Flach and C.R. Willis, Phys. Rep. **295**, 181 (1998)
32. R. Lai and A.J. Sievers, Phys. Rep. **314**, 148 (1999)
33. P.G. Kevrekidis, B.A. Malomed, and Y.B. Gaididei, Phys. Rev. E **66**, 016609 (2002)
34. H.S. Eisenberg, Y. Silberberg, R. Morandotti, and J.S. Aitchison, Phys. Rev. Lett. **85**, 1863 (2000)
35. T. Pertsch, T. Zentgraf, U. Peschel, A. Bräuer, and F. Lederer, Phys. Rev. Lett. **88**, 093901 (2002)
36. D.H. Dunlap and V.M. Kenkre, Phys. Rev. B **34**, 3625 (1986)
37. F. Bloch, Z. Phys. **52**, 555 (1928)
38. U. Peschel, T. Persch, and F. Lederer, Opt. Lett. **23**, 1701 (1998)
39. T. Pertsch, P. Dannberg, W. Elflein, A. Bräuer, and F. Lederer, Phys. Rev. Lett. **83**, 4752 (1999)
40. R. Morandotti, U. Peschel, J.S. Aitchison, H.S. Eisenberg, and Y. Silberberg, Phys. Rev. Lett. **83**, 4756 (1999)
41. G. Lenz, R. Parker, M.C. Wanke, and C.M. de Sterke, Opt. Communications **218**, 87 (2003)
42. S. Longhi, Opt. Lett. **30**, 2137 (2005)
43. S. Longhi, M. Marangoni, M. Lobino, R. Ramponi, P. Laporta, E. Cianci, and V. Foglietti, Phys. Rev. Lett. **96**, 243901 (2006)
44. R. Iyer, J.S. Aitchison, J. Wan, M.M. Dignam, and C.M. de Sterke, Opt. Express **15**, 3112 (2007)
45. G. Floquet, Ann. École Norm. Sup. **12**, 47 (1883)
46. R. de L. Kronig and W.G. Penney, P. Roy. Soc. Lond. A Mat. **814**, 499 (1931)
47. J.C. Slater, Phys. Rev. **87**, 807 (1952)
48. D. Mandelik, H.S. Eisenberg, Y. Silberberg, R. Morandotti, and J.S. Aitchison, Phys. Rev. Lett. **90**, 053902 (2003)
49. N.K. Efremidis, J. Hudock, D.N. Christodoulides, J.W. Fleischer, O. Cohen, and M. Segev, Phys. Rev. Lett. **91**, 213906 (2003)
50. C. Zener, P. Roy. Soc. Lond. A Mat. **145**, 523 (1934)
51. H. Trompeter, T. Pertsch, F. Lederer, D. Michaelis, U. Streppel, and A. Bräuer, Phys. Rev. Lett. **96**, 023901 (2006)
52. H. Trompeter, W. Krolikowski, D.N. Neshev, A.S. Desyatnikov, A.A. Sukhorukov, Y.S. Kivshar, T. Pertsch, U. Peschel, and F. Lederer, Phys. Rev. Lett. **96**, 053903 (2006)
53. O. Cohen, B. Freedman, J.W. Fleischer, M. Segev, and D.N. Christodoulides, Phys. Rev. Lett. **93**, 103902 (2004)
54. O. Manela, M. Segev, and D.N. Christodoulides, Opt. Lett. **30**, 261 (2005)
55. M.J. Ablowitz and J.F. Ladik, J. Math. Phys. **17**, 1011 (1976)

56. Y.S. Kivshar and D.K. Campbell, Phys. Rev. E **48**, 3077 (1993)
57. S. Darmanyan, A. Kobyabov, and F. Lederer, JETP **86**, 682 (1998)
58. P.G. Kevrekidis, K.Ø. Rasmussen, and A.R. Bishop, Int. J. Mod. Phys. B **15**, 2833 (2001)
59. Y.S. Kivshar, Opt. Lett. **14**, 1147 (1993)
60. R.S. MacKay and S. Aubry, Nonlinearity **6**, 1623 (1994)
61. P.J.Y. Louis, E.A. Ostrovskaya, C.M. Savage, and Y.S. Kivshar, Phys. Rev. A **67**, 013602 (2003)
62. D.N. Neshev, E.A. Ostrovskaya, Y.S. Kivshar, and W. Krolikowski, Opt. Lett. **28**, 710 (2003)
63. D. Mandelik, R. Morandotti, J.S. Aitchison, and Y. Silberberg, Phys. Rev. Lett. **92**, 093904 (2004)
64. A.A. Sukhorukov, D.N. Neshev, W. Krolikowski, and Y.S. Kivshar, Phys. Rev. Lett. **92**, 093901 (2004)
65. D.N. Neshev, A.A. Sukhorukov, B. Hanna, W. Krolikowski, and Y.S. Kivshar, Phys. Rev. Lett. **93**, 083905 (2004)
66. Y.S. Kivshar and M. Salerno, Phys. Rev. E **49**, 3543 (1994)
67. J. Meier, G.I. Stegeman, D.N. Christodoulides, Y. Silberberg, R. Morandotti, H. Yang, G. Salamo, M. Sorel, and J.S. Aitchison, Phys. Rev. Lett. **92**, 163902 (2004)
68. M. Stepić, C. Wirth, C.E. Rüter, and D. Kip, Opt. Lett. **31**, 247 (2006)
69. Y.S. Kivshar, W. Krolikowski, and O.A. Chubykalo, Phys. Rev. E **50**, 5020 (1994)
70. R. Morandotti, H.S. Eisenberg, Y. Silberberg, M. Sorel, and J.S. Aitchison, Phys. Rev. Lett. **86**, 3296 (2001)
71. D.N. Christodoulides. and R.I. Joseph, Opt. Lett. **13**, 53 (1988)
72. S. Darmanyan, A. Kobyakov, E. Schmidt, and F. Lederer, Phys. Rev. E **57**, 3520 (1998)
73. J. Meier, J. Hudock, D.N. Christodoulides, G.I. Stegeman, Y. Silberberg, R. Morandotti, and J.S. Aitchison, Phys. Rev. Lett. **91**, 143907 (2003)
74. O. Cohen, T. Schwartz, J.W. Fleischer, M. Segev, and D.N. Christodoulides, Phys. Rev. Lett. **91**, 113901 (2003)
75. A.A. Sukhorukov and Y.S. Kivshar, Phys. Rev. Lett. **91**, 113902 (2003)
76. T. Peschel, U. Peschel, and F. Lederer, Phys. Rev. E **57**, 1127 (1998)
77. R. Iwanow, R. Schiek, G.I. Stegeman, T. Pertsch, F. Lederer, Y. Min, and W. Sohler, Phys. Rev. Lett. **93**, 113902 (2004)
78. A.B. Aceves, G.G. Luther, C. De Angelis, A.M. Rubenchik, and S.K. Turitsyn, Phys. Rev. Lett. **75**, 73 (1995)
79. D. Cheskis, S. Bar-Ad, R. Morandotti, J.S. Aitchison, H.S. Eisenberg, Y. Silberberg, and D. Ross, Phys. Rev. Lett. **91**, 223901 (2003)
80. A.S. Desyatnikov, E.A. Ostrovskaya, Y.S. Kivshar, and C. Denz, Phys. Rev. Lett. **91**, 153902 (2003)
81. H.J. Shin, Phys. Rev. E **69**, 067602 (2004)
82. Y. Kominis, Phys. Rev. E **73**, 066619 (2006)
83. K.G. Makris, S. Suntsov, D.N. Christodoulides, G.I. Stegeman, and A. Hache, Opt. Lett. **30**, 2466 (2005)
84. Y.V. Kartashov, V.A. Vysloukh, and L. Torner, Phys. Rev. Lett. **96**, 073901 (2006)

85. S. Suntsov, K.G. Makris, D.N. Christodoulides, G.I. Stegeman, and A. Haché, R. Morandotti, H. Yang, G. Salamo, and M. Sorel, Phys. Rev. Lett. **96**, 063901 (2006)
86. C.R. Rosberg, D.N. Neshev, W. Krolikowski, A. Mitchell, R.A. Vicencio, M.I. Molina, and Y.S. Kivshar, Phys. Rev. Lett. **97**, 083901 (2006)
87. R. Morandotti, U. Peschel, J.S. Aitchison, H.S. Eisenberg, and Y. Silberberg, Phys. Rev. Lett. **83**, 2726 (1999)
88. M.J. Ablowitz, Z.H. Musslimani, and G. Biondini, Phys. Rev. E **65**, 026602 (2002)
89. T.R.O. Melvin, A.R. Champneys, P.G. Kevrekidis, and J. Cuevas, Phys. Rev. Lett. **97**, 124101 (2006)
90. D.N. Christodoulides, N.K. Efremidis, P. Di Trapani, and B.A. Malomed, Opt. Lett. **29**, 1446 (2003)
91. Y. Lahini, E. Frumker, Y. Silberberg, S. Droulias, K. Hizanidis, R. Morandotti, and D.N. Christodoulides, Phys. Rev. Lett. **98**, 023901 (2007)
92. F. Fedele, J. Yang, and Z. Chen, Opt. Lett. **30**, 1506 (2005)
93. N.K. Efremidis and K. Hizanidis, Opt. Express **13**, 10571 (2005)
94. N.K. Efremidis and D.N. Christodoulides, Phys. Rev. E **67**, 026606 (2003)
95. E.A. Ultanir, G.I. Stegeman, and D.N. Christodoulides, Opt. Lett. **29**, 845 (2004)
96. K. Maruno, A. Ankiewicz, and N. Akhmediev, Opt. Communications **221**, 199 (2003)
97. J. Meier, G.I. Stegeman, Y. Silberberg, R. Morandotti, and J.S. Aitchison, Phys. Rev. Lett. **93**, 093903 (2004)
98. M.J. Ablowitz and Z.H. Musslimani, Phys. Rev. Lett. **87**, 254102
99. G. Assanto, A. Fratalocchi, and M. Peccianti, Opt. Express **15**, 5248 (2007)
100. A.B. Aceves, C. De Angelis, T. Peschel, R. Muschall, F. Lederer, S. Trillo, and S. Wabnitz, Phys. Rev. E **53**, 1172 (1996)
101. H.S. Eisenberg, R. Morandotti, Y. Silberberg, J.M. Arnold, G. Pennelli, and J.S. Aitchison, J. Opt. Soc. Am. B **19**, 2938 (2002)
102. U. Peschel, R. Morandotti, J.M. Arnold, J.S. Aitchison, H.S. Eisenberg, Y. Silberberg, T. Pertsch, and F. Lederer, J. Opt. Soc. Am. B **19**, 2637 (2002)
103. V.K. Mezentsev, S.L. Musher, I.V. Ryzhenkova, and S.K. Turitsyn, JETP Lett. **60**, 829 (1994)
104. S. Flach, K. Kladko, and R.S. MacKay, Phys. Rev. Lett. **78**, 1207 (1997)
105. M.I. Weinstein, Nonlinearity **12**, 673 (1999)
106. N.G. Vakhitov and A.A. Kolokolov, Radiophys. Quantum Electron. **16**, 783 (1973)
107. E.W. Laedke, K.H. Spatschek, and S.K. Turitsyn, Phys. Rev. Lett. **73**, 1055 (1994)
108. Y. Sivan, G. Fibich, N.K. Efremidis, and S. Bar-Ad, Nonlinearity **21**, 509 (2008)
109. J. Yang and Z.H. Musslimani, Opt. Lett. **21**, 2094 (2003)
110. H. Martin, E.D. Eugenieva, Z. Chen, and D.N. Christodoulides, Phys. Rev. Lett. **92**, 123902 (2004)
111. R. Fischer, D. Träger, D.N. Neshev, A.A. Sukhorukov, W. Krolikowski, C. Denz, and Y.S. Kivshar, Phys. Rev. Lett. **96**, 023905 (2006)
112. Y.V. Kartashov, A.A. Egorov, V.A. Vysloukh, and L. Torner, Opt. Express **14**, 4049 (2006)

113. K.G. Makris, J. Hudock, D.N. Christodoulides, G.I. Stegeman, O. Manela, and M. Segev, Opt. Lett. **31**, 2774 (2006)
114. X. Wang, A. Bezryadina, Z. Chen, K.G. Makris, D.N. Christodoulides, and G.I. Stegeman, Phys. Rev. Lett. **98**, 123903 (2007)
115. A. Szameit, Y.V. Kartashov, F. Dreisow, T. Pertsch, S. Nolte, A. Tünnermann, and L. Torner, Phys. Rev. Lett. **98**, 173903 (2007)
116. N.K. Efremidis, D.N. Christodoulides, and K. Hizanidis, Phys. Rev. A **76**, 043839 (2007)
117. J. Yang, I. Makasyuk, A. Bezryadina, and Z. Chen, Opt. Lett. **29**, 1662 (2004)
118. Z. Chen, A. Bezryadina, I. Makasyuk, and J. Yang, Opt. Lett. **29**, 1656 (2004)
119. T. Schwartz, G. Bartal, S. Fishman, and M. Segev, Nature **446**, 52 (2007)
120. B.A. Malomed and P.G. Kevrekidis, Phys. Rev. E **64**, 026601 (2001)
121. B.B. Baizakov, B.A. Malomed, and M. Salerno, Europhys. Lett. **63**, 642 (2003)
122. D.N. Neshev, T.J. Alexander, E.A. Ostrovskaya, Y.S. Kivshar, H. Martin, I. Makasyuk, and Z. Chen, Phys. Rev. Lett. **92** (2004)
123. J.W. Fleischer, G. Bartal, O. Cohen, O. Manela, M. Segev, J. Hudock, and D.N. Christodoulides, Phys. Rev. Lett. **92**, 123904 (2004)
124. G. Bartal, O. Manela, O. Cohen, J.W. Fleischer, and M. Segev, Phys. Rev. Lett. **95**, 053904 (2005)
125. O. Manela, O. Cohen, G. Bartal, J.W. Fleischer, and M. Segev, Opt. Lett. **29**, 2049 (2004)
126. T.J. Alexander, A.A. Sukhorukov, and Y.S. Kivshar, Phys. Rev. Lett. **93**, 063901 (2004)
127. T.J. Alexander, A.S. Desyatnikov, and Y.S. Kivshar, Opt. Lett. **32**, 1293 (2007)
128. H. Buljan, O. Cohen, J.W. Fleischer, T. Schwartz, M. Segev, Z.H. Musslimani, N.K. Efremidis, and D.N. Christodoulides, Phys. Rev. Lett. **92**, 223901 (2004)
129. O. Cohen, G. Bartal, H. Buljan, T. Carmon, J.W. Fleischer, M. Segev, and D.N. Christodoulides, Nature **433**, 500 (2005)
130. G. Bartal, O. Cohen, H. Buljan, J.W. Fleischer, O. Manela, and M. Segev, Phys. Rev. Lett. **94**, 163902 (2005)
131. H. Buljan, G. Bartal, O. Cohen, T. Schwartz, O. Manela, T. Carmon, M. Segev, J.W. Fleischer, and D.N. Christodoulides, Stud. Appl. Math. **115**, 173 (2005)

6

Complex Nonlinear Photonic Lattices: From Instabilities to Control

Jörg Imbrock, Bernd Terhalle, Patrick Rose, Philip Jander, Sebastian Koke, and Cornelia Denz

Institut für Angewandte Physik and Center for Nonlinear Science, Westfälische Wilhelms-Universität Münster, Corrensstr. 2/4, 48149 Münster, Germany
imbrock@uni-muenster.de

6.1 Introduction

Nonlinear periodic structures have become an active area of research due to many exciting possibilities of controlling wave propagation, steering and trapping. Periodicity changes the wave bandgap spectrum and therefore strongly affects propagation and localization, leading to the formation of discrete and gap solitons which have already been studied in several branches of science [1–4].

In optics, a periodic modulation of the refractive index can either be prefabricated as in photonic crystals [5] or optically induced in photorefractive materials [6–9]. Until now, several different approaches for the fabrication of photonic crystals exist [10–12]. Although these mechanisms enable a precise material structuring with periodicities adequate for optical waves, they do not allow for flexible changes of structural parameters (e.g., lattice period or modulation depth). In contrast, the optical induction in photorefractive crystals provides highly reconfigurable, wavelength-sensitive nonlinear structures which can be induced at very low power levels.

When dealing with optically induced photonic lattices in these photorefractive materials, it is crucially important to consider the anisotropic properties of photorefractive crystals. The light-induced refractive index change strongly depends on orientation as well as polarization of the lattice wave [13, 14]. In particular, its orientation with respect to the c-axis of the crystal determines the symmetry of the induced pattern [15]. The shape of the induced refractive index pattern also changes with increasing lattice strength depending on the saturation of the photorefractive nonlinearity. For instance, an ordinarily polarized light pattern created by several interfering plane waves induces a change of the refractive index while propagating linearly along the crystal. The lattice wave does not 'feel' the periodic modulated refractive index during propagation. If the lattice is weak, i.e. it is not affected by the saturation of the photorefractive nonlinearity, the light-induced refractive index follows the light intensity distribution and forms a two-dimensional photonic lattice, being uniform in the direction of propagation. Many exciting features of nonlinear light propagation have been investigated in these lattices and have been presented in chapter 5.

An extraordinarily polarized periodic wave in contrast gets self-trapped if the diffractionless light pattern can propagate without change of its profile, becoming an eigenmode of the self-induced periodic potential. Such a stationary periodic nonlinear wave is a soliton-like lattice. These flexible nonlinear photonic lattices offer many new possibilities for the study of nonlinear effects in periodic systems.

In this chapter we will give an overview of our recent experimental and theoretical results of the properties and features of optically induced strongly nonlinear photonic lattices in photorefractive media. In the following section we will present the experimental techniques to engineer the desired photonic lattices and discuss the underlying theory. Many kinds of lattices with different symmetries can be optically imprinted in a photorefractive crystal. Exploiting the anisotropy and saturation of the photorefractive nonlinearity, allows inducing strongly nonlinear photonic lattices up to self-trapped lattices. The orientation and polarization anisotropy of the lattices is discussed in section 6.3. The periodic modulation of the refractive index changes the wave bandgap spectrum and therefore the light propagation. By flexibly tuning the strength and form of the refractive index change light can be self-trapped in the form of discrete and gap solitons (section 6.4). The combination of weakly and strongly nonlinear features of a lattice leads to hybrid lattices which are presented in section 6.5. The variation of the strength of the nonlinearity represents an excellent tool to create more complex lattices for light propagation. Among them, superimposed lattices, quasiperiodic or random lattices seem to be attractive for studying fundamental laws of non-periodicity. In section 6.6, we show the potential to realize these lattices exploiting techniques of holographic optical storage, especially multiplexing of different lattices at a single location. Section 6.7 is devoted to the next step of complexity: complex beam propagation in complex photonic lattices – as e.g. the propagation of a dipole-mode gap soliton in a complex, triangular lattice. Finally, we discuss the potential of optically induced lattices to serve as control systems to stabilize spatial and temporal instabilities in nonlinear optical feedback systems. We show exemplarily how one- or two-dimensional photonic lattices can be used to stabilize counterpropagating spatial solitons. We demonstrate that spatio-temporal oscillations and chaotic temporal oscillations of two counterpropagating solitons can be successfully 'tamed' by a photonic lattice.

6.2 Optically Induced Lattices in Photorefractive Media

Two-dimensional photonic lattices can be optically induced in photorefractive crystals in different ways, e.g. by interfering a certain number plane waves (see chapter 5), or by amplitude modulation of a partially coherent optical beam (see chapter 7). In our experimental approach, a spatial *phase modulation* of a coherent optical beam is used. We have demonstrated experimentally that an array of in-phase spatial solitons can be produced by amplitude modulation where every soliton of the lattice induces a waveguide [16] as long as the spatial solitons are separated spatially in a sufficient way to prevent interaction. Therefore, the spatial periodicity of these soliton lattices is limited by attractive or repulsive interaction of the neighboring solitons that generates their strong instability (see also Sec. 6.8). In contrast, lattices created by out-of-phase spatial solitons are robust [13]. Therefore, phase-engineering is a powerful tool to generate non-diffracting light patterns. When engineering the light patters in an appropriate way, a variety of different forms and symmetries like square, diamond, hexagonal, or triangular-shaped lattices can be easily generated, opening the door to new insights into light propagation in complex lattices. A typical experimental setup is shown schematically in Fig. 6.1. A beam derived from a frequency-doubled Nd:YAG laser at a wavelength of 532 nm is sent through a combination of half wave plate and polarizing beam splitter in order to adjust the intensities. The desired pure phase modulation of the transmitted beam is achieved by using a programmable spatial light modulator. The modulated beam is then imaged at the input face of a $z = 20$ mm long Cerium doped strontium barium niobate (SBN:Ce) crystal by a high numerical aperture telescope. The crystal is biased by an externally applied electric field and uniformly illuminated with a white-light source to control the dark irradiance. A half wave plate is placed in front of the telescope so that lattices can be induced with ordinarily as well as extraordinarily polarized light.

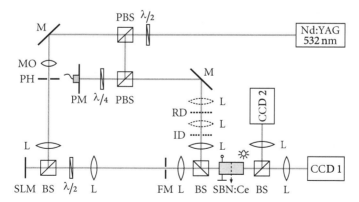

Fig. 6.1. Experimental setup, (P)BS: (Polarizing) beam splitter, MO: Microscope objective, PH: Pinhole, L: Lens, M: Mirror, PM: Piezoelectric mirror, SLM: Spatial light modulator, FM: Fourier mask, ID: Iris diaphragm, RD: Rotating diffuser, CCD1: Real space camera, CCD2: Fourier camera

The structure of the induced photonic lattice can be visualized in different ways. By switching off the modulator, the crystal can be illuminated with a broad plane wave to observe the wave guiding properties of the induced lattice, and hence to visualize the light-induced refractive index structure [17]. To ensure that the light will experience a strong refractive index modulation of the lattice, the beam is extraordinarily polarized. If a weak nonlinear lattice shall be operating, ordinary polarization can also be employed by simply adjusting the polarization in the system. The output of the crystal is analyzed with two CCD cameras. CCD1 monitors the real space output, whereas CCD2 is placed in the focal plane of a lens to visualize the Fourier power spectrum of the light exiting the lattice. Another possibility to analyze the structure is given by the Brillouin zone spectroscopy [15,18]. For this purpose, the second beam, also extraordinarily polarized, is passed through a rotating diffuser and the partially spatially incoherent output of the diffuser is imaged at the front face of the crystal. This results in partially coherent multi-band excitation of the lattice modes and enables a direct visualization of the lattice structure in Fourier space by mapping the boundaries of the extended Brillouin zone, which are defined through the Bragg reflection planes. By removing the rotating diffuser, two lenses and the iris diaphragm (dashed), the setup can be changed to observe the evolution of Bloch waves on the lattice. This is achieved by focusing a Gaussian probe beam onto the front face of the crystal and analyzing the output using CCD1.

To describe the propagation of an extraordinarily polarized beam in a photonic lattice as well as the optical induction process of the lattice, we employ the generalized nonlinear Schrödinger equation

$$i\frac{\partial E}{\partial z} + \nabla^2 E + n(I) E = 0, \tag{6.1}$$

where $I = |E|^2$ is the light intensity. The nonlinear contribution to the refractive index is given by

$$n(I) = \Gamma \frac{\partial \varphi}{\partial x}, \tag{6.2}$$

where $\Gamma = k^2 n_0^2 x_0^2 r_{\text{eff}} E_{\text{ext}}$ is defined through the effective electro-optic coefficient r_{eff}, the externally applied electric field E_{ext}, and $k = 2\pi n_0/\lambda$. The electro-optic coefficient of $Sr_{0.6}Ba_{0.4}Nb_2O_6$ for extraordinarily polarized light ($r_{33} \approx 235\,\text{pm/V}$) is about five times larger than the coefficient for ordinarily polarized light ($r_{13} \approx 47\,\text{pm/V}$). The electro-static potential of the optically induced space charge field pattern satisfies the equation [19]

$$\nabla^2 \varphi + \nabla\varphi \nabla \ln(1+I) = \frac{\partial}{\partial x}\ln(1+I), \tag{6.3}$$

where the gradient operator is $\nabla^2 = \partial^2/\partial x^2 + \partial^2/\partial y^2$ and the intensity I is measured in units of the background illumination intensity.

The desired stationary solutions which have a transverse intensity profile that does not change during propagation can be written as

$$E(x, y, z) = U(x, y) \exp(\mathrm{i} k z). \tag{6.4}$$

Inserting this into (6.1) results in an equation for the field envelope U

$$-kU + \left(\frac{\partial^2}{\partial x^2} + \frac{\partial^2}{\partial y^2}\right) U + \Gamma U \frac{\partial \phi}{\partial x} = 0. \tag{6.5}$$

Following [13, 14], we are looking for phase-modulated periodic waves of the form $U(X, Y) = U(X + 2\pi, Y + 2\pi)$ and the initial ansatz

$$U_0(X, Y) = A \sin X \sin Y \tag{6.6}$$

is used. Equations (6.3) and (6.5) can then be solved numerically using the relaxation technique described in [20].

Numerical simulations reveal that two different families of solutions evolve from the initial ansatz (6.6) depending on the spatial orientation of the lattice wave.

The *diamond lattice* is oriented diagonally at an angle of 45 degrees with respect to the c-axis of the crystal, thus giving $X = (x+y)/\sqrt{2}$ and $Y = (x-y)/\sqrt{2}$. The calculated field and the refractive index for this type of solution are shown in Fig. 6.2 for three different levels of saturation and a

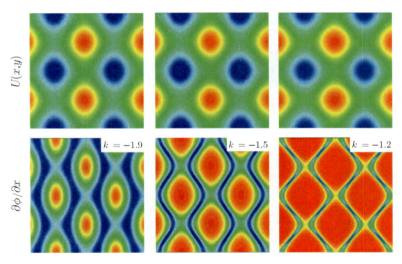

Fig. 6.2. Field $U(x, y)$ and refractive index $\partial_x \phi$ of the diamond lattice shown for three distinct values of the mode amplitude $\max(U(x, y))$ or the propagation constant k from low ($k = -1.9$) to high ($k = -1.2$) saturation

focusing nonlinearity (i.e. $\sigma = +1$). Note that the different levels of saturation are indicated by different values of k. This is due to the specific structure of the used equations. In general, the saturation of the induced refractive index change depends on the intensity I. However, Eq. (6.5) shows that the field envelope U and consequently the intensity I depend on the propagation constant k as well. Therefore, the level of saturation can also be described in terms of this constant [14]. The maximum amplitude $\max(U(x,y))$ vanishes in the limit $k \rightarrow -2$. Hence, $k = -2$ can be considered as the linear limit. Further analysis shows that in the general case $\Gamma \neq 1$ the solutions cover a band of $k = [-2, \Gamma - 2]$.

The second family of solutions is given by the *square lattice* which is essentially given by a 45 degrees tilted diamond lattice. Its corresponding variables are $X = x$ and $Y = y$ and the field and the refractive index are shown in Fig. 6.3 for the same values of the propagation constant k as for the diamond lattice.

Comparing the refractive index structures shown in Fig. 6.2 and Fig. 6.3, a striking difference between the solutions with different orientations can be observed. In the strong saturation regime, the regions of high refractive index are well separated for the diamond lattice but fuse to vertical lines for the square lattice.

Thus, the symmetry of the induced refractive index change strongly depends on the orientation of the lattice wave with respect to the c-axis of the crystal. The diamond lattice induces a truly two-dimensional refractive index lattice whereas the square lattice leads to an effectively one-dimensional refractive index pattern, although the original lattice wave is fully two-dimensional.

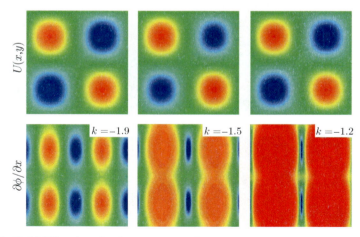

Fig. 6.3. Field $U(x,y)$ and refractive index $\partial_x \phi$ of the square lattice shown for three distinct values of the mode amplitude $\max(U(x,y))$ or the propagation constant k from low ($k = -1.9$) to high ($k = -1.2$) saturation

6.3 Anisotropy in Nonlinear Photonic Lattices

The optically induced refractive index structure in a photonic lattice depends on the orientation of the light pattern as well as on the polarization of the lattice wave. These anisotropies are examined and discussed in the following sections.

6.3.1 Orientation Anisotropy

To investigate the orientation anisotropy, the structure of the induced refractive index change is analyzed with two different methods: guiding of a broad extraordinarily polarized plane wave on the one hand, and Brillouin zone spectroscopy on the other hand.

Fig. 6.4 demonstrates the obtained results using a lattice wave with a period of 22 µm for a diamond (top row) and square pattern (bottom row), respectively. The lattice input is shown in real space (Figs. 6.4a and 6.4d). The intensity distribution of the plane wave guided by the lattice is shown in Figs. 6.4b and e. As the output intensity of the guided wave matches the induced refractive index change, the differences in the refractive indices for the two orientations caused by the anisotropy of the photorefractive crystal is clearly present. The waveguiding output consists of vertical lines for the square pattern (Fig. 6.4e), but keeps its fully two-dimensional structure for the diamond pattern (Fig. 6.4b).

The orientation-dependent structure of the induced refractive index change is clearly demonstrated in the Brillouin zone pictures (Figs. 6.4c and 6.4f), too. For the diamond pattern the two-dimensional structure of the induced refractive index change is revealed by the clearly visible first Brillouin zone of

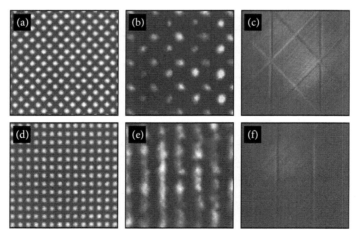

Fig. 6.4. Structure analysis for diamond (top) and square pattern (bottom). (**a**), (**d**) Lattice wave (input), (**b**), (**e**) guided wave, (**c**), (**f**) Brillouin zone

the diamond lattice. The Brillouin zone representation of the square pattern however is dominated by two vertical lines representing the borders of the first Brillouin zone of the corresponding one-dimensional stripe pattern.

6.3.2 Polarization Anisotropy

Due to the electro-optic anisotropy of axial photorefractive crystals like strontium barium niobate (SBN), one can distinguish between linear and nonlinear material response in order to create the desired photonic lattices. An ordinarily-polarized light beam only experiences a negligible nonlinearity due to the small electro-optic coefficient and therefore propagates in an almost linear regime. An extraordinarily polarized light beam on the other hand is influenced by a strong photorefractive nonlinearity and propagates in the nonlinear regime. For both polarizations the symmetry of the induced refractive index change depends on orientation of the lattice wave with respect to the c-axis [13, 14], and the saturation of the refractive index depends on the intensity of the lattice wave. The larger the intensity, the stronger the saturation of the refractive index. However, when the lattice wave is extraordinarily polarized, the self-focusing effect increases the intensity of each spot such that the lattice is effectively induced with higher peak intensity. Consequently, for otherwise same parameters (initial intensity, background illumination, and applied voltage), the refractive index modulation induced with extraordinarily polarized light is stronger than that for ordinarily polarized light. This is illustrated in Fig. 6.5 showing the output of the guided plane wave for square and diamond patterns, respectively [13, 14]. Lattices with a period of 60 µm are induced with either ordinary or extraordinary polarization using a very low power of 3 µW for the whole lattice wave. The polarization of the plane wave can be made ordinary (o) as well as extraordinary (e). As expected, the strongest modulation can be observed when lattice and the probe plane wave are extraordinarily polarized. If the lattice and the plane waves are both ordinarily polarized, no modulation of the guided wave can be observed. Indeed, in the latter case the lattice intensity is too low to induce a significant refractive index change as well as the coupling of the ordinarily polarized plane wave is week due to very small electro-optic coefficient.

6.4 Two-dimensional Discrete Solitons in Nonlinear Photonic Lattices

To study the propagation of an optical beam in a biased photorefractive crystal with an optically induced lattice, the set of equations (6.1)–(6.3) can be solved with a total intensity I that includes the intensity of the lattice wave $V(x, y)$, the intensity of the optical beam $|E|^2$ as well as the background illumination. Thus, the intensity relation

6 Complex Nonlinear Photonic Lattices   109

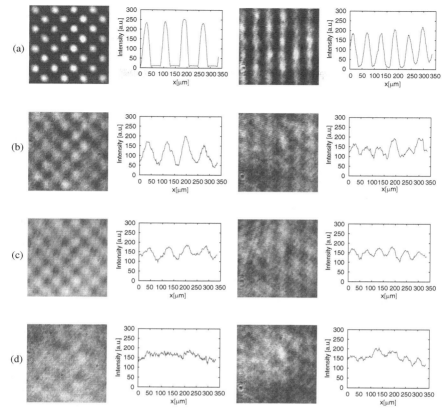

Fig. 6.5. Guided wave output (first and third column) and horizontal intensity profiles (second and fourth column) of diamond pattern (left) and square pattern (right) with varied polarizations for lattice wave and guided wave. (**a**) lattice wave and guided wave e-polarized, (**b**) lattice wave e-polarized, guided wave o-polarized, (**c**) lattice wave o-polarized, guided wave e-polarized, (**d**) lattice wave and guided wave o-polarized

$$I = 1 + V(x,y) + |E|^2 \qquad (6.7)$$

is used. The periodic modulation of the refractive index causes a bandgap spectrum for the transverse components of the wave vectors K_x and K_y and the beam propagation through the lattice is described by two-dimensional Bloch waves [21] of the form

$$E(x,y,z) = U(x,y) \exp\left[i\left(\beta z + K_x x + K_y y\right)\right] \qquad (6.8)$$

with the propagation constant β and the spatially periodic amplitude $U(x,y)$.

To calculate the band structure of a specific lattice type (Fig. 6.6), it is important to consider the anisotropic properties of the refractive index.

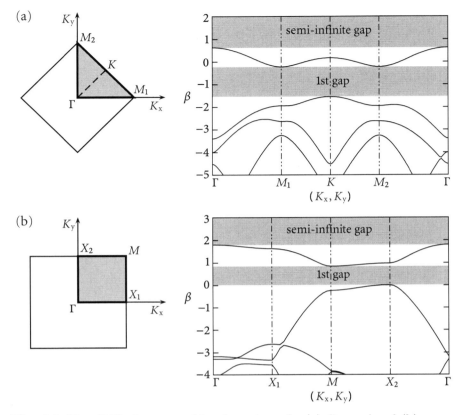

Fig. 6.6. First Brillouin zone and band spectrum for (**a**) diamond and (**b**) square lattice

From a geometrical point of view, there is no qualitative difference between diamond (Fig. 6.6a) and square (Fig. 6.6b) pattern except the spatial orientation. However, it has been demonstrated that different spatial orientations of the lattice wave result in fundamentally different refractive index structures and accordingly in different band structures.

The points in the band spectrum with maximum β are of special importance. In case of a focusing nonlinearity, the nonlinear response increases the beam propagation constant and leads to a shift inside the gaps for modes associated with these points. Once the propagation constant is located inside the gap, the propagation is no longer restricted to the spatially extended Bloch waves and the formation of self-trapped waves becomes possible. In the following experiments, the Γ point at the top of the first band will be investigated. Increasing the propagation constant from this point shifts it inside the semi-infinite total internal reflection gap. Solitary states in this gap are commonly denoted as *discrete solitons*. Additionally, another possible

6 Complex Nonlinear Photonic Lattices   111

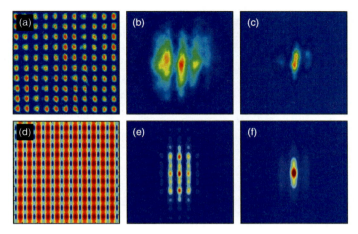

Fig. 6.7. Experimental results (top) and numerical simulations (bottom) for the square pattern. (**a**) Lattice wave intensity and (**d**) refractive index distribution, (**b**), (**e**) diffraction of the probe beam at low intensity, (**c**), (**f**) localized state at high intensity

localized state originates from the point with maximum β of the second band. In this case, the propagation constant is shifted inside the Bragg-reflection gap and the corresponding localized state is called *gap soliton*.

In order to explore the influence of the anisotropy of photonic lattices on the symmetries of discrete solitons, we generate two-dimensional solitons experimentally by focusing an extraordinarily polarized Gaussian beam into a lattice site at the front face of the crystal. The result for the square lattice is shown in Fig. 6.7. Although the probe beam is extraordinarily polarized, we also observe its linear propagation (discrete diffraction) using the slow response of the photorefractive nonlinearity. Indeed, the process of optical induction is much slower than the propagation of light and immediately after launching the probe beam the periodic refractive index induced by the lattice wave is undistorted.

We observed that the discrete diffraction in the diamond lattice follows the dynamics known for a truly two-dimensional square photonic lattice, while corresponding images for the square lattice in Fig. 6.7b are significantly different. The modulation of the beam intensity after discrete diffraction, similar to the guided waves in [Figs. 6.4e and 6.5], is effectively one-dimensional.

To model discrete diffraction (Fig. 6.7e) we solve Eqs. (6.1–6.3) with the total intensity $I = 1 + V$ and the lattice intensity in analogy to (6.6)

$$V(x, y) = I_0 \cos^2 X \cos^2 Y, \tag{6.9}$$

where again $X = (x + y)/\sqrt{2}$ and $Y = (x - y)/\sqrt{2}$ for the diamond lattice while $X = x$ and $Y = y$ for square lattice. The numerical results are in good qualitative agreement with corresponding experimental pictures.

Increasing the power of the probe beam and allowing for sufficient time for self-action effects to take place, we record the formation of discrete solitons in Fig. 6.7c. Numerically, we obtain the profiles of the solitons by solving Eqs. (6.1–6.3) with the ansatz $E = U(x,y)\exp(\mathrm{i}\beta z)$, and the total intensity $I = 1 + V + U^2$. The resulting discrete solitons carry the symmetry of the underlying nonlinear lattice, i.e., the soliton in the square lattice is only modulated in the x-direction.

The observations closely resemble two-dimensional solitons in saturable media propagating in a one-dimensional lattice potential investigated theoretically in [22]. These solitons naturally show a strong anisotropy, making them essentially different from usual two-dimensional solitons. An additional advantage of using a one-dimensional lattice potential is that the remaining free direction offers the possibility of soliton motion, and therefore allows the study of soliton collisions or the formation of bound states [23]. The mobility of two-dimensional solitons was also shown to be strongly anisotropic in two-dimensional lattices [24].

6.5 Hybrid Lattices

An ordinarily polarized lattice wave is not influenced by the refractive index changes it produces due to the small electro-optic coefficient. This is why the ordinarily polarized lattice is often denoted as *fixed lattice*. The extraordinarily polarized lattice wave propagates in the nonlinear regime and also induces the nonlinear refractive index change which will be different for different orientations and different intensities. However, in the nonlinear regime, there are instabilities which may brake the stationary lattice during propagation. Similarly, an additional extraordinarily polarized probe beam can be seen as a perturbation to the lattice such that it gets deformed or even destroyed. Therefore, this lattice is sometimes denoted as *flexible lattice* [25, 26].

Hybrid lattices with mixed polarization include both types of lattices. If the polarization of the lattice wave gets tilted with respect to the c-axis, the projection to the ordinary axis will propagate in an almost linear regime and the projection to the extraordinary axis will propagate in the nonlinear regime. Still, it is the total intensity that is important for saturation of the refractive index, even if the lattice is induced with two parts, fixed and flexible. A two-dimensional flexible lattice is rather unstable to perturbations, and therefore gets easily destroyed. In a hybrid lattice, the flexible part does not propagate in a free medium as the pure flexible lattice but in a periodically modulated media created by the fixed part. As a result, the flexible part is stabilized, and the probe beam 'sees' the stable periodic media while it still able to interact with the flexible part of the hybrid lattice.

This stabilization of the flexible lattice part can easily be observed in experiment (Fig. 6.8). After induction of an ordinarily polarized lattice, an extraordinarily polarized probe beam is launched collinear into the lattice

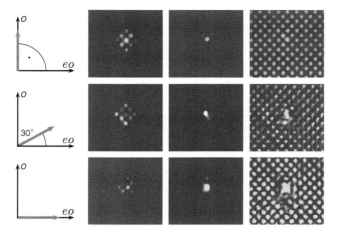

Fig. 6.8. Beam propagation in hybrid lattices with mixed polarization

to excite the Γ-point. The evolution of the probe beam has been observed for increased intensities. Figure 6.8 clearly shows the discrete diffraction and the corresponding discrete soliton. After the discrete soliton had formed, the lattice and the probe beam have been imaged together to check if the lattice remains stable while being probed. As expected, the ordinary polarized lattice is not influenced by the probe beam. Then, while keeping the total intensity constant, the polarization of the lattice wave has been tilted to different angles with respect to the c-axis and the steps mentioned above have been repeated. When the polarization is tilted by 30° with respect to the c-axis, the lattice gets strongly modified or even destroyed in the probed region.

Obviously, the observed localized states look similar for all the chosen polarizations of the lattice wave. A detailed knowledge of the interaction between the hybrid lattice and the probe beam therefore brings an additional degree of freedom in engineering the lattice guiding properties. In particular, tuning the relative amplitudes of both lattice types allows to precisely control the mobility of the localized probe beam with the polarization of the lattice wave being the control parameter. Thus, hybrid lattices are a forthcoming and promising area of research that will lead optically induced photonic lattices to a higher level of complexity.

6.6 Multiperiodic Lattices

Besides the comparatively simple geometries like diamond, square [9, 13, 14] or hexagonal [27] lattices special attention is also paid to more complex photonic structures like modulated waveguide arrays [28], lattice interfaces [29] or double-periodic one-dimensional photonic lattices [30]. In general, such multiperiodic structures are of great interest since they offer many exciting possibilities to engineer the diffraction properties of light by opening additional mini gaps in the transmission spectrum and thereby facilitate the existence of new soliton families in nonlinear media [31].

We have implemented a new approach for all optical induction of multiperiodic superlattices by superposition of several single periodic lattices [32]. The demonstrated method is closely related to the incremental recording in holographic data storage [33] and enables the induction of reconfigurable multiperiodic structures in one and two dimensions. Unfortunately, for the induction of multiperiodic lattices the self-evident idea to use the spatial light modulator for a direct modulation of the lattice wave with a corresponding pattern is not successful. The reason is that lattice waves of different periodicities acquire different phase shifts during propagation and their coherent superposition therefore leads to an intensity modulation in the propagation direction due to interference. Consequently, a method of incoherent superposition is required.

Of course a simple overlay of multiple incoherent interference patterns is feasible but lacks the flexibility benefits offered by the usage of a modulator. A solution is given by the multiplexing technology known from holographic data storage. Several different approaches like wavelength, angular and phase code multiplexing [34–36] allow the superposition of different refractive index patterns inside the volume of a photorefractive crystal and can therefore serve as a basis for the induction of multiperiodic photonic superlattices.

Compared to the commonly used sequential recording scheme, the method of incremental multiplexing [33] offers the possibility to induce the superimposed lattices with varied modulation depths by simply adjusting their relative illumination times. In fact, this enhances the flexibility of the induction process even more.

If only one lattice period is used during the induction process, the Brillouin zone pictures show two dark lines marking the borders of the first Brillouin zone of the corresponding lattice. This is demonstrated in Figs. 6.9d and 6.9e for lattice periods of 15 µm and 24 µm, respectively. The corresponding real space images of the lattice wave are shown in Figs. 6.9a and 6.9b. The induction of a one-dimensional photonic superlattice as a superposition of these two structures is depicted in Figs. 6.9c and 6.9f. The arrows in Fig. 6.9c indicate the alternating sequence of the two single periodic lattice waves. Both waves are sent onto the crystal in an alternating scheme having the same power of 35 µW for two seconds, respectively. It is important to note that this illumination time is at least one magnitude smaller than the typical dielectric relaxation time of the used crystal, which at these intensities typically is in the range of tens of seconds. The total illumination time of the induction process is about 60 s. In this case, (Fig. 6.9f) clearly shows four dark lines corresponding to the Brillouin zone structure of the double periodic one-dimensional superlattice induced by superposition of the two single periodic structures.

In addition to the optical induction of one-dimensional superlattices, the method can easily be extended to achieve multiperiodic structures in

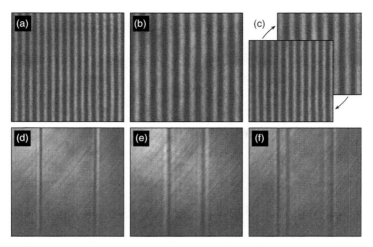

Fig. 6.9. Experimental realization of a *superlattice* by incrementally recording two stripe patterns with different periods. Lattice wave (top), Brillouin-zone spectroscopy (bottom)

two transverse dimensions as well [32]. The only fundamental restriction on the successively multiplexed structures is their diffraction-free propagation through the medium. Therefore, the method of holographic multiplexing may also be extended to induce more sophisticated refractive index structures, for example asymmetric lattices being a superposition of many single periodic lattices of different symmetries. Due to its simplicity and high flexibility, the presented method can serve as a novel tool for the investigation of several fascinating effects of nonlinear wave propagation in multiperiodic photonic lattices.

6.7 Complex Beam Propagation in Complex Lattices

Up to now, in almost all demonstrations the optically induced lattices have been restricted to a fourfold symmetry in a diamond-like orientation. This is due to the fact that in this lattice configuration effects of the anisotropy of the electro-optic properties of photorefractive crystals can be neglected. Recently, in the same spirit, the formation of discrete and gap solitons in hexagonal lattices has also been demonstrated [27, 37]. Again, the orientation of the lattice wave has been chosen to minimize the effect of anisotropy.

To overcome previous limitations of useable lattice configurations the concept of solitons in optically induced lattices can be extended to more complex anisotropic lattices. Triangular lattices represent an example of highly-symmetric patterns that are transformed into lattices with strongly reduced symmetry and they are also an extension of the commonly known fourfold symmetry diamond or square lattices. Triangular lattices are

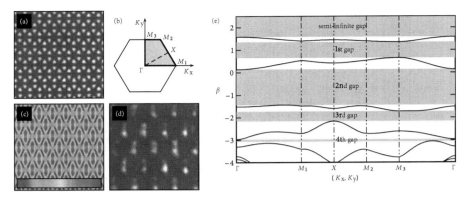

Fig. 6.10. Parallel triangular pattern. (**a**) Lattice wave, (**b**) first Brillouin zone, (**c**) numerical simulation of the light-induced refractive index change, (**d**) guided wave, (**e**) bandgap spectrum

higher-order lattices consisting of dipole structures oriented in a diamond pattern with angles of 60°.

A similar effect of symmetry reduction can also be observed for the hexagonal lattices already implemented. However, the photorefractive anisotropy leads for the triangular lattices to crucially different refractive index structures (refer to Fig. 6.10c). Certainly, this results in different bandgap spectra and hence influences the characteristics of possible solitons dramatically.

In the following, we will expand the analysis to the propagation of dipole-modes [38] and show that triangular lattices in parallel orientation enable the formation of dipole-mode gap solitons [39]. This is a demonstration of a stable dipole structure – a molecule of light – in a highly anisotropic lattice.

To obtain the induced refractive index change, the two equations (6.3) and (6.5) have to be solved numerically with the initial ansatz [14]

$$U_0^\Delta(X,Y) = E \sin\left(2Y/\sqrt{3}\right) \sin\left(Y/\sqrt{3} + X\right) \sin\left(Y/\sqrt{3} - X\right). \quad (6.10)$$

Again, the numerical simulations show that two different families of solutions evolve from the initial ansatz. The parallel orientation with lines of π phase jumps oriented parallel to the c-axis is described by $(X,Y) = (x,y)$ and the perpendicular orientation is given by $(X,Y) = (y,x)$. Similar to the diamond and square pattern, these lattices also show a strong orientation anisotropy in the symmetry of the induced refractive index changes (see section 6.3). Numerical calculations for the parallel orientation show that every two vertically neighboring out-of-phase lobes of the field distribution induce a focusing dipole-island and these islands form essentially a diamond pattern with angles of 60° (Fig. 6.10c). The experimentally observed guided waves (Fig. 6.10d) confirm these numerical results. In order to study the formation of discrete and gap solitons in these lattices, we also calculate the bandgap spectrum for the parallel lattice as shown in Fig. 6.10e.

6 Complex Nonlinear Photonic Lattices   117

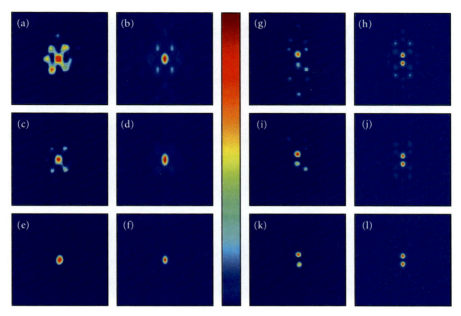

Fig. 6.11. Formation of fundamental discrete solitons in the parallel triangular lattice ((a)–(f)) and formation of dipole-mode gap solitons in the parallel triangular lattice ((g)–(l)). Experimental results (first and third column) and numerical simulations (second and fourth column). (**a**), (**b**) and (**g**), (**h**) diffraction of the probe beam at low power, (**c**), (**d**) and (**i**), (**j**) localized state at moderate power, (**e**), (**f**) and (**k**), (**l**) discrete soliton. Used colormap going from blue (low intensity) to red (high intensity) depicted between the columns

Experimentally, we generate fundamental discrete solitons by focusing an extraordinarily polarized Gaussian beam into one lattice site at the front face of the crystal. The results for the parallel pattern are shown in Fig. 6.11.

It is clearly visible that the diffraction of the probe beam in the parallel-oriented lattice at low power shows a behavior similar to the diamond lattice forming a fully two-dimensional diffraction pattern (Fig. 6.11a). It has also been experimentally verified that the perpendicular lattice shows an effectively one-dimensional diffraction pattern consisting of vertical stripes as observed for the square pattern (Fig. 6.4b).

Increasing the power of the probe beam, we observe the evolution from the described diffraction pattern to the strongly localized discrete solitons (Fig. 6.11e).

In addition to the previously discussed fundamental discrete solitons in triangular photonic lattices, our numerical simulations reveal that the lattice in parallel orientation with its dipole-like islands of high refractive index gives rise to the formation of dipole-mode gap solitons originating from the M_3-point of the irreducible first Brillouin zone (Fig. 6.10b).

To compare these numerical simulations to the experiment, we generate a dipole-like input beam by using the superposition of two counter-rotating vortices in a Mach-Zehnder interferometer and observe the output of the probe beam at the back face of the crystal for different probe beam powers.

The experimental as well as the numerical results are summarized in Figs. 6.11g–l. At low probe beam powers the diffraction pattern consists of a central dipole surrounded by four side lobes each forming a dipole itself (Fig. 6.11g). With increased power the side lobes vanish and a stable dipole-mode gap soliton evolves (Fig. 6.11k).

The existence of these stable solitons in triangular lattices offers novel possibilities to control dipole-like beams. In bulk photorefractive media they are known to experience strong repulsion [40], whereas in the presence of the lattice they are confined in one dipole-like island of high refractive index.

6.8 Controlling Instabilities of Counterpropagating Solitons by Optically Induced Photonic Lattices

Light-induced refractive index changes of photorefractive optical spatial solitons offer the possibility for reconfigurable optical links. Therefore, the interaction of spatial optical solitons is of crucial importance. The interaction of photorefractive solitons can include attraction or repulsion as well as energy exchange and fusion or even a creation of new solitons [41,42]. The temporally constant input of *copropagating* solitons leads to a stationary state after initial transient dynamics [43]. In contrast, *counterpropagating* solitons [44–51] induce a common waveguide. With increasing medium length L or increasing coupling strength Γ this waveguide structure breaks up into a so called bidirectional waveguide structure. Increasing the nonlinear interaction parameter ΓL further, the bidirectional waveguide loses stability via regular limit cycle oscillations until the temporal dynamics become irregular [47–50]. To control and stabilize counterpropagating solitons, optically induced photonic lattices [52] are ideal candidates. The modulation of the refractive index acts as a periodic potential, reducing the mobility of the spatial extension of the beams, and enabling the formation of lattice solitons in the total internal reflection gap. This suppresses the instability of the counterpropagating solitons, therefore allowing for stability of the resulting composite waveguide. In this section we discuss the influence of the lattice depth and period of an optically induced lattice on the dynamics of counterpropagating solitons.

To numerically simulate the situation, we focus on the interaction of two counterpropagating mutually incoherent screening solitons with a one-dimensional model of a saturable nonlinearity. Spatial evolution of the slowly varying field envelopes E_F and E_B of the forward and backward propagating beams is modelled in paraxial approximation:

$$i\partial_z E_F + \partial_x^2 E_F = \Gamma E_{sc} E_F$$
$$-i\partial_z E_B + \partial_x^2 E_B = \Gamma E_{sc} E_B \quad (6.11)$$

where E_{sc} is the space charge field created by the beams inside the photorefractive crystal and $\Gamma = k^2 n_0^2 w_0^2 r_{eff} E_{ext}$ denotes the photorefractive coupling constant. The transverse x-coordinate is scaled to the beam waist w_0 whereas the propagation z-coordinate is scaled with the diffraction length $L_D = 2kw_0^2$, with $k = 2\pi n_0/\lambda$ and λ denoting the laser wavelength [48].

For the temporal evolution of the space charge field we assume relaxation type dynamics

$$\tau(I)\partial_t E_{sc} + E_{sc} = -\frac{I}{1+I}, \quad (6.12)$$

where $I = |E_F|^2 + |E_B|^2$ is the total intensity – again scaled to the background intensity – and E_{sc} is scaled to the applied electrical field. The relaxation time τ depends on the total intensity as $\tau(I) = \tau_0/(1+I)$.

The presence of the index grating created by the interference of two plane waves is modelled by including the additional term $V(x) = A\cos^2(\pi x/p)$ in the expression for the total light intensity $I = |E_F|^2 + |E_B|^2 + V$, where A and p determine the peak intensity and periodicity of the lattice, respectively. As the lattice-forming waves are ordinarily polarized, they do not interact directly with solitons, and the lattice intensity distribution remains stationary throughout the crystal.

In our investigations, we assume the incident waves to be stationary in a form of two identical Gaussian beams with beam waists 1 and peak intensity $I = 1$ each. We simulate their head-on collision by launching both beams at the same lateral position perpendicularly to the crystal face. For the chosen parameter values, the soliton solutions in bulk media already exhibited very irregular temporal oscillations [50].

In order to study the influence of the induced photonic lattice on the soliton dynamics, we vary the peak intensity and periodicity of the lattice wave and determine the degree of instability in terms of the *level of dynamics* [50] defined as

$$\text{lod}^N = \sum_{t=1}^{N}\sum_x \left[\frac{|E_F(x,L,t) - E_F(x,L,t-1)|^2}{|E_F(x,L,t)|^2} + \frac{|E_B(x,0,t) - E_B(x,0,t-1)|^2}{|E_B(x,0,t)|^2}\right]. \quad (6.13)$$

This parameter represents the time and space integrated modulus of the differences in the field envelopes at either exit face between two successive simulated time-steps. It follows directly from Eq. (6.13) that a decrease in the intensity variation of the solitons (i.e. a weaker instability) leads to a lower value of lod^N ($\text{lod}^N = 0$ for stationary solutions).

As can be clearly seen from Fig. 6.12, the rate of decrease of the temporal dynamics of counterpropagating solitons launched on-site strongly depends on

120 J. Imbrock et al.

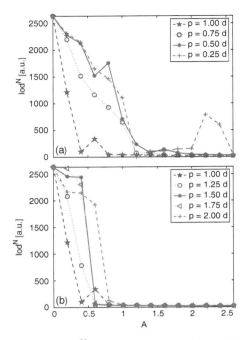

Fig. 6.12. Level of dynamics lod^N for counterpropagating solitons launched on-site as a function of lattice intensity A for different lattice period p. (Here and in the following on-site will denote the case when the solitons maximal intensity coincides with the maximum of the lattice.) (**a**) period smaller than the beam diameter, (**b**) lattice period larger than the beam diameter. Non-monotonous decrease of lod^N for $p = 1\,d$ and $A = 0.4$–0.8, is due to long lasting transient dynamics

lattice strength and its period. The influence of the lattice on soliton dynamics is strongest for period $p = 1\,d$, where $d = 2$ is the beam diameter of the incident Gaussian beams. This can be explained as follows: for small lattice periodicity, the self-trapped beam covers many lattice sites. As a result, the effect of the lattice is weaker, and in the limit $p \to 0$ the medium can be regarded as homogeneous with higher refractive index. If the periodicity is comparable with the beam diameter, the soliton experiences maximal guiding by the lattice induced refractive index modulation. For larger periodicity, the region in which the refractive index change is negligible increases, so that in the limit $p \to \infty$ one ends up again with a homogeneous medium. Consequently the influence of the lattice decreases again. Furthermore, our simulations show that the decrease of the dynamics is not as rapid for smaller periodicity as it is for larger one.

Figure 6.13 depicts the numerical solution of the temporal evolution of the intensity distribution for $p = 1\,d$ for three different values of the lattice intensity A. For weaker lattices the output oscillates irregularly. These oscillations are similar to those occurring in a bulk continuous medium. The presence

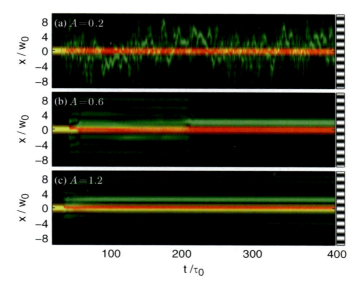

Fig. 6.13. Temporal evolution of the intensity distribution at one crystal face with increasing lattice peak intensity at the plane of incidence of E_F. The constant line (red) marks the input of E_F while the intensity distribution of E_B oscillates (green) since it is the output plane of beam E_B. Depicted is the evolution along the curve for periodicity equal to beam diameter. Qualitatively similar results are obtained at the plane of incidence of E_B for the transmitted beam E_F

of the lattice leads to solitons residing more frequently at the lattice sites, which is reflected in the appearance of faint horizontal lines in Fig. 6.13a. For a stronger lattice, we still observe some transient dynamics with oscillation periods which are quite short (Fig. 6.13b). During these oscillations, the output couples to the neighboring lattice sites. However, after the initial oscillations the output becomes stationary and resides at the lattice site closest to the input waveguide. For even larger lattice intensity the range of time transient dynamics shortens. Notice that in the stable state a small fraction of the soliton is trapped at the input lattice channel while the majority of soliton power is confined in the neighboring site.

The impact of a one-dimensional optically induced lattice on the dynamics of soliton interaction can be investigated experimentally by interfering two plane waves of equal power inside the photorefractive crystal. The periodicity of the lattice is chosen to be comparable with the beam diameter, for which the soliton stabilization is expected to be strongest.

The experimental results are shown in Fig. 6.14 which depicts temporal evolution of one of the beams at the exit facet of the crystal. It is evident that soliton dynamics, i.e. spatio-temporal oscillations are suppressed with the increasing strength of the optical lattice. These results agree qualitatively well with the numerical simulations. Figure 6.14a shows irregular oscillations forming out at low lattice power. These oscillations are comparable to those

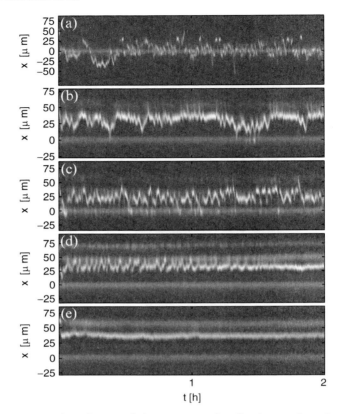

Fig. 6.14. Temporal evolution of the intensity distribution projected onto the x-axis (parallel to the c-axis) for different lattice powers of (**a**) $250\,\mu\mathrm{W}$, (**b**) $1.0\,\mathrm{mW}$, (**c**) $1.5\,\mathrm{mW}$, (**d**) $2.0\,\mathrm{mW}$, and $2.5\,\mathrm{mW}$. The faint horizontal lines at $x = 0\,\mu\mathrm{m}$ mark the reflected beam at this crystal face which acts as a reference. Again the results for the other crystal face show similar behavior

emerging in a bulk medium under the same set of parameters. For higher lattice beam power (Figs. 6.14b and 6.14c) the oscillation amplitude starts to decrease and within the irregularity sequences of regular oscillations are observed. For even higher power of the lattice waves we see long lasting transient dynamics tending to a stationary state (Fig. 6.14d). Finally, a stationary state over the observation period of two hours is seen for lattice beam power of $2.5\,\mathrm{mW}$ (see Fig. 6.14e).

So far we have considered the dynamics of counterpropagating solitons in a one-dimensional lattice, emphasizing the arrest of instability in a single transverse direction only. For application of the counterpropagating solitons as a self-adjusting bi-directional waveguide, it is necessary, however, to ensure stabilization in both transverse directions. This can be achieved in two-dimensional optical lattices [52]. The dependence of beam dynamics on the lattice period qualitatively matched those obtained for one-dimensional optical lattice. For a small lattice period the potential induced by the lattice is too weak to arrest the instability of the counterpropagating beams but with increasing lattice period the instability is practically removed for a certain range of lattice intensities. Without the lattice, both beams overlap weakly and their individual propagation is strongly affected by the beam self-focusing and self-bending. The introduction of the lattice suppresses the oscillatory instability observed in homogenous crystals due to the induced periodic potential and therefore reduces the mobility of solitons. In the presence of the lattice both beams align well with each other and the oscillatory motion is suppressed. With the increase of the lattice power the instability dynamics of the counterpropagating beams is reduced and the solitons join steadily at both sides of the crystal.

In conclusion, the dynamics of counterpropagating solitons can be suppressed by optically induced lattices of proper period and strength. For a one-dimensional lattice the period should be comparable to the beam diameter and the lattice strength should exceed a certain value. The stabilization of the dynamics of counterpropagating solitons in two transverse direction can be achieved with a two-dimensional square lattice.

6.9 Summary

We have presented an overview of the properties and applications of strongly nonlinear optically induced photonic lattices in photorefractive media. By exploiting the anisotropic nature and the saturation of the photorefractive nonlinearity, the strong nonlinearity allows designing strongly complex lattices and variably tuning the strength and shape of the periodic refractive index change. With nonlinear photonic lattices, we are able to influence the propagation of light beams in many different ways. Self-trapped light beams in form of discrete and gap solitons can be generated inside the lattice, and counterpropagating solitons can be stabilized by photonic lattices. As light-induced optical lattices provide a much greater control of the grating parameters than fabricated waveguide arrays, we believe that these results open many new possibilities for the study of various nonlinear effects in more complex optically induced lattices, including lattices with mixed polarization and superlattices with different spatial periodicities.

Acknowledgement

Many parts of this work have been inspired or realized together with excellent collaboration partners. Among them, Anton Desyatnikov and Yuri S. Kivshar, Nonlinear Physics Centre, Australian National University, Canberra contributed many important numerical simulations on strongly nonlinear lattices,

Tobias Richter and Friedemann Kaiser, Institut für Angewandte Physik, Technische Universität Darmstadt, Germany, supported the investigation of complex beams in complex lattices and Milan Petrovic and Milivoj Belic, Institute of Physics, University of Belgrade discussed many aspects of instabilities of counterpropagating solitons. The activities with our collaboration partners have mainly been support by the German Academic Research Association, by the binational exchange program as well as a binational dissertation grant.

References

1. A.S. Davydov, J. Theor. Biol. **38**, 559 (1973)
2. W.P. Su, J.R. Schrieffer, and A.J. Heeger, Phys. Rev. Lett. **42**, 1698 (1979)
3. A. Trombettoni and A. Smerzi, Phys. Rev. Lett. **86**, 2353 (2001)
4. D.N. Christodoulides and R.I. Joseph, Opt. Lett. **13**, 794 (1988)
5. Y.S. Kivshar and G.P. Agrawal, *Optical Solitons: From Fibers to Photonic Crystals* (Academic, San Diego 2003), p. 540 ff
6. N.K. Efremidis, S. Sears, D.N. Christodoulides, J.W. Fleischer, and M. Segev, Phys. Rev. E **66**, 046602 (2002)
7. J.W. Fleischer, T. Carmon, M. Segev, N.K. Efremidis, and D.N. Christodoulides, Phys. Rev. Lett. **90**, 023902 (2003)
8. D. Neshev, E. Ostrovskaya, Y.S. Kivshar, and W. Krolikowski, Opt. Lett. **28**, 710 (2003)
9. J.W. Fleischer, M. Segev, N.K. Efremidis, and D.N. Christodoulides, Nature **422**, 147 (2003)
10. U. Grüning, V. Lehmann, S. Ottow, and K. Busch, Appl. Phys. Lett. **68**, 747 (1996)
11. T.F. Krauss, R. De La Rue, and S. Brand, Nature **383**, 699 (1996)
12. J. Serbin, A. Ovsianikov, and B. Chichkov, Opt. Expr. **12**, 5221 (2004)
13. A.S. Desyatnikov, D.N. Neshev, Y.S. Kivshar, N. Sagemerten, D. Träger, J. Jägers, C. Denz, and Y.V. Kartashov, Opt. Lett. **30**, 869 (2005)
14. A.S. Desyatnikov, N. Sagemerten, R. Fischer, B. Terhalle, D. Träger, D.N. Neshev, A. Dreischuh, C. Denz, W. Krolikowski, and Y.S. Kivshar, Opt. Expr. **14**, 2851 (2006)
15. B. Terhalle, A.S. Desyatnikov, C. Bersch, D. Träger, L. Tang, J. Imbrock, Y.S. Kivshar, and C. Denz, Appl. Phys. B **86**, 399 (2007)
16. J. Petter, J. Schröder, D. Träger, and C. Denz, Opt. Lett. **28**, 438 (2003)
17. B. Terhalle, D. Träger, L. Tang, J. Imbrock, and C. Denz, Phys. Rev. E **74**, 057601 (2006)
18. G. Bartal, O. Cohen, H. Buljan, J.W. Fleischer, O. Manela, and M. Segev, Phys. Rev. Lett. **94**, 163902 (2005)
19. A.A. Zozulya and D.Z. Anderson, Phys. Rev. A **51**, 1520 (1995)
20. A.A. Zozulya, D.Z. Anderson, A.V. Mamaev, and M. Saffman, Phys. Rev. A **57**, 522 (1998)
21. D. Träger, R. Fischer, D.N. Neshev, A.A. Sukhorukov, C. Denz, W. Krolikowski, and Y.S. Kivshar, Opt. Expr. **14**, 1913 (2006)
22. B.B. Baizakov, B.A. Malomed, and M. Salerno, Phys. Rev. A **70**, 053613 (2004)
23. T. Mayteevarunyoo and B.A. Malomed, Phys. Rev. E **73**, 036615 (2006)

24. R. Fischer, D. Träger, D.N. Neshev, A.A. Sukhorukov, W. Krolikowski, C. Denz, and Y.S. Kivshar, Phys. Rev. Lett. **96**, 023905 (2006)
25. A.S. Desyatnikov, E.A. Ostrovskaya, Y.S. Kivshar, and C. Denz, Phys. Rev. Lett. **91**, 153902 (2003)
26. A. Bezryadina, D.N. Neshev, A. Desyatnikov, J. Young, Z. Chen, and Y.S. Kivshar, Opt. Expr. **14**, 8317 (2006)
27. C.R. Rosberg, D.N. Neshev, A.A. Sukhorukov, W. Krolikowski, and Y.S. Kivshar, Opt. Lett. **32**, 397 (2007)
28. C.R. Rosberg, I.L. Garanovich, A.A. Sukhorukov, D.N. Neshev, W. Krolikowski, and Y.S. Kivshar, Opt. Lett. **31**, 1498 (2006)
29. S. Suntsov, K.G. Makris, D.N. Christodoulides, G.I. Stegemann, A. Hach, R. Morandotti, H. Yang, G. Salamo, and M. Sorel, Phys. Rev. Lett. **96**, 063901 (2006)
30. E. Smirnov, C.E. Rüter, D. Kip, K. Shandarova, and V. Shandarov, Appl. Phys. B **88**, 359 (2007)
31. P.J.Y Louis, E.A. Ostrovskaya, and Y.S Kivshar, Phys. Rev. A **71**, 023612 (2005)
32. P. Rose, B. Terhalle, J. Imbrock, and C. Denz, J. Phys. D: Appl. Phys. **41**, 224004 (2008)
33. Y. Taketomi, J.E. Ford, H. Sasaki, J. Ma, Y. Fainman, and S.H. Lee, Opt. Lett. **16**, 1774 (1991)
34. G.A. Rakuljic, V. Leyva, and A. Yariv, Opt. Lett. **17**, 1471 (1992)
35. F.H. Mok, Opt. Lett. **18**, 915 (1993)
36. C. Denz, G. Pauliat, G Roosen, and T. Tschudi, Opt. Commun. **85**, 171 (1991)
37. O. Peleg, G. Bartal, B. Freedman, O. Manela, M. Segev, and D.N. Christodoulides, Phys. Rev. Lett. **98**, 103901 (2007)
38. J. Yang, I. Makasyuk, A. Bezryadina, and Z. Chen, Opt. Lett. **29**, 1662 (2004)
39. P. Rose, T. Richter, B. Terhalle, J. Imbrock, F. Kaiser, and C. Denz, Appl. Phys. B **89**, 521 (2007)
40. W. Krolikowski, E.A. Ostrovskaya, C. Weilnau, M. Geisser, G. McCarthy, Y.S. Kivshar, C. Denz, and B. Luther-Davies, Phys. Rev. Lett. **85**, 1424 (2000)
41. G.I. Stegeman and M. Segev, Science **286**, 1518 (1999)
42. W. Krolikowski, B. Luther-Davies, and C. Denz, IEEE J. Sel. Top. Quantum Electron. **39**, 3 (2003)
43. C. Weilnau, M. Ahles, J. Petter, D. Träger, J. Schröder, and C. Denz, Ann. Phys. **11**, 573 (2002)
44. O. Cohen, R. Uzdin, T. Carmon, J. W. Fleischer, M. Segev, and S. Odoulov, Phys. Rev. Lett. **89**, 133901 (2002)
45. D. Kip, C. Herden, and M. Wesner, Ferroelectrics **274**, 135 (2002)
46. P. Jander, J. Schröder, C. Denz, M. Petrovic, and M.R. Belic, Opt. Lett. **30**, 750 (2005)
47. K. Motzek, P. Jander, A. Desyatnikov, M. Belic, C. Denz, and F. Kaiser, Phys. Rev. E **68**, 066611 (2003)
48. M.R. Belic, P. Jander, A. Strinic, A. Desyatnikov, and C. Denz, Phys. Rev. E **68**, 025601 (2003)
49. M. Belic, P. Jander, K. Motzek, A. Desyatnikov, D. Jovic, A. Strinic, M. Petrovic, C. Denz, and F. Kaiser, J. Opt. B **6**, 190 (2004)
50. P. Jander, J. Schröder, T. Richter, K. Motzek, F. Kaiser, M. R. Belic, and C. Denz, Proc. SPIE **6255**, 62550A (2006)

51. M. Haelterman, A.P. Sheppard, and A.W. Snyder, Opt. Commun. **103**, 145 (1993)
52. S. Koke, D. Träger, P. Jander, M. Chen, D.N. Neshev, W. Krolikowski, Y.S. Kivshar, and C. Denz, Opt. Expr. **15**, 6279 (2007)

7

Light Localization by Defects in Optically Induced Photonic Structures

Jianke Yang[1], Xiaosheng Wang[2], Jiandong Wang[1], and Zhigang Chen[2,3]

[1] Department of Mathematics and Statistics, University of Vermont, VT 05401, USA
jyang@cems.uvm.edu, jwang@cems.uvm.edu
[2] Department of Physics and Astronomy, San Francisco State University, CA 94132, USA
gxkren@gmail.com, zchen@stars.sfsu.edu
[3] TEDA Applied Physical School, Nankai University, Tianjin 300457, China

7.1 Introduction

In the past ten years, there has blossomed an interest in the study of collective behavior of wave propagation in periodic waveguide arrays and photonic lattices [1–3]. The unique bandgap structures of these periodic media, coupled with nonlinear effects, give rise to many types of novel soliton structures [1–26]. On the other hand, it is well known that one of the unique and most interesting features of photonic band-gap structures is a fundamentally different way of waveguiding by defects in otherwise uniformly periodic structures. Such waveguiding has been demonstrated with an "air-hole" in photonic crystal fibers (PCF) for optical waves [27, 28], in an isolated defect in two-dimensional arrays of dielectric cylinders for microwaves [29–31], and recently in all-solid PCF with a lower-index core [32, 33]. In addition, laser emission based on photonic defect modes has been realized in a number of experiments [34–38]. In one-dimensional (1D) fabricated semiconductor waveguide arrays, previous experiments have investigated nonlinearity-induced escape from a defect state [39] and interactions of discrete solitons with structural defects [40] (see also [41]).

Despite the above efforts, theoretical understanding on defect guiding was still limited, and experimental demonstrations of defect guiding was still scarce. In addition, when nonlinear effects are significant, how defect guiding is affected by nonlinearity is largely an open issue. Recently, in a series of theoretical and experimental studies, we optically induced 1D, 2D and ring-like photonic lattices with single-site negative defects in photorefractive crystals, and investigated their linear and nonlinear light guiding properties [42–48]. This work will be reviewed in this Chapter. In addition, we present the first experimental demonstration of nonlinear defect modes which undergoes nonlinear propagation through the defects. Our work not only has a direct link to technologically important systems of periodic structures such as PCF, but also brings about the possibility for studying, in an optical setting, many novel phenomena in periodic systems beyond optics such as edge dislocation, defect healing, eigenmode splitting, and nonlinear mode coupling which have been intriguing scientists for decades [49–51].

7.2 Optically Induced Lattices and Defects

With today's nano-fabrication technology, creation of a closely-spaced uniform 1D waveguide array on a substrate material is not a difficult task. For instance, such waveguides have been fabricated with AlGaAs semiconductor materials or LiNbO$_3$ crystals. Yet, it has been a challenge to create or fabricate 2D or 3D waveguide arrays in bulk media. In Ref. [9], 2D photonic lattices were successfully created by sending multiple interfering beams into a crystal. This interference method has some disadvantages, such as its sensitivity to ambient perturbation, and its inability to generate more complicated lattice structures with single-site defects. In view of that, we used a different method of optical induction which is based on the amplitude modulation of a partially coherent optical beam.

The experimental setup for our study is illustrated in Fig. 7.1. The experiments are performed in a biased SBN:60 (strontium barium niobate) photorefractive crystal (typically, $r_{33} \sim 280\,\mathrm{pm/V}$ and $r_{13} \sim 24\,\mathrm{pm/V}$) illuminated by a laser beam (either Coherent argon ion laser $\lambda = 488\,\mathrm{nm}$ or solid-state laser $\lambda = 532\,\mathrm{nm}$) passing through a rotating diffuser and an amplitude mask. The biased crystal (bias field can be varied from -2.0 to 6.0 kV/cm) provides

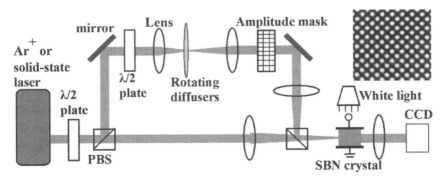

Fig. 7.1. Experimental setup for optical induction of waveguide lattices in a biased photorefractive crystal by amplitude modulation of a partially coherent beam. PBS: polarizing beam splitter, SBN: strontium barium niobate. Top path is the lattice beam, and bottom path is the probe beam (either a Gaussian beam or a vortex beam if a vortex mask is inserted). The right insert shows a typical experimental picture of 2D uniform lattice created by optical induction

7 Light Localization by Defects in Optically Induced Photonic Structures   129

a self-focusing or defocusing noninstantaneous nonlinearity [52]. The rotating diffuser turns the laser beam into a partially spatially incoherent beam with controllable degree of spatial coherence, as first introduced in experiments with incoherent optical solitons [53–55]. The amplitude mask provides spatial modulation after the diffuser on the otherwise uniform beam, which exhibits a periodic intensity pattern at the input face of the crystal [56,57]. This partially coherent and spatially modulated beam is used as our lattice beam. Another beam, either split from the same laser or emitted from a different laser and not passing through the diffuser and the mask, is used as our probe beam, propagating along with the lattice. In our experiments on defect modes, the lattice beam has its polarization close to being o-polarized, thus the lattice beam induces a weak periodic index variation to form the waveguide arrays. The probe beam, on the other hand, is always a coherent e-polarized beam, but its intensity and/or wavelength can be adjusted so it can undergo linear propagation (for study of linear guidance or linear defect modes) or nonlinear propagation (for study of nonlinear trapping or nonlinear defect modes) as detailed in later sections. The two beams are monitored separately with CCD cameras at the input and output facets of the crystal. In addition, a white-light background beam illuminating from the top of the crystal is normally used for fine-tuning the photorefractive nonlinearity [52–60].

In our experiments, the periodic lattice must stay stationary during its quasi-linear propagation through the crystal. In order for this to happen, we need to understand how to eliminate the *Talbot effect*. The Talbot effect is a phenomenon of coherent light propagation in a homogeneous media with spatially-periodic initial conditions [61,62]. Light exhibiting this phenomenon does not propagate stationarily, and it shrinks and expands as it moves along, and its intensity pattern repeats itself periodically along the propagation direction. Our lattice beam travels in a homogeneous crystal (as it does not feel the probe beam), and its initial condition on the input face of the crystal is periodic (due to the amplitude mask). Because of the Talbot effect, it can not form a stationary lattice. To overcome this difficulty, our idea is to use frequency filtering to remove half of the spatial frequencies in the initial conditions. The filtered lattice beam, when slightly tilted, can propagate stationarily along the crystal; thus, the Talbot effect is eliminated. Our experimentally created 1D and 2D stationary lattices are presented in Fig. 7.2. A theoretical understanding for the elimination of the Talbot effect by frequency filtering and beam tilting can be found in [63].

In our studies, we need to optically create stationary periodic lattices with a local defect akin to an "air defect" in photonic crystals. To explore this possibility, we prepare an initial periodic lattice with a single-site negative defect using amplitude masks. Under linear propagation, however, we find that the frequency filtering and beam tilting techniques are not enough to maintain the defect and keep the lattice stationary. The defect tends to be washed out at the exit face of the crystal. In order to maintain the defect, we employ two additional techniques. One is to introduce a small amount of nonlinearity

130   J. Yang et al.

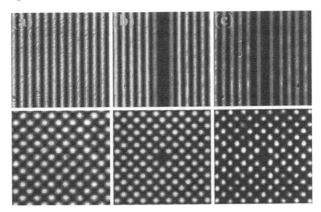

Fig. 7.2. Optically induced 1D and 2D lattices with a single-site negative defect, as obtained from our experiments. 1D results (*top*), 2D results (*bottom*). (**a**) uniform lattice at output, lattice spacing 42 µm (1D) and 27 µm (2D), (**b**) lattice with defect at input, (**c**) lattice with defect at output maintained by weak nonlinearity

into the lattice beam (by setting its polarization to contain a small amount of e-polarized component), and the other one is to introduce partial incoherence into the lattice beam (by letting the lattice beam go through a rotating diffuser). With the combined use of these techniques, we have successfully created 1D and 2D single-site defects in the otherwise uniform lattice which remains nearly stationary throughout the crystal (length varies from 10 to 20 mm). Typical examples are presented in Fig. 7.2.

7.3 Linear Defect Modes in 1D Lattices

When a periodic lattice has a local defect, this defect can affect the propagation of a probe beam significantly and in a way fundamentally different from linear propagation in continuous media. For instance, if the defect is negative (repulsive), i.e., the lattice intensity at the defect is lower than that in neighboring lattice sites, the defect can guide a linear localized mode (defect mode). This is quite counter-intuitive. The physical mechanism for this unusual light guiding is the repeated Bragg reflections, rather than the conventional total internal reflections, analogous to light transmission in air-hole photonic crystal fibers.

To understand the linear light-guiding property of a negative defect, a theoretical analysis is performed first for our present physical system [42,43]. The non-dimensionalized model is [7–9]

$$iU_z + U_{xx} - \frac{E_0}{1+I_\mathrm{L}(x)}U = 0 , \qquad (7.1)$$

7 Light Localization by Defects in Optically Induced Photonic Structures   131

where U is the envelope function of the probe beam, E_0 is the applied bias field, $I_L(x) = I_0 \cos^2(x)[1 - f_D(x)]$ is the lattice intensity containing a defect, I_0 is the lattice peak intensity, and $f_D(x) = \exp(-x^8/128)$ accounts for the single-site negative defect. If we take $I_0 = 3$, this defective lattice is shown in Fig. 7.3i. A surprising fact is that this negative defect supports localized defect modes (DMs) of the form

$$U(x,z) = u(x)e^{-i\mu z}, \qquad (7.2)$$

where μ is the propagation constant (DM eigenvalue). These eigenvalues versus E_0 are shown in Fig. 7.3ii. It is seen that these eigenvalues all lie in the gaps between Bloch bands. None of them exists in the semi-infinite bandgap (total internal reflection region). As E_0 increases, these modes disappear from lower bandgaps, and appear in higher bandgaps. A typical DM profile in the second gap is shown in Fig. 7.3iii. This mode has its intensity maximum inside the negative defect, with double-peaks in each lattice spacing, and its neighboring intensity peaks are out of phase with each other [42, 43].

The above results on DMs are confirmed experimentally, and our experimental results are shown in Fig. 7.3a–d. Here, Fig. 7.3a is the input of the 1D lattice with a defect (lattice spacing about 42 µm), the polarization angle

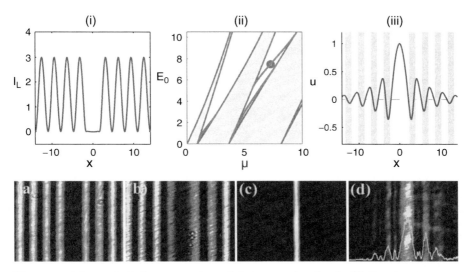

Fig. 7.3. Theoretical demonstration of defect modes (*top*). (**i**) Lattice intensity profile, (**ii**) DM branches in the (μ, E_0) plane, (**iii**) a DM in the second bandgap marked by a circle in (**ii**). Experimental observation of DMs (*bottom*), shown are transverse intensity patterns of the lattice beam at crystal input (**a**) and output (**b**) with a single-site defect, and those of the probe beam at input (**c**) and output (**b**) after 20 mm propagation through the defect channel. Lattice spacing 42 µm, Bias field 1.1 kV/cm (after Ref. [42, 43, 46])

is about 8% relative to the o-axis, and the propagation distance is 20 mm. At the bias field of 1.1 kV/cm, the output of the lattice is shown in Fig. 7.3b. It is seen that the defect is well maintained throughout propagation. After such a lattice is "fabricated", its light guiding property can be studied. To do so, we launch a low-intensity e-polarized probe beam into the defect. The experimental result is shown in Figs. 7.3c and 7.3d. It can be seen that after 20 mm propagation, most of the probe-beam energy is still confined inside the negative defect. This is remarkable, as without the defect, the probe beam would strongly scatter to nearby lattice sites in case of strong coupling. The experimentally observed defect mode (Fig. 7.3d) closely resembles the theoretical one in Fig. 7.3iii.

7.4 Linear Defect Modes in 2D Square Lattices

Guiding light by defects in 2D periodic lattices is even more interesting. Using experimental techniques similar to those for 1D DMs, we have demonstrated 2D defect guiding as well. The experimental results are shown in Fig. 7.4. Here a 2D lattice with a single-site negative defect is first created in a 20 mm crystal as shown in Fig. 7.2. Then we launch a Gaussian probe beam into the defect (Fig. 7.4a). Under different lattice conditions, we observed different guided structures as shown in Fig. 7.4b–d. At lattice spacing of 27 μm and bias field of 2.8 kV/cm, the Gaussian beam evolves into a DM, with most of its energy concentrated in the defect site (Fig. 7.4b). At spacing 42 μm and bias field of 3.0 kV/cm, the tails along the principal axes of the square lattice (which are diagonally oriented) are more prominent, and they show interesting vortex-array-like structures (Fig. 7.4c). Figure 7.4d shows a typical interferogram corresponding to intensity pattern of Fig. 7.4c, where the locations of vortices are indicated by arrows. It is seen that the vortex cells have different sign of topological charge along two diagonal "tails" [45].

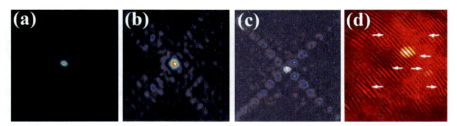

Fig. 7.4. Experimental observations on 2D defect guidance. (**a**) Input probe beam, (**b**), (**c**) intensity patterns of the output probe beam under different lattice conditions, (**d**) zoom-in interferogram of (c) with a tilted plane wave where arrows indicate location of vortices. The brightest spot corresponds to the defect site (after Ref. [45])

7 Light Localization by Defects in Optically Induced Photonic Structures   133

To theoretically analyze these various 2D DM structures, we use the 2D model equation

$$iU_z + U_{xx} + U_{yy} - \frac{E_0}{1 + I_L(x,y)} U = 0 ,\qquad(7.3)$$

where

$$I_L(x,y) = I_0 \cos^2\left(\frac{x+y}{\sqrt{2}}\right)\cos^2\left(\frac{x-y}{\sqrt{2}}\right)\left\{1 + \varepsilon \exp\left[-\frac{(x^2+y^2)^8}{128}\right]\right\}\qquad(7.4)$$

is the 2D defective lattice, and ε is the defect depth. Localized DMs in the form of $U(x,y,z) = u(x,y)\exp(-i\mu z)$ are sought for both attractive ($\varepsilon > 0$) and repulsive ($\varepsilon < 0$) defects. We first examine the dependence of defect-modes on the defect strength ε by fixing $E_0 = 15$, $I_0 = 6$, and varying ε from -1 to 1. When the defect is weak, i.e., $\varepsilon \ll 1$, such dependence can be derived analytically by asymptotic methods, and we find that the distance between μ and a Bloch-band edge μ_c decreases exponentially with the defect strength ε, i.e. [48],

$$\mu = \mu_c + Ce^{-\beta/|\varepsilon|} ,\qquad \varepsilon \ll 1 ,\qquad(7.5)$$

where C and $\beta > 0$ are constants. Fig. 7.5 shows the analytical results (dashed lines) as well as our numerical results (solid lines) for both weak and strong defects. When ε is small, they are in very good agreement. Fig. 7.5 also shows that there is one DM branch bifurcating from each band: DMs in an attractive defect ($\varepsilon > 0$) bifurcate from the left edge of each band, and DMs in a repulsive defect ($\varepsilon < 0$) bifurcate from the right edge of each band. As the defect strength ε varies, branches of attractive defect-modes stay inside their respective gaps, while repulsive DM branches march to higher Bloch bands, disappear when reaching the edge of the band, and reappear in higher gaps. Fig. 7.6 shows typical DM profiles corresponding to the letter-marked points in Fig. 7.5. DMs on branches "i", "a" and "b" are symmetric in both x and

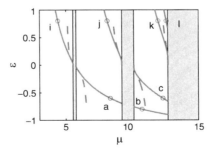

Fig. 7.5. Bifurcation of DMs with the defect lattice (7.4) at $E_0 = 15$ and $I_0 = 6$. Solid lines: numerical results, dashed lines: analytical results. The shaded regions are the Bloch bands. Profiles of defect modes at the circled points are displayed in Fig. 7.6 (after Ref. [48])

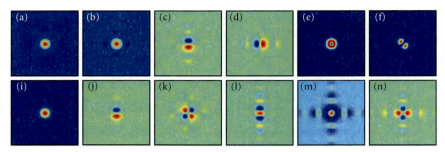

Fig. 7.6. (a)-(c) and (i)-(l): profiles of defect modes at the circled points in Fig. 7.5. Repulsive defect modes (*top*), attractive defect modes (*bottom*). (d) the co-existing mode of mode (c), (e) vortex mode derived from the superposition of mode (c) and (d) with $\pi/2$ phase delay, i.e., in the form of $u(x,y) + iu(y,x)$, (f) dipole mode derived from the superposition of mode (c) and (d) without phase delay, i.e., in the form of $u(x,y) + u(y,x)$, (m) and (n) modes derived from the superposition of mode (l) and its co-existing mode in the form of $u(x,y)+u(y,x)$ and $u(x,y)-u(y,x)$ respectively (after Ref. [48])

y, with a dominant hump at the defect site, and satisfy the symmetry relation $u(x,y) = u(y,x)$. They can be called fundamental defect modes. DMs on the branches "j" and "c" are dipole-like and do not satisfy the above symmetry relation. Note that there is one branch, i.e., "l", which bifurcates not from a band edge, but rather from an interior point in the third Bloch band. On DM branches with asymmetric modes $u(x,y)$, such as "j", "c" and "l", another linearly independent DM $u(y,x)$ co-exists. For instance, at point "c" in Fig. 7.5, the DM is a dipole along the vertical direction (Fig. 7.6c), while its co-existing DM is a dipole along the horizontal direction (Fig. 7.6d). Due to the linear property of the model equation (7.3), we can get other co-existing DMs by arbitrarily superimposing these two DMs $u(x,y)$ and $u(y,x)$. This linear superposition can produce more interesting DM patterns. For instance, a vortex mode, shown in Fig. 7.6e, can be derived from a superposition of modes "c" and "d" in the form of $u(x,y) + iu(y,x)$. A dipole mode oriented along the diagonal direction, shown in Fig. 7.6f, can be derived as well by the superposition $u(x,y) + u(y,x)$. Similarly, Fig. 7.6m and 7.6n are derived from the superposition of mode "l" and its co-existing mode in the form of $u(x,y) + u(y,x)$ and $u(x,y) - u(y,x)$, respectively.

Comparing the experimental results of Fig. 7.4 with the theoretical results of Figs. 7.5 and 7.6, we can see that Fig. 7.4b closely resembles a fundamental defect mode of Fig. 7.6b. The tails of vortex arrays in Figs. 7.4c and 7.4d should be related to the vortex DM of Fig. 7.6e when this DM gets close to a Bloch band [45, 64]. Further analysis of tail structures in Figs. 7.4c and 7.4d as well as observations of various DMs in Figs. 7.6 are currently underway.

7.5 Linear Defect Modes in 2D Ring Lattices

Ring-like photonic-lattice structures with a negative defect (a low-index core) are of particular interest as they have direct connections with Bragg fibers and PCFs. Yet it poses a challenge to make such structures by optical induction. Recently, we have succeeded in creating such structures in a *self-defocusing* photorefractive crystal, and subsequently demonstrated defect guiding in 2D ring lattices [44].

The crystal used for this experiment is a 10 mm long SBN:61. The experimental setup is similar to that used for generation of discrete ring lattice solitons reported in [65], except that the Bessel-like lattices [66, 67] are induced with a *self-defocusing* nonlinearity, so the center of the lattice is a low index core. With proper filtering, the mask gives rise to a Bessel-like intensity pattern at the crystal input, which remains nearly invariant during propagation throughout the 10 mm long crystal even under a negative bias field of 2 kV/cm (Fig. 7.7A). Starting from the first ring, the measured spacing between adjacent rings in Fig. 7.7A is about 20 µm. We note that the ring pattern created this way is somewhat different from the true Bessel pattern, since the intensity of rings here decreases more slowly (along the radial direction) than in a true Bessel lattice. Under a negative bias field, the crystal has a *self-defocusing* nonlinearity [52, 68, 69]. This means that the locations of the ring waveguides correspond to the dark (low intensity) areas of the lattice beam, while the center (high-intensity) corresponds to an anti-guide. Thus the ring pattern in Fig. 7.7A induces a periodic ring waveguide lattice with a low-index core.

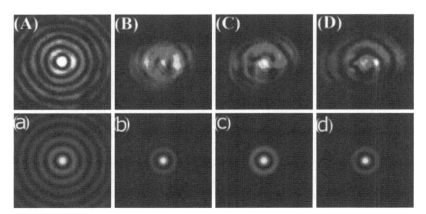

Fig. 7.7. Experimental observation of defect guidance in a ring lattice (*top*). (**A**) a ring lattice (20 µm spacing) established in experiments, (**B**)–(**D**) output of the probe beam in a ring lattice with 37 µm spacing as the negative bias is increased gradually. Theoretical demonstration of quasi-localized defect modes (*bottom*). (**a**) a Bessel-like ring lattice, (**b**) a guided mode in (a), (**c**) the lattice of (a) with outer rings removed, (**d**) a guided mode in (c) (after Ref. [44])

To investigate the waveguiding property in such a ring lattice, a Gaussian-like probe beam (FWHM: 14 µm) is launched directly into the core and propagates collinearly with the lattice. The probe beam is e-polarized but has a wavelength of 632.8 nm that is nearly photo-insensitive for our crystal, so that nonlinear self-action of the probe beam is negligible [52]. Since the index at the center of the lattice is lower than that at its surrounding, the probe beam tends to escape from the center and couple into the surrounding ring waveguides due to evanescent coupling. However, under appropriate conditions, guiding of the probe beam into the core is observed. Typical experimental results are presented in Fig. 7.7B–D for a ring lattice with 37 µm spacing. They show the output patterns of the probe beam as the negative applied dc field is set at three different levels (−0.6, −1.4, and −2.0 kV/cm), while the lattice intensity (normalized to background illumination) is fixed. When the bias field is low, the probe beam tends to diffract away from the core (Fig. 7.7B), but as the bias field increases, the probe beam undergoes a transition from discrete diffraction to central guidance (Fig. 7.7C). At even higher bias field, the guidance starts to deteriorate (Fig. 7.7D) because the experimental condition deviates from that for the formation of the defect mode.

To better understand the experimental results in Fig. 7.7, we use the model equation
$$iU_z + U_{xx} + U_{yy} - \frac{E_0}{1 + I_0|J_0(r)|^{3/2}} U = 0 , \tag{7.6}$$

where $J_0(r)$ is the Bessel function, and $r = \sqrt{x^2 + y^2}$. Here x and y are normalized by the spacing (pitch) of the lattice far away from the center, and normalizations for I_0, E_0 and z are the same as in [42,43]. The Bessel function $|J_0(r)|^{3/2}$ was chosen for the ring lattice since the first four peaks of this function decay as 1.00, 0.25, 0.16, 0.12, closely resembling those in experiments. Numerical simulations under experimental conditions produce results qualitatively similar to those in Fig. 7.7B–D. Furthermore, we have also searched for guided modes of the above model in the form of $U(x, y, z) = \exp[-i\mu z]u(r)$ with normalized parameters $E_0 = -15$, and $I_0 = 750$ (corresponding to experimental parameters). At $\mu = 0.97$, we found solutions $u(r)$ which have a high central peak and weak oscillatory tails. One such solution is shown in Fig. 7.7b, which resembles those observed in Fig. 7.7C. Note that in our ring lattice, the intensity decays along the radial direction, and thus bandgaps do not really open in the above model. Thus, the solution in Fig. 7.7b can not be a truly localized defect mode, but has tails decaying very slowly like a Bessel function, and its power is infinite. We can call it a quasi-localized mode. If we keep only the central beam and the first ring of the lattice (see Fig. 7.7c), we find that quasi-localized modes as in Fig. 7.7b persist (see Fig. 7.7d). This finding indicates that the guidance observed in Fig. 7.7C may not be attributed to the repeated Bragg reflections of outer rings, but rather it is dominated by the first high-index ring in our lattice. This guidance seems analogous to that in antiresonant reflecting optical waveguides [70], and certainly merit further investigation. For instance, one of the subjects in our future research is to see if such low-index core can create any "coloring" effect as that occurred in photonic crystal fibers.

7.6 Nonlinear Defect Modes

In previous sections, we discussed linear DMs and bandgap guidance in different types of photonic lattices containing defects, for which the probe beam propagates in the linear regime. On the other hand, the probe beam can also propagate in the nonlinear regime, where light can also be trapped as localized nonlinear defect modes (defect solitons). This issue has not received much attention before in photonic crystals or PCFs because the nonlinearity there is very weak. In nonlinear waveguide arrays and photonic lattices, however, the nonlinearity is high [1–18]. Thus far, there has been only limited theoretical work on defect solitons in waveguide lattices [41, 47, 51]. In this section, we investigate how nonlinearity affects the formation of 1D defect modes both theoretically and experimentally. The theoretical work is an extension of our earlier results in [47], while the experimental work is new.

We first consider defect solitons in a negative defect under focusing nonlinearity. The experimental setup is similar to the one illustrated in Fig. 7.1. The linear 1D lattice with a single-site negative defect as created in our experiment is shown in Figs. 7.8a and 7.8b. Here the lattice spacing is about 42 μm, the peak intensity (normalized to background illumination) is about 0.65, and the propagation distance is 20 mm. It is seen that the defect is maintained quite well. Next, a probe stripe beam with peak intensity about 0.23 (normalized to background illumination) is launched into this defect. At bias field of 1.1 kV/cm (for self-focusing nonlinearity), the output of linear propagation (taken instantaneously) is shown in Fig. 7.8c. It is seen that the probe beam evolves into a linear defect mode as we have demonstrated in Fig. 7.3d. However, under nonlinear propagation, the output (taken after a steady state has reached) has evolved into a defect soliton shown in Fig. 7.8d. The nonlinear output is similar to the linear output, except that the central stripe is more localized due to self-focusing nonlinearity. These observations demonstrate that defect guiding is quite robust, and sustains under nonlinear effects. This phenomenon is quite different from the nonlinearity-induced escape from defect sites of waveguide arrays as reported in [39].

Theoretically, we employ the model equation similar to (7.1) except that nonlinearity is involved:

$$iU_z + U_{xx} - \frac{E_0}{1 + I_L(x) + |U|^2}U = 0 . \tag{7.7}$$

Defect soliton solutions of (7.7) are sought in the form of (7.2). Corresponding to experimental conditions, we take $E_0 = 20$, and $I_0 = 0.65$. At these parameter values, the lowest linear defect mode is at $\mu = 20.40$, which lies

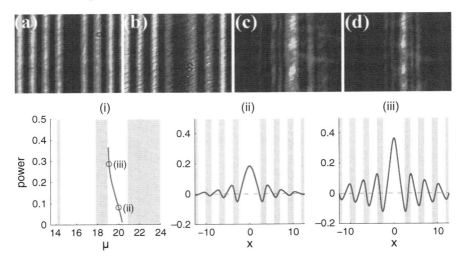

Fig. 7.8. Experimental demonstration of defect solitons in a repulsive defect under focusing nonlinearity (*top*). (**a**) lattice input, (**b**) lattice output, (**c**) probe linear output, (**d**) probe nonlinear output. Theoretical demonstration of defect solitons (*bottom*). (**i**) the power diagram (shaded regions are Bloch bands), (**ii**), (**iii**) profiles of defect solitons at μ values marked by circles in (i). The shaded stripes denote locations of high intensities in the defective lattice

in the second bandgap (between the second and third Bloch bands). Using asymptotic methods, we have shown that bifurcating from every linear defect mode, a family of nonlinear defect solitons will arise [47]. We have computed this family of defect solitons using the modified squared-operator method developed in [71], and its power curve is shown in Fig. 7.8i. We see that at higher powers, this soliton branch moves left toward the lower Bloch band. Two representative mode profiles are displayed in Figs. 7.8ii and 7.8iii. They have symmetric shapes, with peak intensities lying inside the repulsive defect. The profile of Fig. 7.8ii has lower power, and it is very close to the linear defect mode. As the power rises, the central peak becomes narrower (due to nonlinear self-focusing), while the side band mini-peaks become more pronounced (see Fig. 7.8iii). These findings agree well with the experimental observations in Figs. 7.8c and 7.8d.

Next, we present our work on defect solitons in a single-site attractive (rather than repulsive) defect. For this purpose, we only need to turn the nonlinearity from self-focusing to self-defocusing by reversing the sign of the bias field E_0 in the experiment of Fig. 7.8 and in the theoretical model of Eq. (7.7). The question is what types of defect solitons can exist in an attractive defect under self-defocusing nonlinearity. To answer this question, we take $E_0 = -8$, $I_0 = 3$ in Eq. (7.7), and numerically find two families of defect solitons, one in the semi-infinte bandgap, and the other one in the first

7 Light Localization by Defects in Optically Induced Photonic Structures 139

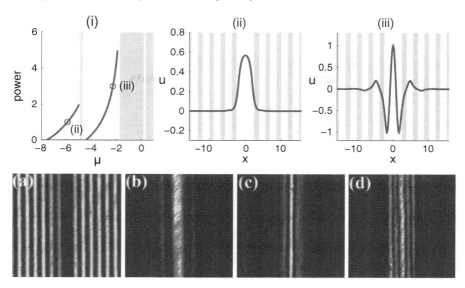

Fig. 7.9. Theoretical demonstration of defect solitons in an attractive defect under defocusing nonlinearity (*top*, $E_0 = -8$, $I_0 = 3$). (**i**) the power diagrams, (**ii**), (**iii**) defect solitons in the semi-infinite and first bandgaps respectively. Experimental observation of defect solitons (*bottom*). (**a**) lattice input, (**b**) defect soliton excited with a single-stripe input at $E_0 = -1.4\,\mathrm{kV/cm}$, corresponding to (ii), (**c**) three-stripe probe input, (**d**) defect soliton excited with (c) at $E_0 = -1.6\,\mathrm{kV/cm}$, corresponding to (iii)

bandgap. The power curves of these families are shown in Fig. 7.9i. We can see that both families bifurcate from the linear defect modes in the low power limit. On the family of the semi-infinite bandgap, a typical soliton profile is shown in Fig. 7.9ii, which consists of a dominant hump inside the defect. Note that this soliton exists under defocusing nonlinearity. It is possible because the negative bias field makes the surrounding lattice sites having lower refractive indices than the center defect, thus due to total internal reflection, such solitons are permitted. For the family in the first bandgap, a typical soliton profile is shown in Fig. 7.9iii. This soliton has three dominant intensity peaks inside the defect site, flanked by weaker peaks nearby. All adjacent peaks are out of phase with each other. This is a higher-order nonlinear defect mode (defect gap soliton) supported by the attractive defect.

Experimentally, these two types of defect solitons under defocusing nonlinearity are also observed. The experimental results are summarized in Fig. 7.9a–d. First, a photonic lattice with a single-site defect is maintained (see Fig. 7.9a). Then we launch a single-stripe probe beam into the defect channel. At bias field of $-1.4\,\mathrm{kV/cm}$, the output of the probe beam is shown in Fig. 7.9b. It can be seen clearly that the probe beam is well guided inside the defect channel despite the fact that a self-defocusing nonlinearity is employed.

This guided mode closely resembles the theoretical defect soliton shown in Fig. 7.9ii. When the input probe beam contains three out-of-phase stripes (Fig. 7.9c), it still remains somewhat localized at the bias field of $-1.6\,\text{kV/cm}$ (Fig. 7.9d). Recalling that such a probe beam would defocus and diverge dramatically should the defect be absent, the formation of such defect gap soliton is again attributed to combined effects of nonlinearity and lattice impurity. When comparing this output with the theoretical solution in Fig. 7.9iii, we see that the outer two peaks in the experiment are more pronounced, but the theoretical and experimental results are in qualitative agreement. These studies of nonlinear defect modes and defect gap solitons can readily be extended to the 2D domain.

7.7 Summary

In summary, we have successfully fabricated 1D and 2D defective photonic lattices by the method of amplitude modulation together with several other techniques such as frequency filtering, partial spatial coherence, and polarization-controlled index variation. We have also studied light guiding in these defective lattices, and shown that these defects can guide a wide array of defect modes in both linear and nonlinear regimes. Our results pave the way for further studies of new phenomena in photonic structures with built-in defects, as well as for exploring potential applications in beam shaping and light routing with reconfigurable lattices. Since defects exist in a wide array of other periodic linear and nonlinear systems, our work may prove to be relevant to the studies of defect-mediated phenomena in other branches of physics and nonlinear sciences.

Acknowledgements

This work was supported in part by NSF, AFOSR, and the 973 program.

References

1. D.N. Christodoulides, F. Lederer, and Y. Siberberg, Nature **424**, 817 (2003)
2. Y.S. Kivshar and G.P. Agrawal, *Optical solitons*, Academic Press, New York (2003)
3. D. Campbell, S. Flach, and Y.S. Kivshar, Phys. Today **57**, 43 (2004)
4. D.N. Christodoulides and R.I. Joseph, Opt. Lett. **13**, 794 (1988)
5. H.S. Eisenberg, Y. Silberberg, R. Morandotti, A.R. Boyd, and J.S. Aitchison, Phys. Rev. Lett. **81**, 3383 (1998)
6. R. Morandotti, H.S. Eisenberg, Y. Silberberg, M. Sorel, and J.S. Aitchison, Phys. Rev. Lett. **86**, 3296 (2001)
7. N.K. Efremidis, S. Sears, D.N. Christodoulides, J.W. Fleischer, and M. Segev, Phys. Rev. E **66**, 046602 (2002)

7 Light Localization by Defects in Optically Induced Photonic Structures   141

8. J.W. Fleischer, T. Carmon, M. Segev, N.K. Efremidis, and D.N. Christodoulides, Phys. Rev. Lett. **90**, 023902 (2003)
9. J.W. Fleischer, M. Segev, N.K. Efremidis, and D.N. Christodoulides, Nature **422**, 147 (2003)
10. D. Neshev, E. Ostrovskaya, Y. Kivshar, and W. Krolikowski, Opt. Lett. **28**, 710 (2003)
11. H. Martin, E.D. Eugenieva, Z. Chen, and D.N. Christodoulides, Phys. Rev. Lett. **92**, 123902 (2004)
12. Z. Chen, H. Martin, E.D. Eugenieva, J. Xu, and A. Bezryadina, Phys. Rev. Lett. **92**, 143902 (2004)
13. B.A. Malomed and P. G. Kevrekidis, Phys. Rev. E **64**, 026601 (2001)
14. J. Yang and Z. H. Musslimani, Opt. Lett. **28**, 2094 (2003)
15. Z. Musslimani and J. Yang, J. Opt. Soc. Am. B **21**, 973 (2004)
16. J. Yang, New Journal of Physics **6**, 47 (2004)
17. D.N. Neshev, T.J. Alexander, E.A. Ostrovskaya, Y.S. Kivshar, H. Martin, I. Makasyuk, and Z. Chen, Phys. Rev. Lett. **92**, 123903 (2004)
18. J.W. Fleischer, G. Bartal, O. Cohen, O. Manela, M. Segev, J. Hudock, and D.N. Christodoulides, Phys. Rev. Lett. **92**, 123904 (2004)
19. R. Iwanow, R. Schiek, G.I. Stegeman, T. Pertsch, F. Lederer, Y. Min, W. Sohler, Phys. Rev. Lett. **93**, 113902 (2004)
20. T. Pertsch, U. Peshchl, J. Kobelke, K. Schuster, H. Bartelt, S. Nolte, A. Tünnermann, and F. Lederer, Phys. Rev. Lett. **93**, 053901 (2004)
21. A. Fratalocchi, G. Assanto, K.A. Brzdakiewicz, and M.A. Karpierz, Opt. Lett. **29**, 1530 (2004)
22. Y.S. Kivshar, Opt. Lett. **18**, 1147 (1993)
23. D. Mandelik, R. Morandotti, J.S. Aitchison, and Y. Silberberg, Phys. Rev. Lett. **92**, 093904 (2004)
24. D. Neshev, A.A. Sukhorukov, B. Hanna, W. Krolikowski, and Y.S. Kivshar, Phys. Rev. Lett. **93**, 083905 (2004)
25. F. Chen, M. Stepic, C. Rter, D. Runde, D. Kip, V. Shandarov, O. Manela, and M. Segev, Opt. Express **13**, 4314 (2005)
26. C. Lou, X. Wang, J. Xu, Z. Chen, and J. Yang, Phys. Rev. Lett. **98**, 213903 (2007)
27. J.D. Joannopoulos, R.D. Meade, J.N. Winn, *Photonic Crystals: Molding the Flow of Light*, Princeton University Press, New Jersey (1995)
28. P. Russell, Science **299**, 358 (2003)
29. S.L. McCall, P.M. Platzman, R. Dalichaouch, D. Smith, and S. Schultz, Phys. Rev. Lett. **67**, 2017 (1991)
30. E. Yablonovitch, T.J. Gmitter, R.D. Meade, A.M. Rappe, K.D. Brommer, and J.D. Joannopoulos, Phys. Rev. Lett. **67**, 3380 (1991)
31. M. Bayindir, B. Temelkuran, and E. Ozbay, Phys. Rev. Lett. **84**, 2140 (2000)
32. F. Luan, A.K. George, T.D. Hedley, G.J. Pearce, D.M. Bird, J.C. Knight, and P.St.J. Russell, Opt. Lett. **29**, 2369 (2004)
33. A. Argyros, T.A. Birks, S.G. Leon-Saval, C.B. Cordeiro, F. Luan, and P.St.J. Russell, Opt. Express **13**, 309 (2005)
34. J. Schmidtke, W. Stille, and H. Finkelmann, Phys. Rev. Lett. **90**, 083902 (2003)
35. J.S. Foresi, P.R. Villeneuve, J. Ferrera, E.R. Thoen, G. Steinmeyer, S. Fan, J.D. Joannopoulos, L.C. Kimerling, H.I. Smith, and E.P. Ippen, Nature **390**, 143 (1997)

36. S. Fan, P.R. Villeneuve, J.D. Joannopoulos, and H.A. Haus, Phys. Rev. Lett. **80**, 960 (1998)
37. X. Wu, A. Yamilov, X. Liu, S. Li, V.P. Dravid, R.P.H. Chang, and H. Cao, Appl. Phys. Lett. **85**, 3657 (2004)
38. O. Painter, R.K. Lee, A. Scherer, A. Yariv, J.D. O'Brien, P.D. Dapkus, and I. Kim, Science **284**, 1819 (1999)
39. U. Peschel, R. Morandotti, J.S. Aitchison, H.S. Eisenberg, and Y. Silberberg, Appl. Phys. Lett. **75**, 1348 (1999)
40. R. Morandotti, H.S. Eisenberg, D. Mandelik, Y. Silberberg, D. Modotto, M. Sorel, C.R. Stanley, and J.S. Aitchison, Opt. Lett. **28**, 834 (2003)
41. A.A. Sukhorukov and Y.S. Kivshar, Phys. Rev. Lett. **87**, 083901 (2001)
42. F. Fedele, J. Yang, and Z. Chen, Opt. Lett. **30**, 1506 (2005)
43. F. Fedele, J. Yang, and Z. Chen, Stud. Appl. Math. **115**, 279 (2005)
44. X. Wang, Z. Chen and J. Yang, Opt. Lett. **31**, 1887 (2006)
45. I. Makasyuk, Z. Chen and J. Yang, Phys. Rev. Lett. **96**, 223903 (2006)
46. X. Wang, J. Yang, Z. Chen, D. Weinstein, and J. Yang, Opt. Express **14**, 7362 (2006)
47. J. Yang and Z. Chen, Phys. Rev. E **73**, 026609 (2006)
48. J. Wang, J. Yang, and Z. Chen, Phys. Rev. A **76**, 013828 (2007)
49. G. Bartal, O. Cohen, H. Buljan, J.W. Fleischer, O. Manela, and M. Segev, Phys. Rev. Lett. **94**, 163902 (2005)
50. B. Freedman, R. Lifshitz, J.W. Fleischer, and M. Segev, Nature **440**, 1166 (2006)
51. M.J. Ablowitz, B. Ilan, E. Schonbrun, and R. Piestun, Phys. Rev. E **74**, 035601 (2006)
52. M. Shih, Z. Chen, M. Mitchell, and M. Segev, J. Opt. Soc. Am. B **14**, 3091 (1997)
53. M. Mitchell, Z. Chen, M. Shih, and M. Segev, Phys. Rev. Lett. **77**, 490 (1996)
54. Z. Chen, M. Mitchell, M. Segev, T.H. Coskun, and D.N. Christodoulides, Science **280**, 889 (1998)
55. Z. Chen, M. Segev, and D.N. Christodoulides, J. Opt. A **5**, S389 (2003)
56. Z. Chen and K. McCarthy, Opt. Lett. **27**, 2019 (2002)
57. Z. Chen, K. McCarthy, and H. Martin, Optics and Photonic News, December 2002
58. J. Petter, J. Schröder, D. Träger, and C. Denz, Opt. Lett. **28**, 438 (2003)
59. M. Petrovic, D. Träger, A. Strinic, M. Belic, J. Schröder, and C. Denz, Phys. Rev. E **68**, 055601 (2003)
60. D.N. Neshev, Y.S. Kivshar, H. Martin, and Z. Chen, Opt. Lett. **29**, 486 (2004)
61. H.F. Talbot, Philos. Mag. **9**, 401 (1836)
62. R. Iwanow, D.A. May-Arrioja, D.N. Christodoulides, G.I. Stegeman, Y. Min, and W. Sohler, Phys. Rev. Lett. **95**, 053902 (2005)
63. Z. Chen and J. Yang, "Controlling light in reconfigurable photonic lattices", chapter in H.A. Abdeldayem and D.O. Frazier (ed.), *Nonlinear Optics and Applications*, pp. 103–150, Research Signpost, Kerala, India (2007)
64. Z. Shi and J. Yang, Phys. Rev. E **75**, 056602 (2007)
65. X. Wang, Z. Chen, and P.G. Kevrekidis, Phys. Rev. Lett. **96**, 083904 (2006)
66. Y.V. Kartashov, V.A. Vysloukh, and L. Torner, Phys. Rev. Lett. **93**, 093904 (2004)
67. Z. Xu, Y.V. Kartashov, L. Torner, and V.A. Vysloukh, Opt. Lett. **30**, 1180 (2005)

68. Z. Chen, H. Martin, A. Bezryadina, D. Neshev, Y.S. Kivshar, and D.N. Christodoulides, J. Opt. Soc. Am. B **22**, 1395 (2005)
69. Z. Chen, H. Martin, E.D. Eugenieva, J. Xu, J. Yang, and D.N. Christodoulides, Opt. Express **13**, 1816 (2005)
70. N.M. Litchinitser, A.K. Abeeluck, C. Headley, and B.J. Eggleton, Opt. Lett. **27**, 1592 (2002)
71. J. Yang and T.I. Lakoba, Stud. Appl. Math. **118**, 153 (2007)

8

Polychromatic Light Localisation in Periodic Structures

Dragomir N. Neshev, Andrey A. Sukhorukov, and Wieslaw Z. Krolikowski, and Yuri S. Kivshar

Nonlinear Physics Centre and Laser Physics Centre, Research School of Physics and Engineering, Australian National University, Canberra, 0200 ACT, Australia
dragomir.neshev@anu.edu.au

8.1 Introduction

Photonic structures with a periodic modulation of the optical refractive index open novel possibilities for designing the fundamental aspects of optical wave dynamics [1]. They offer important opportunity for spatial beam control, including manipulation of beam refraction and diffraction. The physics of beam propagation in periodic photonic structures is governed by the scattering of waves from the high refractive index regions and the subsequent interference of the scattered waves. This is a resonant process, which is sensitive to both the frequency and the propagation angle. In photonic crystals, for example, there appear sharp spectral features where the propagation of optical signals is highly sensitive to the wavelength. As such, most of the demonstrated effects of spatial beam manipulation in periodic structures are primarily optimised for a narrow-frequency range. In many practical cases, including ultra-broad bandwidth optical communications, manipulation of ultra-short pulses or supercontinuum radiation, the bandwidth of the optical signals can span over a wide frequency range. This motivates the studies on the propagation of broad-bandwidth optical beams in periodic photonic structures.

The strong dispersion of waves in periodic structures can lead to enhanced spatial separation of the spectral components of the incident light, an effect known as superprism [2,3]. Importantly, the wave dispersion and diffraction can be balanced or enhanced in materials with nonlinear optical response, thus opening opportunities for all-optical spectral management [4]. The nonlinear spatial control of polychromatic light represents an intriguing physical problem, and is the main scope of our studies. In this review we present, what we believe is the first experimental demonstration of nonlinear control of broadband and supercontinuum light in spatially-extended periodic photonic structures, as well as provide theoretical description of the observed phenomena.

146   D.N. Neshev et al.

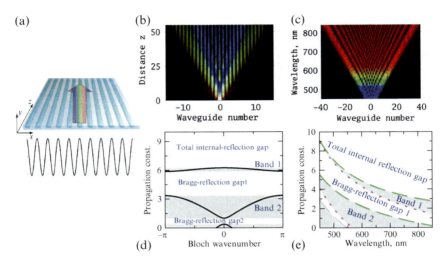

Fig. 8.1. (a) Schematic of the waveguide array structure (*top*) and the characteristic transverse refractive index profile (*bottom*), (b), (c) numerical simulation of polychromatic beam diffraction, (b) image of beam evolution inside the array and (c) the spectrally-resolved output profile, (d) Bloch-wave dispersion at 530 nm wavelength, (e) dependence of photonic bands on wavelength. Numerical simulations correspond to the parameters of LiNbO$_3$ waveguide arrays [21, 22]

To understand the basic concepts of nonlinear control of polychromatic light, we utilise a combination of broadband coherent light radiation from supercontinuum generation and a one-dimensional periodic structure in the form of a waveguide array (Fig. 8.1a). This type of structures belong to a class of nonlinear photonic lattices [5–16], and feature refractive index modulation in the transverse spatial dimension with a characteristic period of several wavelengths, resembling the periodic cladding of photonic crystal fibers [17]. In such structures the back-scattering of light is absent and transmission coefficients can approach unity simultaneously for all spectral components, making them specifically attractive for manipulation of polychromatic or supercontinuum light. By using this system, we reveal novel possibilities for all-optical spatial switching, spectral reshaping, and localisation of supercontinuum light beams. We demonstrate that the interplay of wave scattering from the periodic structure and nonlinearity-induced interaction of multiple colours allows one to selectively separate or combine different spectral components. Additional flexibility is implemented through interactions with induced defects and boundaries in the structure, where small refractive index changes may enable tunable spectral filtering.

The Chapter is organised as follows: In Sec. 8.2 we present the basic concepts of linear polychromatic light propagation in photonic lattices. In Sec. 8.3 we describe the effect of nonlinear material response on the beam spectral-spatial reshaping. We show how the collective nonlinear interaction of spectral components in a medium with slow nonlinearity leads to the formation of polychromatic gap solitons. In Sec. 8.4 we provide experimental verification of the theoretical predictions and demonstrate how the slow defocusing photorefractive nonlinearity in lithium niobate waveguide arrays can be used for tunable all-optical reshaping of supercontinuum light. We also describe the ability of surfaces to alter the nonlinear propagation of polychromatic light, leading to the formation of polychromatic nonlinear surface states.

8.2 Polychromatic Light in Periodic Structures

In this section we discuss the general features of polychromatic beam diffraction in planar periodic photonic structures (Fig. 8.1a), such as optically-induced lattices or waveguide arrays [5–16]. The physical mechanism of beam diffraction in such structures is based on the coupling between the modes of neighbouring waveguides [9, 18, 19]. When the beam is launched into a single waveguide at the input, it experiences 'discrete diffraction' where most of the light is directed into the wings of the beam. This is in sharp contrast to the diffraction of Gaussian beams in bulk materials where the peak intensity remains at the beam centre at any propagation distance. The light couples from one waveguide to another due to the spatial overlaps of the waveguide modes. Since the mode profile and confinement depend on the wavelength, the discrete diffraction exhibits strong spectral dispersion. The mode overlap at neighbouring waveguides is usually much stronger for red-shifted components [20], which therefore diffract faster than their blue counterparts. This leads to spatial redistribution of the colours of the polychromatic beam which increases along the propagation direction, see Fig. 8.1b. As a result, at the output the red components dominate in the beam wings, while the blue components are dominant in the central region, see Fig. 8.1c.

The mathematical modeling of the diffraction process for optical sources with a high degree of spatial coherence, such as supercontinuum light generated in photonic-crystal fibres, is based on a set of equations for the spatial beam envelopes $A_m(x, z)$ of different frequency components at vacuum wavelengths λ_m. Since we consider beam propagation at small angles along the lattice and the refractive index contrast is usually of the order of 10^{-4} to 10^{-2}, then the general wave equations can be reduced to parabolic equations employing the conventional paraxial approximation [22–24],

$$\mathrm{i}\frac{\partial A_m}{\partial z} + \frac{\lambda_m}{4\pi n_0(\lambda_m)}\frac{\partial^2 A_m}{\partial x^2} + \frac{2\pi}{\lambda_m}\Delta n(x, \lambda_m) A_m = 0 \,, \tag{8.1}$$

where x and z are the transverse and longitudinal coordinates, respectively, and $n_0(\lambda_m)$ is the background refractive index. The function $\Delta n(x, \lambda_m)$ describes the effective refractive index modulation, which depends on the vertical mode confinement in the planar guiding structure. Since the vertical mode profile changes with wavelength, the dispersion of the effective index modulation is defined by the geometry of the photonic structures. For an array of optical waveguides in LiNbO$_3$, the modulation can be approximately described as $\Delta n(x, \lambda) = \Delta n_{\max}(\lambda) \cos^2(\pi x/d)$, where the wavelength dependence of the effective modulation depth $\Delta n_{\max}(\lambda)$ can be calculated numerically or determined experimentally by measuring the waveguide coupling [22, 24]. Even if the material and geometrical dispersion effects are weak, the beam propagation would still strongly depend on its frequency spectrum [23, 25] since the values of λ_m appear explicitly in Eqs. (8.1).

The linear propagation of optical beams through a periodic lattice can be fully characterised by decomposing the input profile in a set of spatially extended eigenmodes, called Bloch waves [26, 27]. The Bloch-wave profiles can be found as solutions of Eqs. (8.1) in the form $A_m(x, z) = \psi_j(x, \lambda_m) \exp[i\beta_j(K_b, \lambda_m)z + iK_b x/d]$, where $\psi_j(x, \lambda_m)$ has the periodicity of the underlying lattice, $\beta_j(K_b, \lambda_m)$ are the propagation constants, K_b are the normalised Bloch wavenumbers, j is the band number, and d is the lattice period. At each wavelength, the dependencies of longitudinal propagation constant (along z) on the transverse Bloch wavenumber (along x) are periodic, $\beta_j(K_b, \lambda_m) = \beta_j(K_b \pm 2\pi, \lambda_m)$, and are fully characterised by their values in the first Brillouin zone, $-\pi \leq K_b \leq \pi$. These dependencies have a universal character [1, 26, 27], where the spectrum consists of non-overlapping bands separated by photonic bandgaps, as shown in Fig. 8.1d. The position and the width of the bands and the gaps, however, are strongly sensitive to the wavelength of light (Fig. 8.1e). As a result, the spatial beam shaping exhibits frequency dispersion. In particular, the rate of beam diffraction is determined by the curvature of the dependencies $\beta_j(K_b, \lambda_m)$. For the input beam coupled to a single waveguide, the first band is primarily excited, and the beam diffraction rate is determined by $\max_{K_b} |\partial^2 \beta_1 / \partial K_b^2|$, which increases at longer wavelengths where the band gets wider. This conclusion is in full agreement with the physical interpretation presented above using the concept of coupling between waveguide modes.

8.3 Nonlinear Localisation of Polychromatic Light

An important task of building a nonlinear photonic device for manipulation of broadband signals requires the ability to tune the spectral transmission in the spatial domain. In this section, we describe an approach to flexible spatial-spectral control of polychromatic light through the effect of nonlinear interaction and selective self-trapping of spectral components inside the individual channels of the waveguide array.

8.3.1 Collective Nonlinear Interactions in Media with Slow Nonlinearity

In order to perform spatial switching and reshaping of polychromatic signals without generating or depleting different spectral regions, the coherent four-wave-mixing processes need to be suppressed. This can be achieved in media which nonlinear response is slow with respect to the scale of temporal coherence [28]. This condition is commonly satisfied for photorefractive materials [29, 30] or liquid crystals [31], where the optically-induced refractive index change is defined by the time-averaged light intensity of different spectral components. The slow nonlinearities thus provide a fundamentally different regime compared to dynamics of light with supercontinuum spectrum in multi-core photonic-crystal fibers with fast nonlinear response [32–34], where nonlinearly-induced spatial mode reshaping is inherently accompanied by the spectral transformations.

We model the nonlinear propagation and interaction of the spectral components by including into Eqs. (8.1) a nonlinearly-induced refractive index change,

$$\mathrm{i}\frac{\partial A_m}{\partial z} + \frac{\lambda_m}{4\pi n_0(\lambda_m)}\frac{\partial^2 A_m}{\partial x^2} + \frac{2\pi}{\lambda_m}\left[\Delta n(x, \lambda_m) + \Delta n_{\mathrm{nl}}(x, z)\right] A_m = 0 \ . \quad (8.2)$$

We write the nonlinear term in Eq. (8.2) in a Kerr-type form: $\Delta n_{\mathrm{nl}}(x, z) = \gamma M^{-1} \sum_{j=1}^{M} \sigma(\lambda_j)|A_j|^2$, which approximately describes the LiNbO$_3$ photovoltaic nonlinearity [30, 35] in the regime of weak saturation. Here γ is the nonlinear coefficient, and M is the number of frequency components included in the numerical modeling. Whereas nonlocality and saturation may affect the soliton properties [36], these effects were weak under our experimental conditions [21, 22, 24, 37]. An important characteristic of this model is the spectral nonlinear response $\sigma(\lambda_j)$.

The physical mechanism of the photovoltaic nonlinearity in lithium niobate arises due to charge excitations by light absorption and corresponding separation of these charges due to diffusion. The spectral response of this type of nonlinearity depends on the crystal doping and stoichiometry and may vary from sample to sample. In general, however, light sensitivity exists in a wide spectral range with a maximum for the blue spectral components [38]. The sensitivity, however, extends well in the near infra-red region [39] resulting in a broad-band nonlinear response. In our analysis, we therefore approximate the photosensitivity dependence by a Gaussian function $\sigma(\lambda) = \exp[-\ln(2)(\lambda - \lambda_\mathrm{b})^2/\lambda_\mathrm{w}^2]$ with $\lambda > \lambda_\mathrm{b} = 400\,\mathrm{nm}$ and $\lambda_w = 150\,\mathrm{nm}$.

8.3.2 Polychromatic Gap Solitons

Multiple frequency components of an optical beam can undergo self-trapping process and propagate in a common direction, when they are nonlinearly coupled together and form a polychromatic spatial soliton. Such solitons are self-trapped polychromatic beams which do not diffract. In bulk media polychromatic or white-light solitons can only be supported in materials with self-focusing nonlinearity [29, 30]. In periodic structures, however, polychromatic solitons were predicted to occur in media with either self-focusing [40] or self-defocusing [23,25] nonlinearities. Below we consider the most intriguing case of polychromatic soliton formation in lattices with defocusing nonlinearity, when localisation would not be possible for bulk crystals and can only occur due to resonant Bragg confinement from the periodic structure.

The numerical simulations based on the system of Eqs. (8.2) (with $\gamma = -1$ for defocusing nonlinear response) show that the input beam experiences self-trapping above a critical power level [21–23, 25]. In this regime, the spectral components become spatially localised and form a polychromatic soliton, which propagates without broadening in the photonic structure, see Fig. 8.2a. In order to identify the physical mechanism of beam localisation and soliton formation, we calculate the spectrum of the propagation constants, which is presented as density plot (white colour corresponds to larger amplitudes) in Fig. 8.2b. This figure shows that the propagation constants are entirely

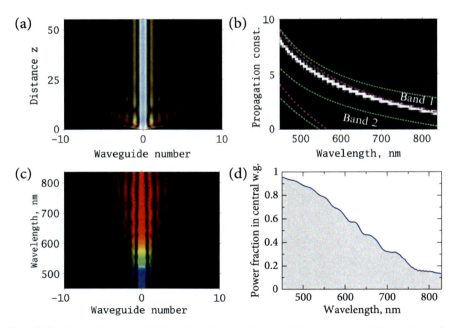

Fig. 8.2. Numerical simulation of nonlinear beam self-trapping and formation of a polychromatic gap soliton [21,22]. (**a**) Soliton excitation inside the waveguide array, (**b**) propagation constants of the soliton beam vs. wavelength, dashed and dashed-dotted lines mark the band edges as indicated in Fig. 8.1e, (**c**) spectrally resolved output intensity profile, (**d**) the fraction of output power remaining at the central waveguide for different spectral components

located inside the Bragg-reflection gap due to the nonlinear decrease of the refractive index in the soliton region. This effect, analogous to the band-gap guiding in hollow-core photonic crystal fibres [17], allows for the spatial self-trapping of all spectral components in an ultra-wide spectral bandwidth. Thus, such self-trapped states can be termed *polychromatic gap solitons*. Note that the spectrum for blue components is shifted deeper inside the gap, whereas the red components have spectra very close to the gap edge. This explains the weaker localisation of red components as shown in Figs. 8.2c and 8.2d, and results in colouring of the soliton profile (Fig. 8.2c) such that the soliton has a blue centre and red tails.

8.4 Experimental Studies of Polychromatic Self-trapping

8.4.1 Experimental Setup

In this section we describe how the nonlinear effects of polychromatic light reshaping and self-trapping can be realised and observed experimentally. The key for such experimental realisation is the combination of a periodic structure featuring a broadband nonlinear response and polychromatic light with a broad frequency spectrum, high-spatial coherence, and high optical intensity. The natural choice of such light source is provided by the effect of supercontinuum generation. In this process, spectrally narrow laser pulses are converted into the broad supercontinuum spectrum through several processes [17, 41, 42], including self-phase modulation, soliton formation due to the interplay between anomalous dispersion and Kerr-nonlinearity, soliton break-up due to higher order dispersion, and Raman shifting of the solitons, leading to non-solitonic radiation in the short-wavelength range. Such supercontinuum radiation has proven to be an excellent tool for characterisation of bandgap materials [43], it posses high spatial coherence [42], as well as high brightness and intensity required for nonlinear effects [44]. In our experiments (Fig. 8.3a), we use a supercontinuum light beam generated by femtosecond laser pulses (140 fs at 800 nm from a Ti:Sapphire oscillator) coupled into 1.5 m of highly nonlinear photonic crystal fiber (Crystal Fiber NL-2.0-740 with engineered zero dispersion at 740 nm) [24]. The spectrum of the generated supercontinuum is shown in Fig. 8.3b, and it spans over a wide frequency range (typically more than an optical octave). After re-collimation and attenuation, the supercontinuum is spectrally analysed by a fiber spectrometer and is refocused by a microscope objective ($\times 20$) to a single channel of a waveguide array (see Fig. 8.3c).

The optical waveguides are fabricated by indiffusion of a thin (100 Å) layer of titanium in a X-cut, 50 mm long mono-crystal lithium niobate wafer [15]. The waveguides are single-moded for all spectral components of the supercontinuum. Arrays with different periodicity and index contrast were fabricated. The choice of $LiNbO_3$ as an experimental platform was dictated by its strong

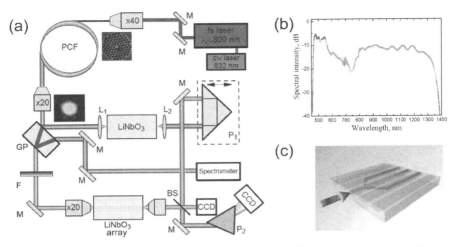

Fig. 8.3. (a) Schematic of the experimental setup. Supercontinuum radiation is generated in a photonic crystal fibre (PCF) and then injected into a single channel of the LiNbO$_3$ waveguide array. The array output is imaged onto a colour CCD camera. A prism can be inserted between the sample and the camera to achieve spectral resolution of the output beam. A fiber spectrometer is used to analyse the supercontinuum spectrum, and a reference beam from a dispersion compensated interferometer is used for interferometric analysis. M: mirrors, L: lenses, P: prisms, and GP: glass plate. (b) Spectrum of the generated supercontinuum radiation. (c) Illustration of the excitation scheme, showing the separation of the colours inside the array of optical waveguides [24]

nonlinear optical response at micro-Watt laser powers due to photorefraction [45, 46]. The photovoltaic photorefractive nonlinearity [35] in LiNbO$_3$ is of the defocusing type, meaning that an increase of the light intensity leads to a local decrease in the material refractive index.

After coupling to the array, its output is imaged by a microscope objective (×5) onto a colour CCD camera, where a dispersive 60° (glass SF-11) prism could be inserted between the imaging objective and the camera in order to spectrally resolve all components of the supercontinuum. Alternatively, the output spectral distribution can be obtained with a high accuracy (resolution 0.3 nm) by a fiber spectrometer, which integrates over the whole transverse mode of each waveguide and provides individual spectra at each waveguide position. Furthermore, a reference supercontinuum beam is used for interferometric measurement of the phase structure of the output beam [22]. To compensate for the pulse delay and pulse spreading inside the LiNbO$_3$ waveguides, this reference beam is sent through a variable delay-line, implemented in a dispersion compensated interferometer, including 5 cm long bulk LiNbO$_3$ crystal (to equalise the material dispersion). In this way, interferometric measurements are possible for ultra-wide spectral range. The experiments were

8 Polychromatic Light Localisation in Periodic Structures   153

performed with two samples designed for weak and strong waveguide coupling, thus allowing to distinguish between different nonlinear localisation scenarios.

8.4.2 Nonlinear Spectral-spatial Reshaping

As a first step we characterise the effect of nonlinearity on the propagation of supercontinuum beam in an array with a weak waveguide coupling (waveguide period of 19 µm). The coupling in this sample has been designed such that the blue spectral components remain entirely in the central waveguide, while other colours couple to the neighbouring waveguides. The spectral distribution at the output of a low power (17 µW) beam (measured by a spectrometer) is shown in Fig. 8.4a. This graph represents a linear distribution of colours with the waveguide number. This can be considered as a simple scheme for colour sorting/separation of broadband radiation into different waveguide channels, an effect similar to wavelength division demultiplexers. This natural colour separation is clearly visible in Fig. 8.4b, where we plot the output spectrum at different waveguides of the array, normalised to the input supercontinuum spectrum. The spectra show a dominant spectral peak, followed by decaying oscillations. Clearly, in each waveguide of the array, in the dominant peak, we select a narrow spectral band from the entire supercontinuum spectrum. The maximum of this transmission peak, however drops rapidly at longer

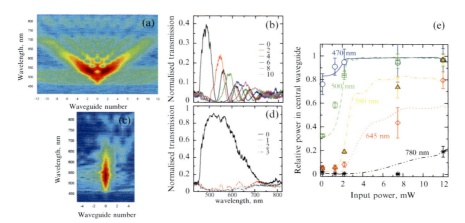

Fig. 8.4. (a) Spectrum measured by a spectrometer for low input power (17 µW) at the output of a waveguide array with periodicity of 19 µm, (b) normalised spectral transmission for the central, second, fourth, sixth, eighth, and the tenth waveguides, respectively, (c), (d) output spectrum for nonlinear localisation at power 7.5 mW and normalised transmission (central, first, second, and third waveguides), respectively, (e) measured (points) and calculated (lines) relative spectral power in the central waveguide as a function of the input power for five different spectral components

wavelengths due to the increased waveguide coupling and the associated beam broadening.

The effect of the nonlinear self-action of the beam can dramatically change the transmission spectrum in each waveguide. In Fig. 8.4c we show the combined pictures for the whole array in the nonlinear regime (power of 7.5 mW). It is clear that at this power the spectral spreading at longer wavelengths is compensated by the nonlinear self-trapping even in the case of defocusing nonlinearity. The normalised spectra of the first three waveguides are shown in Fig. 8.4d. The normalised transmission spectrum for the central waveguide has a well defined bell shape with a steep front at shorter wavelengths and a gradually decaying slope at longer wavelengths, revealing weaker trapping in the latter case. Not surprisingly, at the neighbouring waveguides only a small fraction of light is being transmitted at this laser power.

To quantify the process of nonlinear spectral reshaping we plot in Fig. 8.4e, the amount of spectral power trapped into the central waveguide for five different wavelengths as measured experimentally (points) and confirmed numerically (lines). Our result demonstrates that the localisation happens at lowest powers for the blue spectral components (470 nm), which experience weakest diffraction. The threshold localisation power is higher at longer wavelengths (500, 580, 645, and 780 nm) as larger nonlinearity-induced index change is necessary to balance the stronger diffraction at these wavelengths. The observed gradual localisation of the spectral components with power, together with strong nonlinear confinement in the spatial domain, suggest important possibilities for nonlinear active control over the transmitted supercontinuum radiation. Such active control is possible due to the balance of wave scattering and nonlinear spatial localisation in the $LiNbO_3$ waveguide array.

8.4.3 Generation of Polychromatic Gap Solitons

In the second type of experiments we investigate the nonlinear process of polychromatic beam localisation in an array with periodicity of 10 μm. Accordingly, the waveguide coupling is stronger compared with the sample used in Sec. 8.4.2 where the waveguide separation is larger. The image in Fig. 8.5a depicts the spectrally resolved discrete diffraction of the supercontinuum beam (measured by a spectrometer) when the input beam is focused to a single waveguide. Again the diffraction of the beam is weakest for the blue spectral components, which experience weaker coupling. However the shortest wavelength component (460 nm) now couples to more than eight neighbouring waveguides. The spectrally resolved discrete diffraction provides visual illustration of the colour distribution at the output, but it also allows for exact determination of the waveguide coupling in the waveguide array. In our sample we measure that the discrete diffraction length (defined as the length at which the beam expands with two extra waveguides) varies from 1 cm, for blue (480 nm), to less than 0.2 cm, for red (800 nm) spectral components. These values correspond to a total propagation distance of 5.5 and 27.5 discrete

8 Polychromatic Light Localisation in Periodic Structures   155

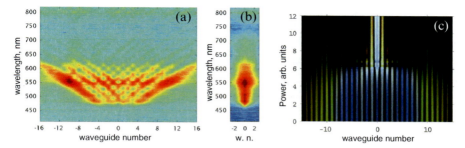

Fig. 8.5. Polychromatic light propagation in a waveguide array of a 10 μm period [21,22]. (**a**) Polychromatic discrete diffraction output profile at low laser power (0.01 mW), (**b**) nonlinear localisation and formation of polychromatic gap soliton at a higher (6 mW) power, (**c**) calculated dependence of the output beam profile with increasing input power

diffraction lengths for the blue and red components, respectively. The propagation of few diffraction lengths for all spectral components is advantageous for nonlinear experiments, and is crucially important for the formation of solitons [47, 48]. This large effective propagation also facilitates strong spectral transformations in the nonlinear regime.

To study the nonlinear beam self-action we increase the power of the input beam while monitoring the output profile. We observe that the broad output beam profile weakly changes up to 150 μW power level, but then quickly transforms into a localised state within a narrow range of input beam powers. Further increase of the beam power leads to gradual trapping of its long-wavelength spectral components. A typical spectrally-resolved output profile is shown in Fig. 8.5b for a supercontinuum average power of 6 mW. A distinguishable characteristic of the observed localisation process is the fact that it traps simultaneously all wavelength components of the supercontinuum spectrum (from blue to red). Numerical simulations of the power dependence of the output beam profile (Figs. 8.5c) indeed reveal that the polychromatic light localisation appears with a sharp power threshold. As such this localisation process results in suppression of the spatial dispersion in the nonlinear regime through the formation of a *polychromatic gap soliton*.

Taking advantage of the high spatial coherence of the supercontinuum light, we also performed interferometric measurement of the phase structure of the localised output profile. Such measurement is important for determining if the localisation is associated with Bragg confinement inside the Bragg reflection gap. The observed interference pattern reveals that all spectral components in the adjacent waveguides are out-of-phase. This fact provides a proof that the localised beam is indeed a polychromatic gap soliton, which characteristic feature is the staggered phase structure [21,22]. We note that this type of localisation is physically different from the gap solitons generated by two-colour coupling in arrays with quadratic nonlinearity [12], where the second harmonic field has a plane phase in contrast to the staggered phase of the fundamental wave component.

8.4.4 Interaction with an Induced Defect

A specific characteristic of the localisation process in LiNbO$_3$ is the slow time response. The response time is inversely proportional to the input laser power, and it can vary from a few seconds to several minutes at low light intensity. The advantage of this slow time response, however, is that once the refractive-index modulation is written it can be preserved in the structure for a long period of time, providing the sample is not exposed to strong light illumination [45]. This opens a novel possibility for dynamic writing of defects with arbitrary geometry [46,49,50]. We demonstrate that defects can play a role of spectral filters for the supercontinuum light. We generate a localised supercontinuum state in an array with periodicity of 19 µm at power of 12 mW and several hours later, we probe this induced defect with a low power supercontinuum beam. To obtain a detailed insight into the spectral distribution at the array output, we resolve the individual spectral components by a prism (P$_2$ in Fig. 8.3a) and acquire a single shot two-dimensional image providing spatial resolution in one (horizontal) direction and spectral resolution in orthogonal (vertical) direction. This technique enables precise determination of the spectral distribution at the array output. We note that the spectral scale in this type of measurements is not linear (Fig. 8.6) due to the prism dispersion.

When the light is injected into the waveguide adjacent to the defect (Fig. 8.6a) we observe reflection of all spectral components below a threshold wavelength value (approximately 800 nm in our case). On the other hand, spectral components of longer wavelengths can tunnel to the left-hand side of the defect. When the input beam is injected into the second or third waveguide away from the induced defect, we observe complex spectral reshaping of

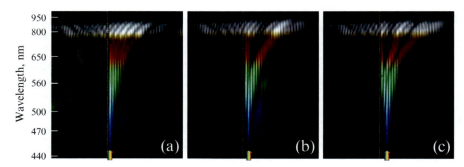

Fig. 8.6. Spectrally-resolved reflection and transmission of a low-power probe beam (0.01 mW) from an optically-induced defect when the array ($d = 19$ µm) is probed (**a**) next to the defect, (**b**) one waveguide away, (**c**) two waveguides away [24]

the output transmission due to reflection of the supercontinuum from the defect state. We note that the nonlinear refractive index change in the detuned waveguide is only of the order of 10^{-4}, but it is sufficient to modify significantly the spectrum of the supercontinuum radiation. This effect is somehow similar to surface manipulation of the supercontinuum light discussed below in Sec. 8.4.5, but with the ability for dynamical reconfiguration and all-optical tuning of the defect properties.

8.4.5 Spatial-spectral Reshaping by Interaction with a Surface

Important flexibility in tailoring beam shaping can be realised when the optical beam interacts with a boundary of the periodic structure [23, 51–54]. Such boundaries or surfaces may support linear localised modes that generalise the so-called surface Tamm states known to exist in solid state physics and other periodic photonic structures. The nonlinearities of the material allow for light localisation even in the cases when linear surface states are absent [55–62]. By making the surface waveguide to be slightly different (see an example in Fig. 8.7a), it is possible to perform spectrally-selective control of linear waves. The surface defect plays the role of an optical waveguide when the refractive index change exceeds a certain threshold, such that the mode eigenvalue can be shifted outside the photonic band [63, 64]. Since the bandgap structure of the photonic lattice depends strongly on frequency, as discussed in Sec. 8.2, the critical change of the refractive index becomes also wavelength-dependent. Only when the optical wavelength is shorter than a certain threshold value, $\lambda < \lambda_{\text{th}}$, where λ_{th} depends on the strength of the surface defect, the optical waves can be localised at the surface [63, 64]. For longer wavelengths the light is reflected from the surface and these spectral components experience

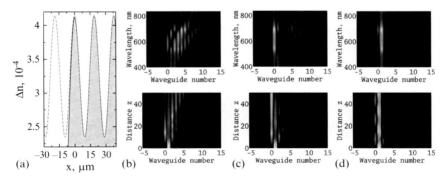

Fig. 8.7. (a) Characteristic refractive index profile in waveguide arrays with a surface defect, dashed line shows, for comparison, the refractive index in an infinite periodic structure, (b) numerically simulated linear and (c), (d) nonlinear dynamics inside the array (*bottom*) and output spectra (*top*) for increasing powers of the input beam coupled to the second waveguide [37]

modified discrete diffraction [55]. For the refractive index profile shown in Fig. 8.7a, the defect can trap blue spectral components, whereas the longer wavelength components are reflected.

In our studies [37] we demonstrate that by coupling light in the second waveguide, next to the surface defect, it is possible to perform power-dependent spatial-spectral reshaping. In the linear regime, the tunneling of short-wavelength components with $\lambda < \lambda_{\rm th}$ to the first waveguide is suppressed almost completely. On the other hand, light at longer wavelengths can penetrate in the first waveguide, see Fig. 8.7b. When the light intensity is increased, the refractive index at the location of the input beam keeps decreasing, approaching gradually the effective refractive index of the surface waveguide. When the mismatch between the two waveguides is reduced, shorter wavelength components start tunneling to the first waveguide. As this happens, nonlinearity acts to increase the mismatch, and light switches permanently to the first waveguide, as shown in Fig. 8.7c. For even higher input powers, the refractive index of the second waveguide decreases to values below the index of the neighbouring waveguides, such that light remains trapped at the input location, see Fig. 8.7d.

The suggested method of beam manipulation was also realised experimentally [37]. As the input power into the second waveguide is increased, we observe enhanced coupling of red, green, and blue components to the surface waveguide (Figs. 8.8a–d) and the formation of polychromatic nonlinear surface modes. At even higher powers the second waveguide is fully detuned from the neighbouring ones and we observe light trapping entirely in the second waveguide (Figs. 8.8e). In the latter case, nonlinearity strongly reduces the influence of the surface on beam propagation. These results demonstrate that collective spatial switching of multiple spectral components can be realised through the nontrivial interplay between the effects of fabricated and self-induced nonlinear defects in photonic lattices.

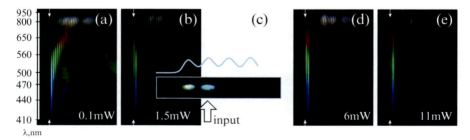

Fig. 8.8. Experimental demonstration of interaction with the surface defect of a supercontinuum light beam coupled to the second waveguide. (**a**), (**b**), (**d**), (**e**) Spectrally resolved spatial distributions at the output for increasing laser power, arrows show the position of the surface waveguide, (**c**) output intensity distribution and a schematic plot of the refractive index profile in the waveguide array [37]

8.5 Summary

We have presented an overview of the basic theoretical studies and experimental observations of spatio-spectral control of polychromatic light in periodic photonic structures. We have described theoretically new types of self-trapped states in the form of polychromatic gap solitons and surface waves, all possessing nontrivial phase structure and spectral features. We have presented the first observation of polychromatic gap solitons in periodic photonic structures with defocusing nonlinearity; such solitons can be generated due to simultaneous spatio-spectral localisation of supercontinuum radiation inside the photonic bandgaps. We anticipate that many of the theoretically predicted and experimentally demonstrated effects can be useful for tunable control of the wavelength dispersion for ultra-broad spectrum pulses offering additional functionality for broadband optical systems and devices.

Acknowledgements

We would like to thank our collaborators and research students for their substantial and valuable contributions to the results summarised in this Chapter. Especially, we are indebted to J. Bolger, A. Dreischuh, B.J. Eggleton, R. Fischer, S. Ha, A. Mitchell, and K. Motzek. The work has been supported by the Australian Research Council.

References

1. J.D. Joannopoulos, R.D. Meade, and J.N. Winn, *Photonic Crystals: Molding the Flow of Light*, Princeton University Press, Princeton (1995)
2. H. Kosaka, T. Kawashima, A. Tomita, M. Notomi, T. Tamamura, T. Sato, and S. Kawakami, J. Lightwave Technol. **17**, 2032 (1999)
3. J. Serbin and M. Gu, Adv. Mater. **18**, 221 (2006)
4. N.C. Panoiu, M. Bahl, and R.M. Osgood, Opt. Lett. **28**, 2503 (2003)
5. D.N. Christodoulides and R.I. Joseph, Opt. Lett. **13**, 794 (1988)
6. H.S. Eisenberg, Y. Silberberg, R. Morandotti, A.R. Boyd, and J.S. Aitchison, Phys. Rev. Lett. **81**, 3383 (1998)
7. J.W. Fleischer, T. Carmon, M. Segev, N.K. Efremidis, and D.N. Christodoulides, Phys. Rev. Lett. **90**, 023902 (2003)
8. D.N. Neshev, E. Ostrovskaya, Y.S. Kivshar, and W. Krolikowski, Opt. Lett. **28**, 710 (2003)
9. D.N. Christodoulides, F. Lederer, and Y. Silberberg, Nature **424**, 817 (2003)
10. A.A. Sukhorukov, Y.S. Kivshar, H.S. Eisenberg, and Y. Silberberg, IEEE J. Quantum Electron. **39**, 31 (2003)
11. A. Fratalocchi, G. Assanto, K.A. Brzdakiewicz, and M.A. Karpierz, Opt. Lett. **29**, 1530 (2004)
12. R. Iwanow, R. Schiek, G.I. Stegeman, T. Pertsch, F. Lederer, Y. Min, and W. Sohler, Phys. Rev. Lett. **93**, 113902 (2004)

13. F. Chen, M. Stepic, C.E. Rüter, D. Runde, D. Kip, V. Shandarov, O. Manela, and M. Segev, Opt. Express **13**, 4314 (2005)
14. J.W. Fleischer, G. Bartal, O. Cohen, T. Schwartz, O. Manela, B. Freedman, M. Segev, H. Buljan, and N.K. Efremidis, Opt. Express **13**, 1780 (2005)
15. M. Matuszewski, C.R. Rosberg, D.N. Neshev, A.A. Sukhorukov, A. Mitchell, M. Trippenbach, M.W. Austin, W. Krolikowski, and Y.S. Kivshar, Opt. Express **14**, 254 (2006)
16. D.N. Neshev, A.A. Sukhorukov, W. Krolikowski, and Y.S. Kivshar, J. Nonlinear Opt. Phys. Mater. **16**, 1 (2007)
17. P.S.J. Russell, Science **299**, 358 (2003)
18. A.L. Jones, J. Opt. Soc. Am. **55**, 261 (1965)
19. S. Somekh, E. Garmire, A. Yariv, H.L. Garvin, and R.G. Hunsperger, Appl. Phys. Lett. **22**, 46 (1973)
20. R. Iwanow, D.A. May-Arrioja, D.N. Christodoulides, G.I. Stegeman, Y. Min, and W. Sohler, Phys. Rev. Lett. **95**, 053902 (2005)
21. A.A. Sukhorukov, D.N. Neshev, A. Dreischuh, R. Fischer, S. Ha, W. Krolikowski, J. Bolger, B.J. Eggleton, A. Mitchell, M.W. Austin, and Y.S. Kivshar, in *Nonlinear Photonics Topical Meeting*, p. JMB5, Optical Society of America, Washington DC (2007)
22. A.A. Sukhorukov, D.N. Neshev, A. Dreischuh, W. Krolikowski, J. Bolger, B.J. Eggleton, L. Bui, A. Mitchell, and Y.S. Kivshar, Opt. Express **16**, 5991 (2008)
23. K. Motzek, A.A. Sukhorukov, and Y.S. Kivshar, Opt. Express **14**, 9873 (2006)
24. D.N. Neshev, A.A. Sukhorukov, A. Dreischuh, R. Fischer, S. Ha, J. Bolger, L. Bui, W. Krolikowski, B.J. Eggleton, A. Mitchell, M.W. Austin, and Y.S. Kivshar, Phys. Rev. Lett. **99**, 123901 (2007)
25. K. Motzek, A.A. Sukhorukov, Y.S. Kivshar, and F. Kaiser, in *Nonlinear Guided Waves and Their Applications*, p. WD25, Optical Society of America, Washington DC (2005)
26. P. Yeh, *Optical Waves in Layered Media*, John Wiley & Sons, New York (1988)
27. P.S.J. Russell, T.A. Birks, and F.D. Lloyd-Lucas, in *Confined Electrons and Photons*, ed. by E. Burstein, C. Weisbuch, pp. 585–633, Plenum, New York (1995)
28. D.N. Christodoulides, S.R. Singh, M.I. Carvalho, and M. Segev, Appl. Phys. Lett. **68**, 1763 (1996)
29. M. Mitchell and M. Segev, Nature **387**, 880 (1997)
30. H. Buljan, T. Schwartz, M. Segev, M. Soljacic, and D.N. Christodoulides, J. Opt. Soc. Am. B **21**, 397 (2004)
31. A. Alberucci, M. Peccianti, G. Assanto, A. Dyadyusha, and M. Kaczmarek, Phys. Rev. Lett. **97**, 153903 (2006)
32. A.B. Fedotov, A.N. Naumov, I. Bugar, D. Chorvat, D.A. Sidorov-Biryukov, D. Chorvat, and A.M. Zheltikov, IEEE J. Sel. Top. Quantum Electron. **8**, 665 (2002)
33. A.B. Fedotov, I. Bugar, A.N. Naumov, D. Chorvat jun., D.A. Sidorov-Biryukov, D. Chorvat, and A.M. Zheltikov, Pis'ma Zh. Éksp. Teor. Fiz. **75**, 374 (2002), in Russian; English translation, JETP Lett. **75**, 304 (2002)
34. A. Betlej, S. Suntsov, K.G. Makris, L. Jankovic, D.N. Christodoulides, G.I. Stegeman, J. Fini, R.T. Bise, and J. DiGiovanni, Opt. Lett. **31**0, 1480 (2006)

35. G.C. Valley, M. Segev, B. Crosignani, A. Yariv, M.M. Fejer, and M.C. Bashaw, Phys. Rev. A **50**, R4457 (1994)
36. Z.Y. Xu, Y.V. Kartashov, and L. Torner, Opt. Lett. **31**, 2027 (2006)
37. A.A. Sukhorukov, D.N. Neshev, A. Dreischuh, R. Fischer, S. Ha, W. Krolikowski, J. Bolger, A. Mitchell, B.J. Eggleton, and Y.S. Kivshar, Opt. Express **14**, 11265 (2006)
38. R.R. Shah, D.M. Kim, T.A. Rabson, and F.K. Tittel, J. Appl. Phys. **47**, 5421 (1976)
39. F. Jermann, M. Simon, and E. Krätzig, J. Opt. Soc. Am. B **12**, 2066 (1995)
40. R. Pezer, H. Buljan, G. Bartal, M. Segev, and J.W. Fleischer, Phys. Rev. E **73**, 056608 (2006)
41. J.K. Ranka, R.S. Windeler, and A.J. Stentz, Opt. Lett. **25**, 25 (2000)
42. J.M. Dudley, G. Genty, and S. Coen, Rev. Mod. Phys. **78**, 1135 (2006)
43. M.H. Qi, E. Lidorikis, P.T. Rakich, S.G. Johnson, J.D. Joannopoulos, E.P. Ippen, and H.I. Smith, Nature **429**, 538 (2004)
44. M. Balu, J. Hales, D.J. Hagan, and E.W. van Stryland, Opt. Express **13**, 3594 (2005)
45. A. Ashkin, G.D. Boyd, J.M. Dziedzic, R.G. Smith, A.A. Ballman, J.J. Levinstein, and K. Nassau, Appl. Phys. Lett. **9**, 72 (1966)
46. D. Kip, Appl. Phys. B **67**, 131 (1998)
47. G.I. Stegeman and M. Segev, Science **286**, 1518 (1999)
48. Y.S. Kivshar and G.P. Agrawal, *Optical Solitons: From Fibers to Photonic Crystals*, Academic Press, San Diego (2003)
49. A. Guo, M. Henry, G.J. Salamo, M. Segev, and G.L. Wood, Opt. Lett. **26**, 1274 (2001)
50. P. Zhang, D.X. Yang, J.L. Zhao, and M.R. Wang, Opt. Eng. **45**, 074603 (2006)
51. K.G. Makris, J. Hudock, D.N. Christodoulides, G.I. Stegeman, O. Manela, and M. Segev, Opt. Lett. **31**, 2774 (2006)
52. K. Motzek, A.A. Sukhorukov, and Y.S. Kivshar, Opt. Lett. **31**, 3125 (2006)
53. M.I. Molina and Y.S. Kivshar, Phys. Lett. A **362**, 280 (2007)
54. S. Suntsov, K.G. Makris, D.N. Christodoulides, G.I. Stegeman, R. Morandotti, M. Volatier, V. Aimez, R. Ares, C.E. Rüter, and D. Kip, Opt. Express **15**, 4663 (2007)
55. K.G. Makris, S. Suntsov, D.N. Christodoulides, G.I. Stegeman, and A. Hache, Opt. Lett. **30**, 2466 (2005)
56. S. Suntsov, K.G. Makris, D.N. Christodoulides, G.I. Stegeman, A. Hache, R. Morandotti, H. Yang, G. Salamo, and M. Sorel, Phys. Rev. Lett. **96**, 063901 (2006)
57. Y.V. Kartashov, V.A. Vysloukh, and L. Torner, Phys. Rev. Lett. **96**, 073901 (2006)
58. G.A. Siviloglou, K.G. Makris, R. Iwanow, R. Schiek, D.N. Christodoulides, G.I. Stegeman, Y. Min, and W. Sohler, Opt. Express **14**, 5508 (2006)
59. E. Smirnov, M. Stepic, C.E. Rüter, D. Kip, and V. Shandarov, Opt. Lett. **31**, 2338 (2006)
60. C.R. Rosberg, D.N. Neshev, W. Krolikowski, A. Mitchell, R.A. Vicencio, M.I. Molina, and Y.S. Kivshar, Phys. Rev. Lett. **97**, 083901 (2006)
61. I.L. Garanovich, A.A. Sukhorukov, Y.S. Kivshar, and M. Molina, Opt. Express **14**, 4780 (2006)
62. Y.V. Kartashov, F.W. Ye, and L. Torner, Opt. Express **14**, 4808 (2006)
63. P. Yeh, A. Yariv, and C.S. Hong, J. Opt. Soc. Am. **67**, 423 (1977)
64. P. Yeh, A. Yariv, and A.Y. Cho, Appl. Phys. Lett. **32**, 104 (1978)

Part III

Periodic Structures for Matter Waves: From Lattices to Ratchets

9

Bose-Einstein Condensates in 1D Optical Lattices: Nonlinearity and Wannier-Stark Spectra

Ennio Arimondo, Donatella Ciampini, and Oliver Morsch

CNR-INFM and CNISM, Dipartimento di Fisica E. Fermi, Università di Pisa, Via Buonarroti 2, 56127 Pisa, Italy
arimondo@df.unipi.it, ciampini@df.unipi.it, morsch@df.unipi.it

9.1 Introduction

The development of powerful laser cooling and trapping techniques has made possible the controlled realization of dense and cold gaseous samples, thus opening the way for investigations in the ultracold temperature regimes not accessible with conventional techniques. A Bose-Einstein condensate (BEC) represents a peculiar gaseous state where all the particles reside in the same quantum mechanical state. Therefore BECs exhibit quantum mechanical phenomena on a macroscopic scale with a single quantum mechanical wavefunction describing the external degrees of freedom. That control of the external degrees of freedom is combined with a precise control of the internal degrees. The BEC investigation has become a very active area of research in contemporary physics. The BEC study encompasses different subfields of physics, i.e., atomic and molecular physics, quantum optics, laser spectroscopy, solid state physics. Atomic physics and laser spectroscopy provide the methods for creating and manipulating the atomic and molecular BECs. However owing to the interactions between the particles composing the condensate and to the configuration of the external potential, concepts and methods from solid state physics are extensively used for BEC description.

Quantum mechanical BECs within the periodic potential created by interfering laser waves ("optical lattices") have attracted a strongly increasing interest [1–4]. In particular, the formal similarity between the wavefunction of a BEC inside the periodic potential of an optical lattice and electrons in a crystal lattice has triggered theoretical and experimental efforts alike. BECs inside optical lattices share many features with electrons in solids, but also with light waves in nonlinear materials and other nonlinear systems. However, the experimental control over the parameters of BEC and of the periodic potential make it possible to enter regimes inaccessible in other systems. Many phenomena from condensed matter physics, such as Bloch oscillations and Landau-Zener (LZ) tunneling have since been shown to be observable also in optical lattices. BEC in an optical lattice even made possible the observation of a quantum phase transition that had, up to then, only been theoretically predicted for condensed matter systems [5]. However an important difference between electrons in a crystal lattice and a BEC inside the periodic potential of an optical lattice is the strength of the self interaction between the BEC components and hence the magnitude of the nonlinearity of the system. Electrons are almost noninteracting whereas atoms inside a BEC interact strongly. A perturbation approach is appropriate in the former case while in the latter the full nonlinearity must be taken into account. Generally, atom-atom interactions in Bose-Einstein condensates lead to rich and interesting nonlinear effects. Most experiments to date have been carried out in the regime of shallow lattice depth, for which the system is well described by the Gross-Pitaevskii equation, a mean-field equation. Moreover, the nonlinearity induced by the mean-field of the condensate has been shown both theoretically and experimentally to give rise to instabilities in certain regions of the Brillouin zone. These instabilities are not present in the corresponding linear system, i.e. the electron system.

The present text initially describes the construction of the optical lattice periodic potential for cold atoms in Sec. 9.2. The following Section reports the analysis of the condensate interference pattern when released from the optical lattice. In Sec. 9.4 the nonlinear term within the Gross-Pitaevskii equation describing the dynamics of a Bose-Einstein condensation is introduced. The following Sections report experimental results on the Bloch oscillations, on the nonlinear Landau-Zener quantum tunneling and on the resonantly enhanced quantum tunneling. A short conclusion terminates the presentation.

9.2 Optical Lattice

In order to trap a Bose-Einstein condensate in a periodic potential, it is sufficient to exploit the interference pattern created by two or more overlapping laser beams and the light force exerted on the condensate atoms. Optical lattices work on the principle of the ac Stark shift. When an atom is placed in a light field, the oscillating electric field of the latter induces an electric dipole moment in the atom. The interaction between this induced dipole and the electric field leads to an energy shift ΔE of an atomic energy level. When we take two identical laser beams and make them counterpropagate in such a way that their cross sections overlap completely see Fig. 9.1a, we expect the two beams to create an interference pattern, with a distance $d_\mathrm{L} = \lambda/2$ between two neighbouring maxima or minima of the resulting light intensity. The potential seen by the atoms is then

$$V(x) = V_0 \cos^2\left(\frac{\pi x}{d}\right) = \frac{V_0}{2}\left[1 + \cos\left(\frac{2\pi x}{d_\mathrm{L}}\right)\right], \tag{9.1}$$

9 Nonlinearity and Wannier-Stark spectra for BEC in Optical Lattices 167

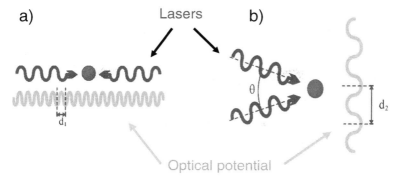

Fig. 9.1. Representation of the laser configuration creating an optical lattice (**a**) in the counterpropagating geometry and (**b**) in the angle tuned geometry

where the lattice depth V_0 is determined by the light shifts ΔE produced by the individual laser beams. The easiest option to create a one-dimensional optical lattice is to take a linearly polarized laser beam and retro-reflect it with a high-quality mirror. If the retro-reflected beam is replaced by a second phase-coherent laser beam as obtained, for instance, by dividing a laser beam in two, another degree of freedom is introduced. It is now possible to have a frequency shift $\Delta \nu_\mathrm{L}$ between the two lattice beams. The periodic lattice potential will now no longer be stationary in space but move at a velocity $v_\mathrm{lat} = d_\mathrm{L} \Delta \nu_\mathrm{L}$. If the frequency difference is varied at a rate $\partial \Delta \nu_\mathrm{L}/\partial t$, the lattice potential will be accelerated with $a_\mathrm{lat} = d_\mathrm{L} \partial \Delta \nu_\mathrm{L}/\partial t$. Clearly, in the rest frame of the lattice there will be a force

$$F = -M a_\mathrm{lat} = -M d_\mathrm{L} \frac{\partial \Delta \nu_\mathrm{L}}{\partial t} \qquad (9.2)$$

acting on the condensate atoms. This gives us a powerful tool for manipulating a BEC inside an optical lattice.

Another degree of freedom of a 1D lattice realized with two laser beams is the lattice constant. The spacing d_L between two adjacent wells of a lattice resulting from two counterpropagating beams can be enhanced by making the beams intersect at an angle $\theta \neq \pi$, see Fig. 9.1b. This will give rise to a periodic potential with lattice constant

$$d_\mathrm{L} = \frac{\lambda}{2 \sin(\theta/2)} \; . \qquad (9.3)$$

To simplify the notation, in the following we shall always denote the lattice constant by d_L and all the quantities derived from it, regardless of the lattice geometry that was used to achieve it. In particular the recoil energy E_R and the recoil frequency ν_R of an atom with mass M are

$$E_\mathrm{R} = h\nu_\mathrm{R} = \frac{\hbar^2 \pi^2}{2 M d_\mathrm{L}^2} \; , \qquad (9.4)$$

and the recoil velocity $v_R = \hbar\pi/(d_L M)$. Naturally, by adding more laser beams one can easily create two- or three-dimensional lattices [4].

The description of the propagation of noninteracting matter waves in periodic potentials is straightforward once one has found the eigenstates and corresponding eigenenergies of the system. The eigenstates are found in by applying Bloch's theorem, which states that the eigenfunctions have the form [6]

$$\phi_{n,q}(x) = e^{iqx} u_{n,q}(x) \, , \tag{9.5}$$

where $\hbar q$ is referred to as quasimomentum and n indicates the band index, the meaning of which will become clear in the following discussion. The quasimomentum q appearing in the Bloch's theorem can always be confined to the first Brillouin zone $(-q_R, q_R)$ with $q_R = \pi/d_L$, because any q' not in the first Brillouin zone can be written as $q' = q + G$, where G is a reciprocal lattice vector and q does lie in the first zone. The eigenenergies $E^n(q)$ of the above eigenstates depend on the potential depth $V_0 = sE_R$ and, additionally, on the quasimomentum q. In Fig. 9.2, we summarize the properties of the eigenbasis for a shallow potential $V_0 = 2E_R$. The eigenenergies form bands that are separated by a gap in the energy spectrum, i.e., certain energies are not allowed. Since the gap energy E_{gap}^n between the nth and $(n+1)$th band scales with V_0^{n+1} in the weak potential limit, it only has appreciable magnitude between the lowest and first excited band. A particle with high energy is very well described as a free particle and the influence of the periodic potential is negligible in this case. It is important to note that for energies near the Brillouin zone edge of the lowest band, the eigenstate probability distribution is a

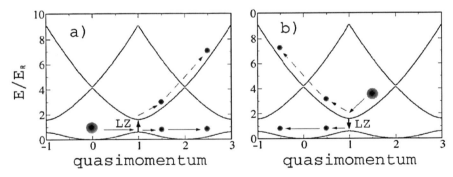

Fig. 9.2. Band structure of a BEC in an optical lattice ($V_0 = 2E_R$) and LZ tunneling, (**a**) ground to excited state and (**b**) excited to ground state. When the BEC is accelerated across the edge of the Brillouin zone (BZ) at quasimomentum $q/q_R = 1$, LZ tunneling occurs. After the first crossing of the edge of the Brillouin zone increasing the lattice depth and decreasing the acceleration leads to a much reduced tunneling rate from the ground state band at successive BZ edge crossings, as shown in (**a**) for the ground to excited state tunneling and in (**b**) for excited to ground state

periodic $\sqrt{2}\sin(2\pi x/2\,d_{\rm L})\exp(\mathrm{i}\pi x/2\,d)$ function, its maxima coinciding with the potential minima and the phase changing by π between adjacent wells. This is the well-known sinusoidal Bloch state at the Brillouin zone edge, in the literature also referred to as a staggered mode. In the limit of deep periodic potentials, also referred to as the tight-binding limit, the eigenenergies of the low lying bands are only weakly dependent on the quasimomentum.

The wave-packet dynamics of a particle in a periodic potential in the presence of an additional external potential, i.e., with an external force, is generally not easy to solve. The problem becomes relatively simple, though, as soon as the width of the wave packet in quasimomentum space is small and thus the wave packet can be characterized by a single mean quasimomentum $q(t)$ at time t. An external force then leads to a time-dependent $q(t)$ via

$$\hbar \dot{q}(t) = F \,. \tag{9.6}$$

In the case of a constant force F, e.g., due to the gravitational field, this results in

$$q(t) = q(t=0) + \frac{Ft}{\hbar} \,. \tag{9.7}$$

In addition, the velocity $v_n(q)$ of the particle in the n band is given by the group velocity of the underlying wavepacket

$$v_n(q) = \frac{1}{\hbar}\frac{\partial E_n(q)}{\partial q} \,. \tag{9.8}$$

The above equations determine that the rate of change of the quasimomentum is given by the external force, but the rate of change of the wavepacket's momentum is given by the total force including the influence of the periodic field of the lattice. In the case of a constant force, the velocity at time t is

$$v_n\left(q(t)\right) = v_n\left(q(t=0) + \frac{Ft}{\hbar}\right) \,. \tag{9.9}$$

Since v_n is periodic in the reciprocal lattice, the velocity is a bounded and oscillatory function of time. Therefore the result of the force is not an acceleration of the wave packet, and instead the wavepacket will show an oscillatory behavior in real space. The velocity oscillatory motion is known as Bloch oscillations [7] and the period as the Bloch time

$$T_{\rm B} = \frac{2\pi\hbar}{F d_{\rm L}} = \frac{1}{F_0 \nu_{\rm R}} \,, \tag{9.10}$$

where we have introduced a dimensionless force $F_0 = F d_{\rm L}/E_{\rm R}$.

In the case of a strong external force acting on matter waves in periodic potentials, transitions into higher bands can occur as schematically represented in Fig. 9.2. In the context of electrons in solids, this is known as the Landau-Zener breakdown [8,9], occurring if the applied electric field is strong enough for the acceleration of the electrons to overcome the gap energy separating the valence and conduction bands. It was shown in [9] that for a given acceleration a_L corresponding to a constant force, one can deduce a tunneling probability across the first-second band gap in the adiabatic limit

$$P_\mathrm{LZ} = \mathrm{e}^{-\pi^2 V_0^2/32 F_0} . \qquad (9.11)$$

The resulting wavepacket dynamics is shown in Fig. 9.2, where LZ tunneling, from $n = 1$ to $n = 2$ band, leads to a splitting of the wave function.

9.3 Analysis of the Interference Pattern

Doing experiments with condensates in optical lattices is useful only if one is able to extract information from the system once the experiment has been carried out. There are essentially two methods for retrieving information from the condensate: in situ and after a time of flight. In the former case, one can obtain information about the spatial density distribution of the condensate, its shape, and any irregularities on it that may have developed during the interaction with the lattice. Also, the position of the center of mass of the condensate can be determined. Looking at a condensate released from a lattice after a time of flight, typically on the order of a few milliseconds, amounts to observing its momentum distribution. When the atomic system is in a steady state, the condensate is distributed among the lattice wells (in the limit of a sufficiently deep lattice in order for individual lattice sites to have well-localized wavepackets). If the lattice is now switched off suddenly, the individual (approximately) Gaussian wavepackets at each lattice site will expand freely and interfere with one another. The resulting spatial interference pattern after a time of flight of t will be a series of regularly spaced peaks with spacing $2 v_\mathrm{R} t$, corresponding to the various diffraction orders. In the case of a condensate that is very elongated along the lattice direction, to a good approximation we initially have an array of equally spaced Gaussians of a width determined by the lattice depth. Figure 9.3 shows a typical time of flight interference pattern of a condensate released from an optical lattice (plus harmonic trap) for a lattice depth $V_0 = 10\, E_\mathrm{R}$. From the spacing of the interference peaks and the time of flight, one can immediately infer the recoil momentum of the lattice and hence the lattice constant d_L. Furthermore, from the relative height of the side peaks corresponding to the momentum classes $\pm 2\, \pi/d_\mathrm{L}$, one can calculate the lattice depth. The top interference pattern was produced by a condensate at rest with zero quasimomentum. Instead the bottom interference pattern was produced by a condensate with quasimomentum at the edge of the Brillouin zone, in the staggered state with the condensate wavefunction changing by π between adjacent wells.

9 Nonlinearity and Wannier-Stark spectra for BEC in Optical Lattices

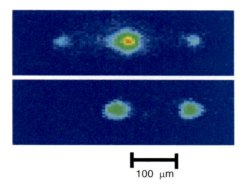

Fig. 9.3. Interference pattern of a Bose-Einstein condensate released from a one-dimensional optical lattice of depth $V_0 = 10E_R$ after a time of flight of 20 ms. In the top pattern the lattice was at rest, whereas in the bottom one the condensate had been accelerated to v_R, i.e., its quasimomentum was at the edge of the Brillouin zone

9.4 Nonlinear Optical Lattice

The motion of a Bose-Einstein condensate in a 1D optical lattice experiencing an acceleration a_L is described by the Gross-Pitaevskii equation

$$i\hbar\frac{\partial\psi}{\partial t} = \frac{1}{2M}\left(-i\hbar\frac{\partial}{\partial x} + Ma_L t\right)^2 \psi + \frac{V_0}{2}\cos\left(\frac{\pi x}{d_L}\right)\psi + \frac{4\pi\hbar^2 a_s}{M}|\psi|^2\psi \, . \tag{9.12}$$

The s-wave scattering length a_s determines the nonlinearity of the system. Equation (9.12) is written in the comoving frame of the lattice, so the inertial force $-Ma_L$ appears as a momentum modification. The wavefunction ψ is normalized to the total number of atoms in the condensate and we define n_0 as the average uniform atomic density. Defining the dimensionless quantities $\tilde{x} = 2\pi x/d_L$, $\tilde{t} = 8E_R t/\hbar$, and rewriting $\tilde{\psi} = \psi/\sqrt{n_0}$, $\tilde{v} = V_0/16E_R$, $\tilde{\alpha} = Ma_L d_L/16E_R\pi$, $C = a_s n_0 d_L^2/\pi$, (9.12) is cast in the following form

$$i\frac{\partial\psi}{\partial t} = \frac{1}{2}\left(-i\frac{\partial}{\partial x} + \alpha t\right)^2 \psi + v\cos(x)\psi + C|\psi|^2\psi \, , \tag{9.13}$$

where we have replaced $\tilde{x}$ with x, etc. In the neighborhood of the Brillouin zone edge we can approximate the wave function by a superposition of two plane waves, assuming that only the ground state and the first excited state are populated. We then substitute in (9.13) a normalized wavefunction

$$\psi(x,t) = a(t)e^{iqx} + b(t)e^{i(q-1)x} \, . \tag{9.14}$$

Projecting on this basis, linearizing the kinetic terms and dropping the irrelevant constant energy, (9.13) assumes the form

$$i\frac{\partial}{\partial t}\begin{pmatrix}a\\b\end{pmatrix} = \left[-\frac{\alpha t}{2}\sigma_3 + \frac{v}{2}\sigma_1\right]\begin{pmatrix}a\\b\end{pmatrix} + \frac{C}{2}\left(|b|^2 - |a|^2\right)\sigma_3\begin{pmatrix}a\\b\end{pmatrix}, \quad (9.15)$$

where σ_i, $i = 1, 2, 3$ are the Pauli matrices. The adiabatic energies of (9.15) have a butterfly structure at the band edge of the Brillouin zone for $C \geq v$ [10], but in the present work we only consider a regime where $C \ll v$, hence that structure plays no role.

In the linear regime ($C = 0$), evaluating the transition probability in the adiabatic approximation, we find the linear LZ formula for the tunneling probability P_{LZ} given by (9.11). Therefore for $C = 0$ the tunneling probability is the same for both tunneling directions whereas for $C \neq 0$ the two rates are different. In the nonlinear regime, as the nonlinear parameter C grows, the lower to upper tunneling probability grows as well until an adiabaticity breakdown occurs at $C = v$ [10]. The upper to lower tunneling probability, on the other hand, decreases with increasing nonlinearity.

The asymmetry in the tunneling transition probabilities can be explained qualitatively as follows: The nonlinear term of the Schrödinger equation acts as a perturbation whose strength is proportional to the energy level occupation. If the initial state of the condensate in the lattice corresponds to a filled lower level of the state model, then the lower level is shifted upward in energy while the upper level is left unaffected. This reduces the energy gap between the lower and upper level and enhances the tunneling. On the contrary, if all atoms fill the upper level then the energy of the upper level is increased while the lower level remains unaffected. This enhances the energy gap and reduces the tunneling. The overall balance leads to an asymmetry between the two tunneling processes.

The nonlinear regime may be reinterpreted by writing (9.15) as

$$i\frac{\partial}{\partial t}\begin{pmatrix}a\\b\end{pmatrix} = \left[-\frac{\alpha t}{2}\sigma_3 + \frac{v}{2}\sigma_1\right]\begin{pmatrix}a\\b\end{pmatrix} - \frac{C}{2}\begin{pmatrix}|a|^2 & -b^*a\\-a^*b & |b|^2\end{pmatrix}\begin{pmatrix}a\\b\end{pmatrix}. \quad (9.16)$$

The nonlinear off-diagonal terms modify the interaction term v in a way equivalent to a Rabi frequency in the two-level model. In (9.16) we identify an offdiagonal term $v + Ca^*b$ which acts as an effective potential. Thus for small C values we can modify the linear LZ formula (9.11) to include nonlinear corrections, substituting the potential $v = V_0/16\,E_R$ with the effective potential $v_{\text{eff}} = V_{\text{eff}}/16\,E_R \equiv |v + Ca^*b|$ (modulus is needed since a^*b is complex). The expression for v_{eff} is

$$v_{\text{eff}} = v\sqrt{1 \pm \frac{C}{v} + \frac{C^2}{4v^2}}, \quad (9.17)$$

where the upper and lower signs corresponds to initial conditions of excited/ground states.

9.5 Bloch Oscillations

A fundamental property of a quasiparticle in a periodic potential subject to an external static force is its localization by Bragg reflections at the boundary of the Brillouin zone, which leads to temporal and spatial oscillations known as Bloch oscillations [7]. Related fundamental transport phenomena are the nonresonant LZ tunneling into a continuum of states of another Bloch band and the resonant LZ tunneling between anticrossing Wannier-Stark states of neighboring Bloch bands. Bloch oscillations were first observed as time-resolved oscillations of wave packets of photo-excited "hot" electrons in biased semiconductor superlattices. Later Bloch oscillations and LZ tunneling were observed in ensembles of cold atoms [11–13]. During the last decade, there were experimentally realized time-resolved Bloch oscillations of coherent electron wave packets in semiconductor superlattices [14,15] subjected to combined electric and magnetic fields. The progress in the fabrication and investigation of complex optical nanostructures has allowed for direct experimental observations of one-dimensional optical Bloch oscillations of an optical laser field in dielectric structures with a transversely superimposed linear ramp of the refractive index [16,17]. A periodic distribution of the refractive index plays a role of the crystalline potential, and the index gradient acts similarly to an external force in a quantum system. This force causes the laser beam to move across the structure while experiencing Bragg reflections on the high-index and total internal reflection on the low-index side of the structure, resulting in an optical analogue of Bloch oscillations. Bloch oscillations and LZ tunneling from the first to second energy band was also demonstrated experimentally in two-dimensional photonic structures [18]. Most recently, acoustic Bloch oscillations and resonant LZ tunneling of phononic wave packets were observed in perturbed ultrasonic superlattices [19].

We report here results for Bloch oscillations in experiments with Bose-Einstein condensates adiabatically loaded into one-dimensional optical lattices. In particular, we discuss the dynamics of the BEC when the periodic potential provided by the optical lattice is accelerated, leading to Bloch oscillations [20, 21]. The condensate was loaded adiabatically into the (horizontal) optical lattice with lattice constant $d_L = 390$ nm immediately after switching off the magnetic trap. Thereafter, the lattice was accelerated with $a = 9.81$ m/s^2 by ramping the frequency difference $\Delta\nu_L$ between the laser beams forming the optical lattice. After a time the lattice was switched off and the condensate was observed after an additional time of flight. Fig. 9.4a and 9.4c shows the results of these measurements in the laboratory frame. The Bloch oscillations are more evident, however, if one calculates the mean velocity v_m of the condensate as the weighted sum over the momentum components after the interaction with the accelerated lattice, as shown in Fig. 9.4b. When the instantaneous lattice velocity v_{lat} is subtracted from v_m, one clearly sees the oscillatory behaviour of $v_m - v_{lat}$. The added feature of using a Bose-Einstein condensate is that the spatial extent of the atomic cloud is sufficiently

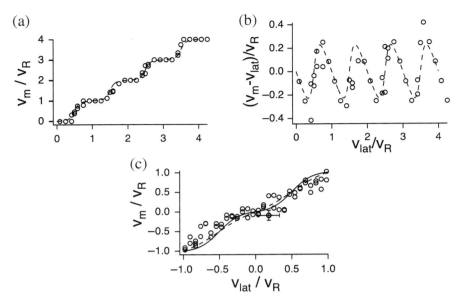

Fig. 9.4. Bloch oscillations of a Bose-Einstein condensate in an optical lattice. (a) Acceleration in the counterpropagating lattice with $d_{\rm L} = 390\,{\rm nm}$, $V_0 \approx 2.3\,E_{\rm R}$ and $a = 9.81\,{\rm m/s}^2$. Dashed line: theory. (b) Bloch oscillations in the rest frame of the lattice, along with the theoretical prediction (dashed line) derived from the shape of the lowest Bloch band. (c) Acceleration in a lattice with $d_{\rm L} = 1.56\,\mu{\rm m}$ and $V_0 \approx 11\,E_{\rm R}$. In this case, the Bloch oscillations are much less pronounced. Dashed and solid lines: theory for $V_0 = 11\,E_{\rm R}$ and $V_{\rm eff} \approx 7\,E_{\rm R}$

small so that after a relatively short time-of-flight the separation between the individual momentum classes is already much larger than the size of the condensate due to its expansion and can, therefore, be easily resolved. Similar observations were reported in [22, 23]. In Ref. [24] by using an optical Bessel beam to form the optical lattice, a very large number of Bloch oscillations of a rubidium condensate was realized and large final velocities were reached.

9.6 Landau-Zener Tunneling

The linear regime of the LZ tunneling in atomic physics was investigated by several authors, using Rydberg atoms in [25], in classical optical systems in [26], for cold atoms in an accelerated optical potential in [27]. We investigated linear and nonlinear LZ tunneling between the two lowest energy bands of a condensate inside an optical lattice in the following way. Initially, the condensate was loaded adiabatically into one of the two bands. Subsequently, the lattice was accelerated in such a way that the condensate crossed the edge of the Brillouin zone once, resulting in a finite probability for tunneling into the other band (higher-lying bands can be safely neglected as their energy separation is much larger than the band gap). Thus, the two bands

9 Nonlinearity and Wannier-Stark spectra for BEC in Optical Lattices   175

had populations reflecting the LZ tunneling probability (assuming only one band exclusively populated initially). In order to experimentally determine the number of atoms in the two bands, we then *increased* the lattice depth and *decreased* the acceleration. Using this experimental sequence we selectively accelerated further that part of the condensate that populated the ground state band, whereas the population of the first excited band was not accelerated further, as shown schematically in Fig. 9.2.

In order to investigate tunneling from the ground state band to the first excited band, we adiabatically ramped up the lattice depth with the lattice at rest and then started the acceleration sequence. The tunneling from the first excited to the ground-state band was investigated in a similar way, except that in this case we initially prepared the condensate in the first excited band by moving the lattice with a velocity of $1.5\,v_R$ (through the frequency difference $\Delta\nu_L$ between the acousto-optic modulators) when switching it on. In this way, in order to conserve energy and momentum the condensate must populate the first excited band at a quasimomentum half-way between zero and the edge of the first Brillouin zone. For both tunneling directions, the tunneling probability is derived from $P_{LZ} = N_{\text{tunnel}}/N_{\text{tot}}$, where N_{tot} is the total number of atoms measured from the absorption picture. For the tunneling from the first excited band to the ground band, N_{tunnel} is the number of atoms accelerated by the lattice, whereas for the inverse tunneling direction, N_{tunnel} is the number of atoms detected in the $v = 0$ velocity class.

For a small value of the interaction parameter C, we verified that the tunneling rates in the two directions are essentially the same and agree well with the linear LZ prediction. By contrast, when C is increased, the two tunneling rates differ greatly, as in Fig. 9.5. We have not yet performed a quantitative

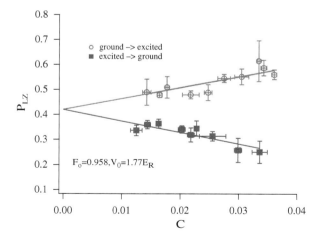

Fig. 9.5. Asymmetric tunneling between the ground state and first excited band of a BEC in an optical lattice as a function of the nonlinear interaction parameter C. In these experiments, $a = 32.1\,\text{m/s}^2$ and the lattice depth was $1.77\,E_R$

comparison of these data with the theoretical predictions of the non-linear LZ model. Previous data published in [28] presented good agreement with the theoretical predictions of the asymmetric nonlinear LZ tunneling.

9.7 Resonantly Enhanced Quantum Tunnelling

Resonantly enhanced tunneling (RET) is a quantum effect in which the probability for tunneling of a particle between two potential wells is increased when the quantized energies of the initial and final states of the process coincide. In spite of the fundamental nature of this effect and the practical interest, it has been difficult to observe experimentally in solid state structures. Quantum tunnelling has found many technological applications, for instance, in scanning tunnelling microscopes and in superconducting squid devices. The most widely application is in tunnelling diodes and related integrated semiconductor devices which go back to the pioneering work of Leo Esaki [29]. The latter also proposed to exploit resonantly enhanced tunnelling (RET) for technical use, and since the 1970s much progress has been made in producing artificial superlattice structures, in which RET of fermionic quasiparticles could be demonstrated.

Here we present our realization of RET using Bose-Einstein condensates held in optically induced potentials. The counter-propagating beams creating the lattice were continuously accelerated such as to mimic a static linear potential in the moving frame of reference. BEC tunneling occurred between the quantised energy levels (the Wannier-Stark levels) in various wells of the potential, see Fig. 9.6. We demonstrated that the tunneling probability is resonantly enhanced and the LZ formula does not give the correct result.

When under the applied external force the quasimomentum explores the Brillouin zone, adiabatic transitions occur at the points of avoided crossings between the adjacent Bloch bands, for example, between the first and second bands in Fig. 9.2. The probability of this transition is given by the LZ

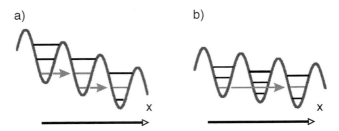

Fig. 9.6. The tunneling of atoms out of a tilted lattice is resonantly enhanced when the tilt induced energy difference $Fd_\mathrm{L}\Delta i$ between lattice wells i and $i+\Delta i$ matches the separation between two quantized energy levels within a well, $\Delta i = 1$ (*left*) and $\Delta i = 2$ (*right*)

9 Nonlinearity and Wannier-Stark spectra for BEC in Optical Lattices   177

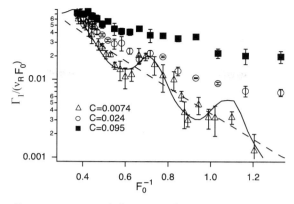

Fig. 9.7. Tunneling resonances of the $n = 1$ lowest energy level for $V_0 = 3.5\,E_R$. The continuous line represents the theoretical RET prediction in the linear regime, and the dashed line the LZ prediction. For the experimental data (at $d_L = 426.1\,\text{nm}$ and $C = 0.007$) and theoretical prediction in the RET linear regime, the $\Delta i = 2$ and $\Delta i = 3$ resonant peaks appear at increasing values of F_0^{-1}, while the $\Delta i = 1$ peak is not completely scanned. In the nonlinear regimes ($d_L = 626.4\,\text{nm}$, $C = 0.024$ and $d_L = 626.4\,\text{nm}$, $C = 0.095$) the resonant peaks are washed out

tunneling formula of (9.11). In a first approximation, one can assume that the adiabatic transition occurs once for each Bloch cycle with the period T_B of (9.10). Then the population of the initial band decreases exponentially with a rate which, for the tunneling out of the ground $n = 1$ band, is given by [30]

$$\Gamma_1 = \nu_R F_0 e^{-\pi^2 V_0^2 / 32\, F_0} \,. \tag{9.18}$$

A plot of this tunneling rate as a function of F_0 in the linear regime is shown in Fig. 9.7. This regime is reached either by choosing small radial dipole trap frequencies or by releasing the BEC from the trap before the acceleration phase and thus letting it expand. In both cases, the density and hence the interaction energy of the BEC is reduced. Superimposed on the overall exponential decay of Γ_1/F_0 with F_0, one clearly sees the resonant tunneling peaks corresponding to $\Delta i = 3, 2$. For this choice of parameters, the $\Delta i = 1$ peak lay outside the region of F values explored in the experiment. In order to highlight the deviation from the LZ prediction, the dashed line represents the prediction of (9.11). The experimental results are in good agreement with numerical solutions obtained by diagonalizing the Hamiltonian of the open decaying system represented by the continuous line.

Resonances in quantum tunneling for atomic motion within an optical lattice were previously observed by few authors. Evidence of RET is apparent in the tunneling measurements on cold atoms of [27]. In a gray optical lattice they appear as a magnetization modulation [31]. In the demonstration of the Mott insulator phase of [5, 32] with each lattice site of Fig. 9.6 occupied by a single atom, RET occurred when the energy difference between neighbouring lattice sites was equal to the on-site atomic interaction energy.

9.8 Summary

In recent years quantum-wave transport phenomena linked to Bloch oscillations and LZ tunneling in a variety of optical lattice configurations have been widely investigated. In addition, Bloch oscillations of ultracold atoms were proposed as a tool for precision measurements of tiny forces with a spatial resolution at the micron level [33]. In [34] Bloch oscillations of ultracold atoms were performed within a few microns from a test mass in order to measure gravity with very large accuracy in order to test deviations from Newton's law. Measurements of the recoil velocity of rubidium atoms based on Bloch oscillations lead to an accurate determination of the fine structure constant [35]. Bloch oscillations of ultracold fermionic atoms have also been proposed as a sensitive measurement of forces at the micrometer length scale, in order to perform a local and direct measurement of the Casimir-Polder force [36]. The use of quantum resonant tunneling that presents a resonant dependence on the external force may improve the accuracy of those measurements.

Acknowledgements

The research work presented here relied on the collaborative effort of M. Anderlini, M. Cristiani, M. Jona-Lasinio, H. Lignier, J.H. Müller, R. Mannella, C. Sias, Y. Singh, S. Wimberger and A. Zenesini. This work was supported by the European Community OLAQUI and EMALI Projects, and by MIUR-PRIN Projects.

References

1. O. Morsch and E. Arimondo, in *Dynamics and Thermodynamics of Systems with Long-Range Interactions*, ed. by T. Dauxois, S. Ruffo, E. Arimondo, and M. Wilkens, p. 312, Springer-Verlag (2002)
2. I. Bloch, J. Phys. B **38**, S629 (2005)
3. I. Bloch, Nat. Phys. **1**, 23 (2005)
4. O. Morsch and M. Oberthaler, Rev. Mod. Phys. **78**, 180 (2006)
5. M. Greiner, O. Mandel, T. Esslinger, T.W. Hänsch, and I. Bloch, Nature **415**, 39 (2002)
6. N. Ashcroft and N.D. Mermin, *Solid State Physics*, International Thomson Publishing, New York (1976)
7. F. Bloch, Z. Phys. **52**, 555 (1929)

8. L. Landau, Phys. Z. Sowjetunion **2**, 46 (1932)
9. G. Zener, Proc. R. Soc. A **137**, 696 (1932)
10. B. Wu and Q. Niu, New J. Phys. **5**, 104 (2003)
11. Q. Niu, X.-G. Zhao, G.A. Georgakis, and M.G. Raizen, Phys. Rev. Lett. **76**, 4504 (1996)
12. M.B. Dahan, E. Peik, J. Reichel, Y.Y. Castin, and C. Salomon, Phys. Rev. Lett. **76**, 4508 (1996)
13. M. Raizen, C. Salomon, and Q. Niu, Phys. Today **50**, 30 (1997)
14. C. Waschke, H.G. Roskos, R. Schwedler, K. Leo, H. Kurz, and K. Köhler, Phys. Rev. Lett. **70**, 3319 (1993)
15. V.G. Lyssenko, G. Valušis, F. Löser, T. Hasche, K. Leo, M.M. Dignam, and K. Köhler, Phys. Rev. Lett. **79**, 301 (1997)
16. T. Pertsch, P. Dannberg, W. Elflein, A. Bräuer, and F. Lederer, Phys. Rev. Lett. **83**, 4752 (1999)
17. R. Morandotti, U. Peschel, J.S. Aitchison, H.S. Eisenberg, and Y. Silberberg, Phys. Rev. Lett. **83**, 4756 (1999)
18. H. Trompeter, W. Krolikowski, D.N. Neshev, A.S. Desyatnikov, A.A. Sukhorukov, Y.S. Kivshar, T. Pertsch, U. Peschel, and F. Lederer, Phys. Rev. Lett. **96**, 053903 (2006)
19. H. Sanchis-Alepuz, Y.A. Kosevich, and J. Sánchez-Dehesa, Phys. Rev. Lett. **98**, 134301 (2007)
20. O. Morsch, J.H. Müller, M. Cristiani, D. Ciampini, and E. Arimondo, Phys. Rev. Lett. **87**, 140402 (2001)
21. M. Cristiani, O. Morsch, J.H. Müller, D. Ciampini, and E. Arimondo, Phys. Rev. A **65**, 063612 (2002)
22. J. Hecker Denschlag, J. Simsarian, H. Häffner, C. McKenzie, A. Browaeys, D. Cho, K. Helmerson, S. Rolston, and W.D. Phillips, J. Phys. B **35**, 3095 (2002)
23. A. Browaeys, H. Häffner, C. McKenzie, S.L. Rolston, K. Helmerson, and W.D. Phillips, Phys. Rev. A **72**, 053605 (2005)
24. S. Schmid, G. Thalhammer, K. Winkler, F. Lang, and J.H. Denschlag, New J. Phys. **8**, 159 (2006)
25. J.R. Rubbmark, M.M. Kash, M.G. Littman, and D. Kleppner, Phys. Rev. A **23**, 3107 (1981)
26. D. Bouwmeester, N.H. Dekker, F.E. van Dorsselaer, C.A. Schrama, P.M. Visser, and J.P. Woerdman, Phys. Rev. A **51**, 646 (1995)
27. C.F. Bharucha, K.W. Madison, P.R. Morrow, S.R. Wilkinson, B. Sundaram, and M.G. Raizen, Phys. Rev. A **55**, R857 (1997)
28. M. Jona-Lasinio, O. Morsch, M. Cristiani, N. Malossi, J.H. Müller, E. Courtade, M. Anderlini, and E. Arimondo, Phys. Rev. Lett. **91**, 230406 (2003); Erratum, Phys. Rev. Lett. **93**, 119903 (2004)
29. L. Esaki, in *Nobel Lectures, Physics 1971-1980*, ed. by S. Lundqvist, World Scientific, Singapore (1992)
30. Q. Niu, X.G. Zhao, G.A. Georgakis, and M.G. Raizen, Phys. Rev. Lett. **76**, 4504 (1996)
31. B.K. Teo, J.R. Guest, and G. Raithel, Phys. Rev. Lett. **88**, 173001 (2002)
32. M. Greiner, *Ultracold quantum gases in three-dimensional optical lattice potentials*, Ph.D. thesis, Ludwig-Maximilians-Universität München (2003)

33. P. Cladé, S. Guellati-Khélifa, C. Schwob, F. Nez, L. Julien, and F. Biraben, Europhys. Lett. **71**, 730 (2005)
34. G. Ferrari, N. Poli, F. Sorrentino, and G.M. Tino, Phys. Rev. Lett. **97**, 060402 (2006)
35. P. Cladé, E. de Mirandes, M. Cadoret, S. Guellati-Khélifa, C. Schwob, F. Nez, L. Julien, and F. Biraben, Phys. Rev. Lett. **96**, 033001 (2006)
36. I. Carusotto, L. Pitaevskii, S. Stringari, G. Modugno, and M. Inguscio, Phys. Rev. Lett. **95**, 093202 (2005)

10

Transporting Cold Atoms in Optical Lattices with Ratchets: Mechanisms and Symmetries

Sergey Denisov[1], Sergej Flach[2], and Peter Hänggi[1]

[1] Institut für Physik, Universität Augsburg, Universitätsstr. 1, 86135 Augsburg, Germany
sergey.denisov@physik.uni-augsburg.de, hanggi@physik.uni-augsburg.de
[2] Max-Planck-Institut für Physik Komplexer Systeme, Nöthnitzer Str. 38, 01187 Dresden, Germany
flach@mpipks-dresden.mpg.de

10.1 Introduction

Thermal fluctuations alone cannot create a steady directed transport in an unbiased system. However, if a system is out of equilibrium, the Second Law of Thermodynamics no longer applies, and then there are no thermodynamical constraints on the appearance of a steady transport [1,2]. A directed current can be generated out of a fluctuating (time-dependent) external field with zero mean. The corresponding *ratchet effect* [3–9] has been proposed as a physical mechanism of a microbiological motility more then a decade ago [4,5]. Later on the ratchet idea has found diverse applications in different areas [6–9], from a molecular nanoscale-machine [10] up to quantum systems and quantum devices [11–17].

When the deviation from an equilibrium regime is small (the case of weak external fields) one may use the linear response theory in order to estimate the answer of the system [18–20]. However, due to the linearization of the response, the current value will be strictly zero since the driving field has zero bias. Therefore, one has to take into account nonlinear corrections and then derive the corresponding nonlinear response functional [20, 21], which may become a very complicated task, if the nonadiabatic regime is to be considered.

To obtain a dc-current, one has to break certain discrete symmetries, which involve simultaneous transformations in space and time. A recently elaborated *symmetry approach* [22, 23] established a clear relationship between the appearance of a directed current and broken space-time symmetries of the *equations of motion*. Thus, the symmetry analysis provides an information about the conditions for a directed transport appearance without the necessity of considering a nonlinear response functional.

Most theoretical and experimental studies have focused on ratchet realizations at a noisy overdamped limit [6–9]. However, systematic studies of the underlying broken symmetries, and the largest possible values of directed currents achieved for different dissipation strength, show that the dc current values typically become orders of magnitude larger in the limit of weak dissipation [24, 25]. The corresponding dynamics is characterized by long space-time correlations which may drastically increase the rectification efficiency [25, 26].

Fast progress in experimental studies of cold atoms ensemble dynamics have provided clean and versatile experimental evidence of a ratchet mechanism in the regime of weak or even vanishing dissipation [27, 28]. The results of the corresponding symmetry analysis for the regime of classical dynamics has already been successfully tested with cold Rubidium and Cesium atoms in optical lattices with a tunable weak dissipation [29–32]. Further decreasing of the dissipation strength leads to the quantum regime [27]. Recent experiments have shown the possibility to achieve an optical lattice with tunable asymmetry in the quantum regime [33]. A Bose-Einstein condensate (BEC) loaded into an optical potential is another candidate for a realization of quantum ratchets in the presence of atom-atom interactions [28]. While there is obvious interest in experimental realizations of theoretically predicted symmetry broken states, another important aspect of the interface between cold atoms and the ratchet mechanism is, that new possibilities for a control of the dynamics of atomic systems by laser fields may be explored [34, 35].

The objective of this work is to provide a general introduction into the symmetry analysis of the rachet effect using a simple, non-interacting one-particle dynamics. Despite its simplicity, this model contains all the basic aspects of classical and quantum ratchet dynamics, and may be used also as a starting point of incorporating atom-atom interactions.

10.2 Single Particle Dynamics

We start with the simple model of an underdamped particle with mass m, moving in a space-periodic potential $U(x) = U(x + \lambda)$ under the influence of the external force $\chi(t)$ with zero mean:

$$m\ddot{x} + \gamma\dot{x} - f(x) - \chi(t) = 0 . \qquad (10.1)$$

Here $f(x) = -U'(x)$, $\int_0^\lambda f(x)\,dx = 0$, and γ is the friction coefficient. Next, we ask whether a directed transport with nonzero mean velocity, $\langle\dot{x}\rangle \neq 0$, may appear in the system (10.1).

If $\chi(t) \equiv \xi(t)$ is a realization of a Gaussian white (i.e. delta-correlated) noise, obeying via its correlation properties the (second) fluctuation-dissipation theorem [20], Eq. (10.1) then describes the thermal equilibrium state of a particle interacting with a heat bath. From the Second Law of Thermodynamics it follows that a directed transport is absent, independently of the particular choice of the periodic potential $U(x)$ [4, 6, 7, 24].

The presence of temporary correlations in $\chi(t)$ may change the situation drastically. A simple way to get such correlations is to use an additive periodic field $E(t)$,

$$\chi(t) = \xi(t) + E(t), \qquad E(t) = E(t+T), \qquad \int_0^T E(t)\,dt = 0. \qquad (10.2)$$

If $\xi(t)$ is a realization of a white noise, then the functions $-\xi(t)$, $\xi(t)$, and $\xi(t+\tau)$ are also realizations of the same white noise, and their statistical weights are equal to the statistical weight of the original realization. For what comes, the noise term $\xi(t)$ will thus not be relevant for the following symmetry analysis. We consider the symmetries of the deterministic differential equation

$$m\ddot{x} + \gamma\dot{x} - f(x) - E(t) = 0. \qquad (10.3)$$

Eq. (10.3) contains two periodic functions, $f(x)$ and $E(t)$, both with zero mean. The properties of the symmetries of the Eq. (10.3) are strongly depending on the symmetry properties of these functions.

10.3 Symmetries

10.3.1 Symmetries of a Periodic Function with Zero Mean

Let us consider a periodic function $g(z+2\pi) = g(z)$ with zero mean, $\int_0^{2\pi} g(z)\,dz = 0$. This function can be expanded into a Fourier series

$$g(z) = \sum_{k=-\infty}^{\infty} g_k \cdot \exp(ikz), \qquad (10.4)$$

where $g_0 \equiv 0$. We will consider real-valued functions; therefore, $g_k = g_{-k}^*$.

The function $g(z)$ may possess three different symmetries. First, it can be *symmetric*, $g(z+z_0) = g(-z+z_0)$, around a certain argument value z_0. For such functions we will use the notation g_s. The Fourier expansion (10.4) contains, after the shift by z_0, only cosine terms, so $g_k(z_0) = g_k \cdot \exp(ikz_0)$ are real numbers, i.e. $g_k(z_0) = g_{-k}(z_0)$.

Second, the function $g(z)$ can be *antisymmetric*, $g(z+z_1) = -g(-z+z_1)$, around a certain value of the argument, z_1. For such functions we will use the notation g_a. The corresponding Fourier expansion (10.4) contains only sine terms (after the shift by z_1), so $g_k(z_0) = g_k \cdot \exp(ikz_0)$ are pure imaginary numbers, and $g_k(z_0) = -g_{-k}(z_0)$.

Finally, the function $g(z)$ can be *shift-symmetric*, $g(z) = -g(z+\pi)$. The Fourier expansion of a shift-symmetric function $g_{\text{sh}}(z)$ contains odd harmonics only, $g_{2m} \equiv 0$.

It is straightforward to show that a periodic function $g(z)$ can have either none of the above mentioned symmetries, or exactly one of them, or all three of them. Let us consider several simple examples. The function $\cos(z)$ possesses all three symmetries. The function $\cos(z)+\cos(3z+\phi)$ always possesses shift-symmetry and in addition may be simultaneously symmetric and antisymmetric for $\phi = 0, \pm\pi$. The function $\cos(z)+\cos(2z+\phi)$ is not shift-symmetric, thus it will either have no other symmetry at all, except for $\phi = 0, \pm\pi$ (symmetric), and $\phi = \pm\pi/2$ (antisymmetric).

10.3.2 Symmetries of the Equations of Motion

The system dynamics in Eq. (10.3) can be described by three first-order autonomous differential equations,

$$\dot{x} = \frac{p}{m}, \qquad \dot{p} = f(x) + E(\tau) - \frac{\gamma}{m}p, \qquad \dot{\tau} = 1. \qquad (10.5)$$

The phase-space dimension is three. We are looking for symmetrytransformations $\hat{S}$, which do not change the equation (10.5), but do change the sign of the velocity $\dot{x}$. Such transformations map the phase space $\{x, p, \tau\}$ onto itself. If we find such a transformation, we then apply it to all points of a given trajectory. We get a new manifold in phase space, which also represents a trajectory, i.e., a solution of the equations (10.5). The original trajectory and its image may coincide (or may not).

Let us assume that we have found such a transformation. Next, we consider the mean velocity, $\bar{v} = \lim_{s\to\infty}(x(t_0+s)-x(t_0))/s$, on the original trajectory. If the trajectory and its image coincide, then $\bar{v} = 0$. If they are different then their velocities have the same absolute value but opposite signs. If, in addition, both the trajectories have the same statistical weights in the presence of a white noise, then we can conclude that the average current in the system (10.3) is equal to zero [22].

There are only two possible types of transformations which change the sign of the velocity $\dot{x}$: they include either a time-reversal operation, $t \to -t$, or a space inversion, $x \to -x$ (but not both operations simultaneously!).

The following symmetries can be identified [22]:

$$\begin{aligned}\hat{S}_a &: \quad x \to -x, \quad t \to t + \frac{T}{2}, \quad \text{if } \{f_a, E_{\text{sh}}\}, \\ \hat{S}_b &: \quad x \to x, \quad t \to -t, \quad \text{if } \{E_s, \gamma = 0\}, \\ \hat{S}_c &: \quad x \to x + \frac{\lambda}{2}, \quad t \to -t, \quad \text{if } \{f_{\text{sh}}, E_a, m = 0\}.\end{aligned} \qquad (10.6)$$

The symmetry $\hat{S}_b$ requires zero dissipation, $\gamma = 0$, i.e., it requires the Hamiltonian regime, and the symmetry $\hat{S}_c$ can be fulfilled in the overdamped limit (i.e. $m = 0$) only. Note that all symmetries require certain symmetry properties of the function $E(t)$. Usually, an experimental setup allows to tune the shape of the time-dependent field $E(t)$ easier than the shape of the spatially periodic potential [36,37]. A proper choice of the force $E(t)$ may break all three symmetries for any coordinate dependence of the force $f(x)$. We restrict the further consideration to the case of a symmetric potential $U(x) = 1 - \cos(x)$ while using a bi-harmonic driving force,

$$E(t) = E_1 \cos(t) + E_2 \cos(2t + \theta) , \tag{10.7}$$

for a symmetry violation. If $\theta \neq 0, \pi/2, \pi, 3\pi/2$ then all three symmetries (10.6) are broken and we may count on a nonzero mean velocity, $v \neq 0$.

10.3.3 The Case of Quasiperiodic Functions

We generalize the symmetry analysis to the case of quasiperiodic driving field $E(t)$ [38, 39].

We consider a quasiperiodic function $g(z)$ to be of the form

$$g(z) \equiv \tilde{g}(z_1, z_2, \ldots, z_N) , \qquad \frac{\partial z_i}{\partial z} = \Omega_i \tag{10.8}$$

where all ratios Ω_i/Ω_j are irrational if $i \neq j$ and $\tilde{g}(z_1, z_2, \ldots, z_i + 2\pi, \ldots, z_N) = \tilde{g}(z_1, z_2, \ldots, z_i, \ldots, z_N)$ for any i. Such a function may have numerous symmetries. With respect to the following symmetry analysis of the equation of motion we will list here only those symmetries of $\tilde{g}$ which are of relevance. It can be symmetric $\tilde{g}_s(z_1, z_2, \ldots, z_N) = \tilde{g}_s(-z_1, -z_2, \ldots, -z_N)$, antisymmetric $\tilde{g}_a(z_1, z_2, \ldots, z_N) = -\tilde{g}_a(-z_1, -z_2, \ldots, -z_N)$. It can be also shift-symmetric for a given set of indices $\tilde{g}_{\text{sh}, \{i,j,\ldots,m\}}$ which means that $\tilde{g}$ changes sign when a shift by π is performed in the direction of each $z_i, z_j, \ldots, z_m$ only, leaving the other variables unchanged.

The relevant symmetry properties of $\tilde{g}$ are thus studied on the compact space of variables $\{z_1, z_2, \ldots, z_N\}$. The irrationality of the frequency ratios guarantees that in the course of evolution of z this compact space is densely scanned by these variables with uniform density in the limit of large z. At the same time we note that it is always possible to find a large enough value Z such that

$$\lim_{\tau \to \infty} \frac{1}{\tau} \int_0^\tau (g(z+Z) - g(z))^2 \, dz < \epsilon \tag{10.9}$$

with (arbitrarily) small absolute value of ϵ. For a given value of ϵ this defines a quasiperiod Z of the function $g(z)$.

In order to make the symmetry analysis of the equation of motion transparent, we rewrite it (skipping the noise term) in the following form [39]:

$$m\ddot{x} + \gamma\dot{x} - f(x) - E(\phi_1, \phi_2, \ldots, \phi_N) = 0 , \tag{10.10}$$

$$\dot{\phi}_1 = \omega_1 ,$$
$$\dot{\phi}_2 = \omega_2 ,$$
$$\vdots$$
$$\dot{\phi}_N = \omega_N .$$

The function $f(x)$ is also assumed to be quasiperiodic with M corresponding spatial frequencies.

The following symmetries can be identified, which change the sign of $\langle \dot{x} \rangle$ and leave (10.10) unchanged:

$$\begin{aligned}
\tilde{S}_\text{a} &: \quad x \to -x, \quad \phi_{i,j,\ldots,m} \to \phi_{i,j,\ldots,m} + \pi, \quad \text{if } \{f_\text{a}, E_{\text{sh},\{i,j,\ldots,m\}}\}, \\
\tilde{S}_\text{b} &: \quad x \to x, \quad t \to -t, \quad \text{if } \{E_\text{s}, \gamma = 0\}, \\
\tilde{S}_\text{c} &: \quad x \to x + \frac{\lambda}{2}, \quad t \to -t, \quad \text{if } \{f_{\text{sh},\{1,2,3,\ldots,M\}}, E_\text{a}, m = 0\}.
\end{aligned} \quad (10.11)$$

The symmetry $\tilde{S}_\text{a}$ is actually a set of various symmetry operations which are defined by the given subset of indices $\{i, j, \ldots, m\}$.

The prediction then is, that if for a given set of parameters any of the relevant symmetries (10.11) is fulfilled, the average current will be zero. If however the choice of functions $f(x)$ and $E(t)$ is such that the symmetries are violated, a nonzero current is expected to emerge.

10.4 Dynamical Mechanisms of Rectification: The Hamiltonian Limit

Let us consider the limit $\gamma = 0$ (Hamiltonian case) [22, 26]. Due to time and space periodicity of the system (10.3) we can map the original three-dimensional phase space (x, p, t) onto a two-dimensional cylinder, $\mathcal{T}^2 = (x \bmod 1, p)$, by using the stroboscopic Poincaré section after each period $T = 2\pi/\omega$. For given initial conditions $\{x(0), p(0)\}$, we integrate the system over time T, and then plot the final point, $\{x(T), p(T)\}$, on the cylinder $\mathcal{T}^2$.

For $E(t) = 0$ the system (10.3) is integrable and there is a separatrix in the phase space which separates oscillating and running solutions. A non-zero field $E(t)$ destroys the separatrix and leads to the appearance of a stochastic layer (see Fig. 10.1). In this part of the phase space the system dynamics is ergodic, i.e., all average characteristics are the same for all trajectories, launched inside the layer. Therefore, the symmetry analysis is valid for all trajectories on this manifold. Numerical studies have confirmed this conclusion [22, 23, 26]. Fig. 10.2 shows several trajectories $x(t)$ from chaotic layers and illustrates the fact that the violation of symmetries causes a directed motion of the particle.

The dynamics within the stochastic layer can be roughly subdivided into two distinct fractions. The first one corresponds to ballistic flights near the layer boundaries. They appear due to a sticking effect [40]. A random diffusion within a chaotic bulk is attributed to the second fraction. A rectification effect appears due to a violation of the balance between ballistic flights in opposite directions [26]. This interpretation supports the view, that even in the presence of damping and noise, the ratchet mechanism relies on harvesting on temporal correlations of the underlying dynamical system. Ballistic flights are just such examples of long temporal correlations on a trajectory which

10 Ratchets with cold atoms 187

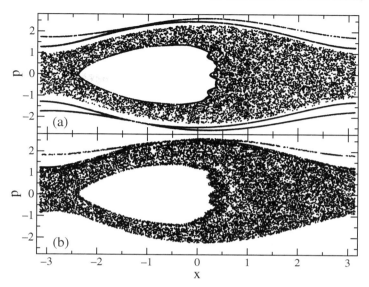

Fig. 10.1. Poincaré map for the system (10.5), (10.7). The parameters are $E_1 = 0.252$, $E_2 = 0.052$, $\gamma = 0$. (a) $\theta = 0$, (b) $\theta = \pi/2$

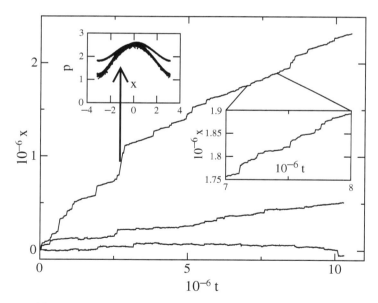

Fig. 10.2. $x(t)$ for $\theta = 0, \pi/5, \pi/2$ (lower, middle and upper curves, respectively). Left upper inset: Poincaré map for a single ballistic flight, $\theta = \pi/2$. Right inset: zoom of $x(t)$ for the case $\theta = \pi/2$

is overall chaotic. Therefore it is not surprising, that the ratchet effect is stronger in the dissipationless limit, since dissipation will introduce finite (and possibly short) time scales which cut the temporal correlations down. The averaged drift velocity can be estimated by using a sum rule [41, 42]. From the corresponding approach, which is based on a statistical argument by the authors of Ref. [41, 42], it follows, that a mixed space, i.e., a stochastic layer with boundaries and embedded regular submanifolds (islands), presents the necessary condition for a directed transport.

The adding of a non-zero dissipation, $\gamma \neq 0$, does not change the situation drastically [23]. The symmetry analysis is still valid for this case. The phase space is shared by different transporting and non-transporting attractors with their corresponding basins of attraction, which are strongly entangled inside the former stochastic layer region. A symmetry violation causes a desymmetrization of basins. Finally, a weak noise leads to a trajectory wandering over different basins, sticking to corresponding attractors, and, finally, to the rectification effect. The long flights which appear at the Hamiltonian limit are damped after a characteristic time which is the shorter, the larger the dissipation strength γ is [23, 26].

A systematic analysis shows that, under the condition of full symmetry violation, the approach of the dissipationless limit leads to a drastic increase of the dc current value [25], which depends on the characteristics of the stochastic layer [26]. It has been shown that, in a full accordance with the symmetry analysis, the dc current disappears near $\theta = 0, \pi$ for the case of weak dissipation, and near $\theta = \pm\pi/2$ at the strong dissipation limit. The value of the phase θ, at which the current becomes zero, is a monotonous function of the dissipation strength γ [25].

An inclusion of a dc-component to the external field, $\tilde{E}(t) = E(t) + E_{\text{dc}}$, may lead to a directed transport against a constant bias E_{dc}, even in the Hamiltonian limit [43].

The abovementioned results have been confirmed in cold atoms experiments, performed in the group of Renzoni [29–31]. In these experiments, atoms of Cs and Rb have been cooled to temperatures of several mK. An optical standing wave, created by a pair of counter-propagating laser beams, formed a periodic potential for the atoms. Finally, a time-dependent force $E(t)$ has been introduced through a periodic modulation of the phase for one of the beams. The results of the above symmetry analysis have been verified by changing the relative phase ϕ and by tuning the effective dissipation strength.

The case of the quasiperiodic driving force $E(t)$ for cold atoms ratchets also has been studied experimentally [32], with a similar outcome.

10.5 Resonant Enhancement of Transport with Quantum Ratchets

A quantum extension of the (dissipationless) system dynamics in Eq. (10.3) can readily be achieved [44,45]. The system evolution can be described by the Schrödinger equation,

$$i\hbar \frac{\partial}{\partial t}|\psi(t)\rangle = H(t)|\psi(t)\rangle \,, \tag{10.12}$$

where the Hamiltonian H is of the form

$$H(x, p, t) = \frac{p^2}{2} + [1 + \cos(x)] - xE(t) \,. \tag{10.13}$$

The system (10.12) describes a cloud of noninteracting atoms, placed into a periodic potential (formed by two counter-propagating laser beams) and exposed to an external ac field (10.7)[1].

Because of the time and space periodicity of the Hamiltonian (10.13), the solutions $|\psi_\alpha(t + t_0)\rangle = U(t, t_0)|\psi_\alpha(t_0)\rangle$ of the Schrödinger equation (10.12) can be characterized by the eigenfunctions of the Floquet operator $U(T, t_0)$ which satisfy the Floquet theorem $|\psi_\alpha(t)\rangle = \exp(-iE_\alpha t/T)|\phi_\alpha(t)\rangle$, $|\phi_\alpha(t + T)\rangle = |\phi_\alpha(t)\rangle$ (here t_0 is the initial time). The quasienergies E_α ($-\pi < E_\alpha < \pi$) and the Floquet eigenstates can be obtained as solutions of the eigenvalue problem of the Floquet operator

$$U(T, t_0)|\phi_\alpha(t_0)\rangle = e^{-iE_\alpha}|\phi_\alpha(t_0)\rangle \tag{10.14}$$

with α denoting the band index and with k being the wave vector [44–47]. An initial state can be expanded over Floquet-Bloch eigenstates, $|\psi(t_0)\rangle = \sum_{\alpha,k} C_{\alpha,k}(t_0)|\phi_{\alpha,k}\rangle$ and the subsequent state's evolution is encoded in the coefficients $\{C_{\alpha,k}\}$. We restrict further consideration to the case $\kappa = 0$ which corresponds to initial states where atoms equally populate all (or many) wells of the spatial potential.

The mean momentum expectation value,

$$J(t_0) = \lim_{t \to \infty} \frac{1}{t} \int_{t_0}^{t} \langle \psi(\tau, t_0)|\hat{p}|\psi(\tau, t_0)\rangle \, d\tau \,, \tag{10.15}$$

measures the asymptotic current. Expanding the wave function over the Floquet states the current becomes

$$J(t_0) = \sum_\alpha \langle p \rangle_\alpha |C_\alpha(t_0)|^2 \,, \tag{10.16}$$

where $\langle p \rangle_\alpha$ is the mean momentum of the Floquet state $|\phi_\alpha\rangle$ [44–46].

---
[1] The dissipation may be included into quantum dynamics by coupling the system (10.13) to a heat bath, $H_{\text{diss}}(x, p, t, \{q\}) = H(x, p, t) + H_B(x, \{q\})$. Here $H_B(x, \{q\})$ describes an ensemble of harmonic oscillators $\{q\}$ at thermal equilibrium interacting with the system [11].

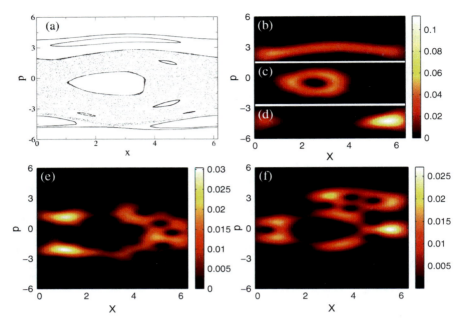

Fig. 10.3. (a) Poincaré section for the classical limit, (10.7), (10.13), (b)–(f) Husimi representations for different Floquet eigenstates for the Hamiltonian (10.13) with $\hbar = 0.2$ (momentum is in units of the recoil momentum, $p_r = \hbar k_L$, with $k_L = 1$). The parameters are $E_1 = E_2 = 2$, $\omega = 2$, $\theta = -\pi/2$ and $t_0 = 0$ for (b)–(e), and $E_1 = 3.26$, $E_2 = 1$, $\omega = 3$, $\theta = -\pi/2$ and $t_0 = 0$ for (f)

The analysis of the transport properties of the eigenstates shows that the quantum system inherits the symmetries of its classical counterpart [44,45]. In particular, the symmetries of the classical equations of motion translate into symmetries of the Floquet operator. The presence of any of these symmetries results in a vanishing the time-averaged expectation value of the momentum operator for each Floquet eigenstate: $\langle p \rangle_\alpha = 0$. Thus, if one of the symmetries, $\tilde{S}_a$, $\tilde{S}_b$ (10.11), holds then $\langle p \rangle_\alpha = 0$ for all α. Consequently $J(t_0) = 0$ in this case.

By using the Husimi representation [48,49] we can visualize different eigenstates in the phase space, $\{x, p, \tau\}$ and establish a correspondence between them and the mixed phase space structures for the classical limit (Fig. 10.3).

Since the Schrödinger equation (10.12) is linear, the system maintains a memory of the initial condition for infinite times [50]. The asymptotic current value depends on the initial time, t_0, and on the initial wave function, $\psi(t_0)$. For a given initial wave function, $|\psi\rangle = |0\rangle$, we can assign a unique current value by performing an averaging over the initial time t_0, $J = 1/T \int_0^T J(t_0)\, dt_0$ [44, 45]. Fig. 10.4 shows the dependence of the average

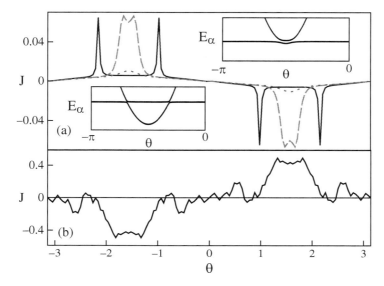

Fig. 10.4. (a) The average current J (in units of the recoil momentum) vs. θ for different amplitude values of the second harmonic, E_2: 0.95 (pointed line), 1 (dashed line) and 1.2 (solid line). Insets: relevant details of the quasienergy spectrum versus θ in the resonance region for $E_2 = 1$ (top right) and $E_2 = 1.2$ (bottom left). The parameters are $E_1 = 3.26$ and $\omega = 3$. (b) The average current J (in units of the recoil momentum) vs. θ for $E_1 = 3$, $E_2 = 1.5$ and $\omega = 1$

current on the asymmetry parameter θ. Sharp resonant peaks for $E_2 = 1.2$ where the current value changes drastically are associated with interactions between two different Floquet eigenstates. The Husimi distributions show that one state locates in the chaotic layer, and another one in a transporting island. Off resonance the initial state mainly overlaps with the chaotic state, which yields some nonzero, yet small, current. In resonance Floquet states mix, and thus the new eigenstates contain contributions both from the original chaotic state as well as from the regular transporting island state. The Husimi distribution of the mixed state is shown in Fig. 10.3f, the strong asymmetry is clearly observed. The regular island state has a much larger current contribution, resulting in a strong enhancement of the current.

To conclude this section, we would like to emphasize the following two points. For both cases, i.e., the classical and the quantum one, the overall, total current over the whole momentum space is zero [41, 42]. Thus, it is essential to have the initial state prepared localized near the line $p = 0$, because for broad initial distributions the asymptotic current tends to zero. However, if the dynamics is restricted to the lowest band of the periodic potential, no current rectification does occur [51].

10.6 Summary

This surveyed symmetry analysis, originally put forward in Refs. [22,23], provides a general toolbox for the prediction of dynamical regimes for which one can (or cannot) obtain the rectification and directed current phenomenon for a given transport dynamics. First, one has to set up the equations of motions and define a observable (current, magnetization, angular velocity, energy flux, etc.) which should become nonzero, in terms of these dynamical variables. Then, one examines whether there exist transformations (symmetries) which change the sign of the observable and at the same time leave the equations of motion invariant. Upon breaking all the symmetries one can expect the emergence of a non-zero, directed current. This strategy has been successfully tested with Josephson junctions (fluxon directed motion) [52,53] and as well with paramagnetic resonance experiments (spin magnetization by a zero-mean field) [54,55].

Herein, we focused only on the one-dimensional case. By use of more laser beams, experimentalists can fabricate two- and three-dimensional optical potentials [28]. By changing the relative phase between lattice beams, 2D- and 3D-potentials with different symmetries and topologies can be achieved [56, 57]. This fact incites for an extension of the present ratchet studies into higher dimensions.

Moreover, for the phenomenon of Bose-Einstein-condensation (BEC) of cold gases, interactions between atoms become essential and nonlinearities start to play an important role [28]. Many features of BEC dynamics are manifestations of general concepts of nonlinear physics, such as soliton creation and propagation. These collective excitations can then themselves be subjected to a ratchet transport mechanism [58].

References

1. M. von Smoluchowski, Phys. Zeitschrift **XIII**, 1069 (1912)
2. R.P. Feynmann, R.B. Leighton, and M. Sands, *The Feyman Lectures on Physics*, 2nd edn., Addison Wesley, Reading, MA (1963), vol. 1, chap. 46
3. M.O. Magnasco, Phys. Rev. Lett. **71**, 1477 (1993)
4. P. Hänggi and R. Bartussek, Lect. Notes. Phys. **476**, 294 (1996)
5. F. Jülicher, A. Ajdari, and J. Prost, Rev. Mod. Phys. **69**, 1269 (1997)
6. P. Reimann and P. Hänggi, Appl. Phys. A **75**, 169 (2002)
7. P. Reimann, Phys. Rep. **361**, 57 (2002)
8. R.D. Astumian and P. Hänggi, Physics Today **55**, 33 (2002)
9. P. Hänggi, F. Marchesoni, and F. Nori, Ann. Phys. **14**, 51 (2005)
10. B. Norden, Y. Zolotaryuk, P.L. Christiansen, and A.V. Zolotaryuk, Phys. Rev. E **65**, 011110 (2002)
11. P. Reimann, M. Grifoni, and P. Hänggi, Phys. Rev. Lett. **79**, 10 (1997)
12. I. Goychuk, M. Grifoni, and P. Hänggi, Phys. Rev. Lett. **81**, 649 (1998); ibid **81**, 2837 (1998) (erratum)

13. I. Goychuk and Hänggi, Europhys. Lett. **43**, 503 (1998)
14. J. Lehmann, S. Kohler, P. Hänggi, and A. Nitzan, Phys. Rev. Lett. **88**, 228305 (2002)
15. M. Grifoni, M.S. Ferreira, J. Peguiron, and J.B. Majer, Phys. Rev. Lett. **89**, 146801 (2002)
16. H. Linke, T.E. Humphrey, A. Löfgren, A.O. Sushkov, R. Newbury, R.P. Taylor, and P. Omling, Science **286**, 2314 (1999)
17. J.B. Majer, J. Peguiron, M. Grifoni, M. Trusveld, and J.E. Mooij, Phys. Rev. Lett. **90**, 056802 (2003)
18. R. Kubo, J. Phys. Soc. Jpn. **12**, 570 (1957)
19. R. Kubo, N. Toda, and N. Hashitsume, *Statistical Physics II*, Springer, Berlin (1985)
20. P. Hänggi and H. Thomas, Phys. Rep. **88**, 207 (1982)
21. N.G. van Kampen, Phys. Norv. **5**, 279 (1971)
22. S. Flach, O. Yevtushenko, and Y. Zolotaryuk, Phys. Rev. Lett. **84**, 2358 (2000)
23. S. Denisov, S. Flach, A.A. Ovchinnikov, O. Yevtushenko, and Y. Zolotaryuk, Phys. Rev. E **66**, 041104 (2002)
24. P. Jung, J.G. Kissner, and P. Hänggi, Phys. Rev. Lett. **76**, 3436 (1996)
25. O. Yevtushenko, S. Flach, Y. Zolotaryuk, and A.A. Ovchinnikov, Europhys. Lett. **54**, 141 (2001)
26. S. Denisov and S. Flach, Phys. Rev. E **64**, 056236 (2001)
27. L. Guidoni and P. Verkerk, J. Opt. B **1**, R23 (1999)
28. O. Morsch and M. Oberthaler, Rev. Mod. Phys. **78**, 179 (2006)
29. M. Schiavoni, L. Sanchez-Palencia, F. Renzoni, and G. Grynberg, Phys. Rev. Lett. **90**, 094101 (2003)
30. P.H. Jones, M. Goonasekera, and F. Renzoni, Phys. Rev. Lett. **93**, 073904 (2004)
31. R. Gommers, S. Bergamini, and F. Renzoni, Phys. Rev. Lett. **95**, 073003 (2005)
32. R. Gommers, S. Denisov, and F. Renzoni, Phys. Rev. Lett. **96**, 240604 (2006)
33. G. Ritt, C. Geckeler, T. Salger, G. Cennini, and M. Weitz, Phys. Rev. A **74**, 063622 (2006)
34. R.J. Gordon and S.A. Rice, Ann. Rev. Phys. Chem. **48**, 601 (1997)
35. S. Denisov, J. Klafter, and M. Urbakh, Phys. Rev. E **66**, 046203 (2002)
36. S. Savel'ev, F. Marchesoni, P. Hänggi, and F. Nori, Europhys. Lett. **67**, 179 (2004)
37. S. Savel'ev, F. Marchesoni, P. Hänggi, and F. Nori, Phys. Rev. E **70**, 066109 (2004)
38. E. Neumann and A. Pikovsky, Eur. Phys. J. B **26**, 219 (1995)
39. S. Flach and S. Denisov, Acta Phys. Pol. B **35**, 1437 (2004)
40. G.M. Zaslavsky, *Physics of chaos in Hamiltonian systems*, Imperial College Press (1998)
41. H. Schanz, M.-F. Otto, R. Ketzmerick, and T. Dittrich, Phys. Rev. Lett. **87**, 070601 (2001)
42. H. Schanz, T. Dittrich, and R. Ketzmerick, Phys. Rev. E **71**, 026228 (2005)
43. S. Denisov, S. Flach, and P. Hänggi, Europhys. Lett. **74**, 588 (2006)
44. S. Denisov, L. Morales-Molina, S. Flach, and P. Hänggi, Phys. Rev. A **75**, 063424 (2007)
45. S. Denisov, L. Morales-Molina, and S. Flach, Europhys. Lett. **79**, 10007 (2007)
46. J. Gong, D. Poletti, and P. Hänggi, Phys. Rev. A **75**, 033602 (2007)

47. M. Grifoni and P. Hänggi, Phys. Rep. **304**, 279 (1998)
48. K. Husimi, Proc. Phys. Math. Soc. Japan **22**, 264 (1940)
49. K. Takahashi and N. Saito, Phys. Rev. Lett. **55**, 645 (1985)
50. F. Haake, *Quantum signature of chaos*, Springer-Verlag, London (1991)
51. I. Goychuk and Hänggi, J. Phys. Chem. B **105**, 6642 (2001)
52. S. Flach, Y. Zolotaryuk, A.E. Miroshnichenko, and M.V. Fistul, Phys. Rev. Lett. **88**, 184101 (2002)
53. A.V. Ustinov, C. Coqui, A. Kemp, Y. Zolotaryuk, and M. Salerno, Phys. Rev. Lett. **93**, 087001 (2004)
54. S. Flach, and A.A. Ovchinnikov, Physica A **292**, 268 (2001)
55. E. Arimondo, Ann. Phys. **3**, 425 (1968)
56. M. Greiner, I. Bloch, O.Mandel, T.W. Hänsch, and T. Esslinger, Phys. Rev. Lett. **87**, 160405 (2001)
57. L. Santos, M.A. Baranov, J.I. Cirac, H.-U. Everts, H. Fehrmann, and M. Lewenstein, Phys. Rev. Lett. **93**, 030601 (2004)
58. A.V. Gorbach, S. Denisov, and S. Flach, Opt. Lett. **31**, 1702 (2006)

11

Atomic Bose-Einstein Condensates in Optical Lattices with Variable Spatial Symmetry

Sebastian Kling, Tobias Salger, Carsten Geckeler, Gunnar Ritt, Johannes Plumhof, and Martin Weitz

Institut für Angewandte Physik, Universität Bonn, Wegelerstr. 8, 53115 Bonn, Germany
kling@iap.uni-bonn.de

11.1 Introduction

Optical lattices for atomic Bose-Einstein condensates raised enormous interest, as they mirror features known from solid state physics to the field of atom optics. In perfect solid state crystals atoms are arranged in a regular array creating a periodic potential for the electrons inside. Felix Bloch was one of the first who investigated in his dissertation (1928) the quantum mechanics of individual electrons in such crystalline solids. In the independent electron approximation interatomic and interelectronic interactions are neglected. Each electron obeys the one electron Schrödinger equation with a periodic potential $V(\boldsymbol{x} + \boldsymbol{a}) = V(\boldsymbol{x})$ with period $\boldsymbol{a}$. According to Bloch's theorem the stationary eigenstates $\psi_{n,q}(\boldsymbol{r})$ are plane waves modulated by a periodic function revealing the periodicity of the atom lattice [1]. With proper periodicity and boundary conditions the eigenstates are quantized, characterized by the band index $n = 0, 1, \ldots$. The plane waves propagate in the direction of the wave vector $\boldsymbol{q}$ with the associated quasimomentum $\hbar\boldsymbol{q}$, which it is sometimes referred to as the crystal or lattice momentum. The energy levels $E_n(\boldsymbol{q})$ are periodic continuous functions of the wave vector $\boldsymbol{q}$ forming the energy bands. Pictures of the energy bands showing the bandstructure are conventionally restricted the first Brillouin-zone of the reciprocal lattice $-\hbar k \leq q \leq \hbar k$. One milestone of Bloch theory and the band structure of particles is the finding of a natural physical explanation of the some 20 orders of magnitude difference in electrical conductivity between an insulator and a good conductor [2].

A realization of the fundamental concept of Bloch theory more recently became also accessable in a field quite different from solid state physics, namely in quantum optics. Due to the breakthrough of creating atomic Bose-Einstein condensates in 1995 [3,4] it is now also possible to investigate condensates confined in periodic optical potentials, so called optical lattices. Ultracold atoms exposed to such periodic potentials exhibits analogies to electrons in solids. In recent years, atoms confined in lattice potentials have allowed for observations of Bloch oscillations and Landau-Zener tunneling [5,6], number squeezing [7], or the Mott-insulator transition [8]. Further, fascinating pictures of cubic clouds depicting the first Brillouin-zone [9] and interference patterns of Bragg scattered atoms were taken [8]. These first experiments demonstrated that optical lattices are an attractive tool for modelling effects known or predicted in solid state physics.

So far, the band structure was investigated only for sinusoidal lattice potentials, in which the lattice periodicity is given by $\lambda/2$, where λ denotes the wavelength of the used laser radiation. Other potential shapes were realized: some authors studied superlattices based on the spatial beating of two neighboring trapping sites [10,11], others used a grey optical lattice configuration to realize asymmetric, dissipative potentials [12]. We here describe work realizing dissipationless lattice potential with spatial periodicity $\lambda/4$, which is realized with a fourth-order Raman process. Building upon this potential, a Fourier-synthesis of lattice potentials for a ^{87}Rb Bose-Einstein condensate is demonstrated. By superimposing the $\lambda/4$ period multiphoton potential with a conventional lattice potential with $\lambda/2$ spatial periodicity of appropriate phase, either symmetric or saw-tooth like asymmetric lattice potentials are synthesized. The scheme is scalable, in principle, to arbitrarily many Fourier components. In subsequent experiments, we have studied quantum transport in such lattices of variable symmetry. We find that the Landau-Zener tunneling rate of atoms between the lowest two excited Bloch bands depends on the shape of the lattice potential. In this way, the band structure of Fourier-synthesized lattices was explored.

The outline is as follows: In Sect. 11.2 we describe our method to generate nonstandard optical lattices. The experimental set up is presented in Sect. 11.3 and experimental results in Sect. 11.4. We conclude this article with Sect. 11.5, where we give an outlook on quantum ratchets.

11.2 Principle of Optical Multiphoton Lattices

A conventional lattice with sinusoidal shape and spatial periodicity $\lambda/2$ is generated by overlapping two counterpropagating off-resonant laser beams with frequency ω forming a standing wave. Due to a spatial varying ac-Stark shift, atoms experience a dipole force depending on the sign of the polarizability, which attracts the atoms to the nodes (for $\omega > \omega_0$) or the anti-nodes (for $\omega < \omega_0$) of the laser intensity, where ω_0 is an atomic resonance frequency. The effective potential for an atom exposed to a standing optical wave may also be described in a quantum picture by the exchange of photons changing the atom's momentum. The atoms here undergo virtual two-photon processes of absorption of one photon from one laser beam and stimulated emission of another photon into a counterpropagating beam, see Fig. 11.1a. An atom undergoing such a two photon process exchanges a momentum amount of $2\hbar k$ with the lattice.

11 BEC in asymmetric lattices 197

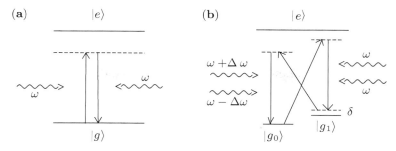

Fig. 11.1. Virtual process in optical lattices. (a) The trapping potential of conventional standing wave lattices is due to a virtual two photon process of absorbing and instantaneously emitting a photon of energy $\hbar\omega$. This yields the well-known spatial periodicity of $\lambda/2$, (b) virtual four photon process contributing to a lattice potential of $\lambda/4$ spatial periodicity

We use a multiphoton Raman technique to generate a lattice potential of periodicity $\lambda/4$, as the first harmonic for a Fourier-synthesis of lattice potentials [13,14]. The scheme uses three-level atoms with two stable ground states $|g_0\rangle$ and $|g_1\rangle$ and the electronically excited state $|e\rangle$. The atoms are irradiated by two optical beams of frequencies $\omega + \Delta\omega$ and $\omega - \Delta\omega$ from the left and by a beam of frequency ω from the right, as shown in Fig. 11.1b. Momentum is transferred to the atoms in units of $4\,\hbar k$ during an induced virtual four-photon processes, being a factor two above the corresponding process in a standing wave. This suggests a spatial periodicity of $\lambda/4$ for the adiabatic light shift potential, which is in agreement with theoretical predictions [13, 14].

This scheme can be extended to generate lattice potentials with higher periodicities $\lambda/2\,n$ by a $2\,n$-th photon process. The high resolution of Raman spectroscopy between two stable ground states over an excited state allows to clearly separate in frequency space the desired $2\,n$-th order process from lower order contributions.

By combining lattice potentials of different spatial periodicities, variable periodic potentials can be synthesized [15]. At present, we experimentally investigate ratchet-type asymmetric and symmetric potentials by combining four-photon potentials based on the scheme of Fig. 11.1b with usual standing wave lattice potentials. The adiabatic lattice has then the form

$$V(z) = V_1 \cos(2\,kz) + V_2 \cos(4\,kz + \phi)\,, \tag{11.1}$$

where V_1 and V_2 are the potential depths of the lattice potentials with spatial periodicities $\lambda/2$ and $\lambda/4$ respectively, and ϕ denotes the relative phase between the spatial harmonics. A constant offset to the potential V is here omitted. Figure 11.2a shows the form of such a lattice potential for typical experimental potential depths for a relative phase between lattice harmonics of $\phi = 0°$ (solid line) and $\phi = 90°$ (dashed line), as an example for a symmetric and asymmetric lattice respectively. Figure 11.2b shows the corresponding

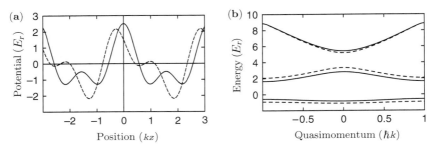

Fig. 11.2. Lattice potential and band structure. (**a**) For the experimentally realized values we show the lattice potential (11.1) for two specific phases ϕ. The solid line corresponds to $\phi = 0°$ and the dashed line to $\phi = 90°$. The depths are $V_1 = 3/2\, E_r$ and $V_2 = 1\, E_r$, where $E_r = \hbar^2 k^2 / 2\, M$ denotes the recoil energy, M the atomic mass. (**b**) Bandstructure corresponding to the lattice potentials shown in (a). It is clearly visible that the energy gap between first and second excited Bloch band depends on the phase ϕ

band structure for those potentials for the lowest Bloch bands. Notably, the band gap between first and second excited Bloch band dependends on the phase ϕ and the potential depths V_1, V_2 of the two lattice harmonics. For given values of the potential depths V_1 and V_2 one finds, that the gap takes its maximum for $\phi = 0°$ and minimum for $\phi = 180°$. For a suitable choice of the lattice depths V_1, V_2 and the phase ϕ between them the gap can even vanish.

11.3 Experimental Approach

Our experimental set-up was described in detail in Ref. [16]. Briefly, a ^{87}Rb Bose-Einstein condensate is produced all-optically by evaporative cooling in a CO_2-laser dipole trap. The used evaporation time is about 10 s, during which an additional magnetic field gradient is activated, to end up with a spin-polarized condensate of about 16 000 atoms in the $|F = 1, m_F = -1\rangle$ ground state is produced.

A magnetic bias field generates a frequency splitting of $\omega_Z \simeq 2\,\pi \times 805$ kHz between neighbouring Zeeman ground states. The direction of the magnetic field forms an angle respectively to the optical beam, so that atoms experience σ^+-, σ^-- and π-polarized light simultaneously. For the multiphoton lattice potential according to the scheme of Fig. 11.1b, the $F = 1$ ground state components $m_F = -1$ and 0 are used as states $|g_0\rangle$ and $|g_1\rangle$, while the $|5\,P_{3/2}\rangle$ manifold serves as the excited state $|e\rangle$. The Raman detuning δ typically is tuned to $2\,\pi \times 50$ kHz. The lattice beams are provided by a tapered diode laser tuned some 2 nm to the red of the rubidium D2-line. The beam is splitted into two, whereafter each of the partial beams pass an acoustooptical modulator. The modulators control the intensities and frequencies of the beams, as is required to generate superpositions of a standing wave and a four-photon potential [15]. The modulators are driven by four phase-locked rf function generators. The optical lattice beams are directed onto the ^{87}Rb condensate under an angle of 49° relatively to axis of gravity via optical fibers in a counterpropagating geometry.

11.4 Measurements and Results

In initial work, we have characterized the generated lattice potentials by diffracting the atomic Bose-Einstein condensate off the variable lattice potentials. Figure 11.3 shows the result of such measurements [15], where atoms were diffracted off a 6 µs long pulse of the lattice potential, where the used free expansion time was 10 ms for the shown time-of-flight images (top figures). For Figs. 11.3a and 11.3b, $\lambda/2$ and $\lambda/4$ periodicity lattice potentials respectively were investigated, where the smaller spatial periodicity of the multiphoton potential results in larger spatial separation of the diffracted peaks in the latter image. Figures 11.3c and 11.3d show the results obtained

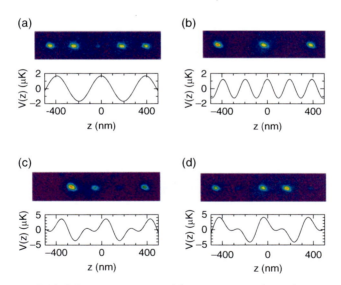

Fig. 11.3. Far-field diffraction images of lattice potentials and corresponding reconstructed spatial structure of the lattice potentials. (**a**) For a two-photon lattice potential with $\lambda/2$ spatial periodicity and (**b**) for a four-photon lattice potential with $\lambda/4$ spatial periodicity. Due to its smaller spatial periodicity, the splitting of the clouds is a factor two above that observed in (a), (**c**) diffractions image for an asymmetric lattice potential realized by superimposing two- and four-photon lattice potentials, (**d**) same as in (c), but with an additional phase shift of 180° for the four-photon lattice potential

when combining lattice potentials of periodicity $\lambda/2$ and $\lambda/4$ for different values of the relative phase between spatial harmonics. The asymmetry of the observed diffraction images is attributed to as evidence for the asymmetry of the Fourier-synthesized lattice potentials. The shown plots below the time-of-flight images give the reconstructed lattice potentials respectively, obtained by fitting the results of the diffraction experiment to the solution of a numerical integration of the momentum-picture Schrödinger equation [15]. More recently, we have experimentally investigated the band structure of Fourier-synthesized lattice potentials by means of quantum transport experiments. One fundamental quantum effect is Landau-Zener tunneling which can occur in accelerated lattices [17]. When the acceleration of the lattice is sufficiently large, transitions between different energy bands become possible, which is called Landau-Zener tunneling. On the other hand, for sufficiently small acceleration the atoms are Bragg diffracted at the Brillouin zone edge and the atomic quasimomentum oscillates within the Bloch bands from $-\hbar k$ to $\hbar k$, which is known as Bloch oscillation. The Landau-Zener transition probability can be estimated as [18, 19]

$$P(a) = \exp(-\pi E_G^2/\hbar^2 k a) \,, \qquad (11.2)$$

where a is the acceleration of the lattice and E_G is the energy gap between two bands. In a series of experiments we have investigated the Landau-Zener tunneling rate between the first and the second excited Bloch band [20]. One of the lattice beams with frequency ω is used both for the standing wave and the fourth order multiphoton potential. This beam was acoustooptically detuned by a small amount δ_{Dopp}, so that the reference frame in which the lattice is stationary moves with a velocity of $v_{\text{rel}} = \lambda \delta_{\text{Dopp}}/4\pi$. The atoms were loaded into the first Bloch band by adiabatically transferring them into a lattice moving at a velocity $v_{\text{rel}} \simeq 1.5\,\hbar k/M$, where M is the atomic mass. To accelerate the lattice we increase the beam detuning δ_{Dopp} with a constant rate, so that an acceleration is achieved which can be larger than the projection of gravity's acceleration on the lattice direction. The tunneling rate depends on the size of the band gap and therefore on the phase ϕ between the two lattice harmonics. The dependence on the phase is measured and depicted in Fig. 11.4 for an acceleration of $6.44\,\text{m/s}^2$ and potential depth of $V_1 \simeq 3/2\,E_r$ and $V_2 \simeq 1\,E_r$, where $E_r = \hbar^2 k^2/2\,M$ denotes the recoil energy. The data fits well to a sinusoidal curve which is fitted to the data (solid line) to guide the eyes. In the course of the experiment we studied also the dependence of the tunneling rate on the depths of the two lattice harmonics for which we refer the reader to [20]. Remarkably, for symmetric potentials $\phi = 0°$ ($\phi = 180°$) the Landau-Zener tunneling rate reaches a minimum (maximum) while for the ratchet type asymmetric potential ($\phi = 90°$) an intermediate value is obtained. This is in contrast to dissipative asymmetric lattices, where maxima and minima tunneling rates are reached for ratchet-like potentials of different symmetry respectively.

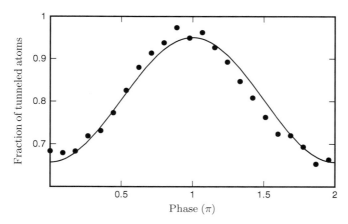

Fig. 11.4. Phase dependence of Landau-Zener tunnel rate. Shown is the ratio $N_{\rm LZ}/N$ of the number of atoms that have undergone a Landau-Zener transition $N_{\rm LZ}$ and the total number of atoms in the condensate N as a function of the phase ϕ between the two lattice harmonics of the potential (11.1). The experimental data (dots) is fitted to a sinusoidal curve (solid line)

11.5 Quantum Ratchets

A mechanical ratchet is a device used to restrict arbitrary motion into one direction. One familiar macroscopic application is the self-winding wristwatch, where the clockwork is rewound by an eccentric weight that rotates to the movement of the wearer's body [21]. Converting microscopic fluctuations, such as Brownian motion, into useful work was suggested by Smoluchowski [22] and later refined in Feynman's Lectures on Physics [23]. Experimental work on ratchet systems is nicely reviewed in [21], see also [24] for recent work on atomic systems. The interest in cold atoms for ratchet systems lies in the here possible realization of a Hamiltonian quantum ratchet system. A quantum mechanical ratchet for cold atoms can be thought of by building a sort of quantum barrier like the asymmetric optical potential described above, see Fig. 11.2a. An nondirected motion can be produced by rocking the ratchet potential forth and back, or alternatively flashing the ratchet on and off. The atoms spread over the whole lattice follow the dragging, but because of the asymmetry of the lattice the probability for tunneling into one direction is preferred. Consequently, the atoms start to move into the preferred direction of tunneling leading to a measurable current of atoms.

By such a periodic rocking no energy is added to the system since the energy contribution vanishes in the time average. The one-dimensional Hamiltonian of such a system is of the form

$$H(z,p,t) = \frac{p^2}{2\,M} + V(z)E(t) \;, \tag{11.3}$$

where p is the atom's momentum, M the atomic mass, and $E(t+T) = E(t)$ is a periodic driving force. A necessary condition to observe the quantum ratchet effect is to break all relevant temporal and spatial symmetries of the system [25]. In a theoretical work concerning the Hamiltonian ratchet (11.3) with the lattice potential (11.1) and an ac-driving force $E(t) = E_1 \cos(2\,\omega_{\mathrm{ac}} t) + E_2 \cos(4\,\omega_{\mathrm{ac}} t + \theta)$, where ω_{ac} is the driving frequency, it was shown that directed transport can here be achieved if $\phi \neq 0, \pm\pi/2$ and $\theta \neq 0, \pm\pi/2$ [26].

Acknowledgement

We acknowledge financial support from the Deutsche Forschungsgemeinschaft.

References

1. F. Bloch, Z. Phys. **52**, 555 (1928)
2. N.W. Ashcroft, N.D. Mermin, *Solid State Physics*, Saunders College Publishing, New York (1976)
3. M.H. Anderson, J.R. Ensher, M.R. Matthews, C.E. Wiemann, and E.A. Cornell, Science **269**, 198 (1995)
4. K.B. Davis, M.-O. Mewes, M.R. Andrews, N.J. van Druten, D.S. Durfee, D.M. Kurn, and W. Ketterle, Phys. Rev. Lett. **75**, 1687 (1995)
5. M.B. Dahan, E. Peik, J. Reichel, Y. Castin, and C. Salomon, Phys. Rev. Lett. **76**, 4508 (1996)
6. M. Cristiani, O. Morsch, J.H. Müller, D. Ciampini, and E. Arimondo, Phys. Rev. A **65**, 63612 (2002)
7. C. Orzel, A.K. Tuchman, M.L. Fenselau, M. Yasuda, and M.A. Kasevich, Science **23**, 291, 2386 (2001)
8. M. Greiner, O. Mandel, T. Esslinger, T.W. Hänsch, and I. Bloch, Nature **415**, 39 (2002)
9. M. Greiner, I. Bloch, O. Mandel, T.W. Hänsch, and T. Esslinger, Phys. Rev. Lett. **87**, 160405 (2001)
10. R. Grimm, J. Söding, and Y.B. Ovchinikov, JETP Lett. **61**, 367 (1995)
11. A. Görlitz, T. Kinoshita, T.W. Hänsch, and A. Hemmerich, Phys. Rev. A **64**, 11401 (2001)
12. C. Mennerat-Robilliard, D. Lucas, S. Guibal, J. Tabosa, C. Jurczak, J.-Y. Courtois, and G. Grynberg, Phys. Rev. Lett. **82**, 851 (1999)
13. P.R. Berman, B. Dubetsky, and J.L. Cohen, Phys. Rev. A **58**, 4801 (1998)
14. M. Weitz, G. Cennini, G. Ritt, and C. Geckeler, Phys. Rev. A **70**, 43414 (2004)
15. G. Ritt, C. Geckeler, T. Salger, G. Cennini, and M. Weitz, Phys. Rev. A **74**, 63622 (2006)
16. G. Cennini, G. Ritt, C. Geckeler, and M. Weitz, Phys. Rev. Lett. **91**, 240408 (2003)
17. C. Zener, Proc. R. Soc. A **137**, 696 (1932)
18. Y. Gefen, E. Ben-Jacob, and A.O. Caldeira, Phys. Rev. B **36**, 2770 (1987)
19. Q. Niu, X.-G. Zhao, G.A. Georgakis, and M.G. Raizen, Phys. Rev. Lett. **76**, 4504 (1996)

20. T. Salger, C. Geckeler, S. Kling, and M. Weitz, Phys. Rev. Lett. **99**, 190405 (2007)
21. P. Reimann, Phys. Rep. **361**, 57 (2002)
22. M. von Smoluchowski, Physik. Zeitschr. **13**, 1069 (1912)
23. R.P. Feynman, R.B. Leighton, and M. Sands, *The Feynman Lectures on Physics*, vol. 1, Addison-Wesley, Reading, (1963)
24. R. Gommers, M. Brown, and F. Renzoni, Phys. Rev. A **75**, 53406 (2007)
25. S. Flach, O. Yevtushenko, and Y. Zolotaryuk, Phys. Rev. Lett. **84**, 2358 (2000)
26. S. Denisov, L. Morales-Molina, S. Flach, and P. Hänggi, Phys. Rev. A **75**, 63424 (2007)

12

Symmetry and Transport in a Rocking Ratchet for Cold Atoms

Ferruccio Renzoni

Department of Physics and Astronomy, University College London, Gower Street, London WC1E 6BT, United Kingdom
f.renzoni@ucl.ac.uk

12.1 Introduction

The ratchet effect [1–3] is the rectification of fluctuations in a periodic potential in the absence of net applied bias forces. In this way directed motion along a macroscopically flat structure is obtained.

The archetypal of a ratchet consists of Brownian particles in a periodic potential. In order to obtain directed motion, two main requirements have to be fulfilled. First, the system has to be driven out of equilibrium, so to overcome the limitations imposed by the second principle of thermodynamics. Second, the relevant symmetries of the system, which would otherwise prevent the generation of a current, have to be broken. Among the different possible implementations of the ratchet effect, we mention here the flashing ratchet [1–3], the rocking ratchet [1–3], and the more recently introduced gating ratchet [4–6].

In the rocking ratchet set-up, Brownian particles in a periodic potential experience an additional time-dependent applied force $F(t)$, which is homogeneous and has zero time-average. The oscillating applied force plays a double role. On one hand, it drives the system out of thermodynamic equilibrium, thus avoiding the restrictions imposed by the second principle of thermodynamics. On the other hand, the temporal symmetry properties of the applied force, together with the spatial symmetry properties of the periodic potential, control the directed motion.

In this work, we review a series of recent experiments with a rocking ratchet for cold atoms in which the relationship between symmetry and transport is investigated. This is quite an unusual system to model a phenomenon of statistical physics, as there is no real thermal bath. The atoms are isolated from the environment, and the laser fields determine both the conservative periodic potential and the applied force, and the dissipative features (damping and fluctuations). The excellent tunability of such a system allows one to precisely investigate the correspondence between symmetry and transport.

This paper is organized as follows. In Section 12.2 we review the theoretical work on the relationship between symmetry and transport in a rocking ratchet [7–10]. Both cases of periodic and quasiperiodic driving are considered. In Section 12.3 we introduce the main features of a near-resonant optical lattice of the type used in the experiments on the rocking ratchet. Section 12.4 reviews the recent experimental work on a cold atom ratchet, which investigates from an experimental point of view the correspondence between symmetry and transport as studied theoretically in the work reviewed in Section 12.2. Finally, Section 12.5 summarizes the work done so far and examines the prospects for future work.

12.2 Symmetries of a Rocking Ratchet

In this Section the symmetries which prevent the generation of a current in a rocking ratchet are identified. We follow closely the treatment of Refs. [7–10].

12.2.1 The Dissipationless Case

We first consider the dissipationless case, which will then be extended to include weak dissipation. In the absence of dissipation, the equation of motion for a particle of mass m is

$$m\ddot{x}(t) = -U'(x) + F(t). \tag{12.1}$$

Here U is a spatially periodic potential of period λ, and F a zero-mean ac driving force of period T.

Now, we are interested in determining the symmetries which forbid directed motion. These correspond to the transformations in x, t which change the sign of the momentum p, i.e., the transformations which map a trajectory $\{x(t; x_0, p_0), p(t; x_0, p_0)\}$, with x_0, p_0 the initial position and momentum, into another one with opposite momentum. These transformations consist of reflections and shifts in time and space, and are

$$\hat{S}_\mathrm{a} \begin{pmatrix} x(t; x_0, p_0) \\ p(t; x_0, p_0) \end{pmatrix} = \begin{pmatrix} -x(t + T/2; x_0, p_0) + 2\chi \\ -p(t + T/2; x_0, p_0) \end{pmatrix}, \tag{12.2}$$

$$\hat{S}_\mathrm{b} \begin{pmatrix} x(t; x_0, p_0) \\ p(t; x_0, p_0) \end{pmatrix} = \begin{pmatrix} x(-t + 2\tau; x_0, p_0) \\ -p(-t + 2\tau; x_0, p_0) \end{pmatrix}, \tag{12.3}$$

with χ and τ constants. If the equation of motion Eq. (12.1) is invariant under $\hat{S}_\mathrm{a}$, $\hat{S}_\mathrm{b}$ directed motion is forbidden. Whether $\hat{S}_\mathrm{a}$, $\hat{S}_\mathrm{b}$ are symmetries of the system depends on the specific form of $U(x)$ and $F(t)$. In the following we consider only the case of a spatially symmetric periodic potential $U(x + \chi) = U(-x + \chi)$, where χ is a constant. This is the case relevant to the experimental realizations reviewed in this work, with the symmetry of the system controlled by the ac driving. Following the notations of Ref. [8], we say that $F(t)$ possesses $\hat{F}_s$ symmetry if $F(t)$ is invariant under time reversal, after some appropriate shift

$$F(t+\tau) = F(-t+\tau). \tag{12.4}$$

Moreover, if $F(t)$ satisfies

$$F(t) = -F(t+T/2), \tag{12.5}$$

we say that F possesses shift-symmetry (F_{sh}).

If the driving is shift-symmetric then the system is invariant under the transformation $\hat{S}_a$, and current generation is forbidden. If the the driving is symmetric under time reversal, then the system is invariant under the transformation $\hat{S}_b$, and once again directed motion is forbidden.

We now carry further the symmetry analysis for a specific form of driving. We consider the case of a bi-harmonic driving force

$$F(t) = A\cos(\omega t) + B\cos(2\omega t + \phi). \tag{12.6}$$

For $A, B \neq 0$ the presence of both an even and an odd harmonic breaks the shift symmetry F_{sh}, independently of the relative value of the phase ϕ. On the other hand, whether the F_s symmetry is broken depends on value of the phase ϕ: for $\phi = n\pi$, with n integer, the symmetry F_s is preserved, while for $\phi \neq n\pi$ it is broken. Therefore for $\phi = n\pi$ current generation is forbidden, while for $\phi \neq n\pi$ it is allowed. Perturbative calculations [7] show that the average current of particles is, in leading order, proportional to $\sin\phi$, in agreement with the above symmetry considerations.

12.2.2 Weak dissipation

We now consider the case of weak, nonzero dissipation [8]. For the sake of simplicity, we restrict our analysis to the case of a bi-harmonic driving of the form of Eq. (12.6). As already mentioned the shift-symmetry is broken as the driving consists both of even and odd harmonics. Consider now the symmetry under time-reversal. For $\phi = n\pi$, with n integer, the driving has F_s symmetry. However, the system is not symmetric under the transformation $\hat{S}_b$ because of dissipation. Therefore the generation of a current is not prevented, despite the symmetry of the driving. It was shown [8] that the generated current I still shows an approximately sinusoidal dependence on the phase ϕ, but acquires a phase lag ϕ_0: $I \sim \sin(\phi - \phi_0)$. Such a phase lag corresponds to the dissipation-induced symmetry breaking.

12.2.3 Quasiperiodic Driving

We now consider the case of quasiperiodic driving [10]. We consider a generic driving with two frequencies ω_1, ω_2. Two different specific forms of driving were examined in the experimental realizations, each probing a different symmetry. Quasiperiodic driving corresponds to an irrational value of the ratio ω_2/ω_1. In order to analyze the relationship between symmetry and transport in the the case of a quasiperiodic driving, the two phases

$$\Psi_1 = \omega_1 t \tag{12.7}$$
$$\Psi_2 = \omega_2 t \tag{12.8}$$

can be treated as *independent* variables [11]. The symmetries valid in the case of a perioding driving can then be generalized to the case of a quasiperiodic ac force [10].

The driving force $F(t)$ is said to be shift-symmetric if it changes sign under one of these transformations

$$\Psi_\alpha \to \Psi_\alpha + \pi, \tag{12.9}$$

where α is any subset of $\{1,2\}$, i.e., the π shift is applied to either any of the two variables, of to both of them. If F is shift-symmetric, then the system is invariant under the transformation

$$\tilde{S}_a : \quad x \to -x, \quad \Psi_\alpha \to \Psi_\alpha + \pi \tag{12.10}$$

and directed motion is forbidden.

The driving is said to be symmetric if

$$F(-\Psi_1 + \chi_1, -\Psi_2 + \chi_2) = F(\Psi_1, \Psi_2) \tag{12.11}$$

with χ_1, χ_2 appropriately chosen constants. If the driving is symmetric, in the dissipationless limit the system is invariant under the transformation

$$\tilde{S}_b : \quad x \to x, \quad \Psi_j \to -\Psi_j + \lambda_j, \quad j = 1, 2 \tag{12.12}$$

and directed transport is forbidden.

The two symmetries are the direct generalization of the symmetries for the periodic case, and control directed motion in the case of a quasiperiodic driving.

12.3 Dissipative Optical Lattices

In this Section we introduce the basic features of dissipative optical lattices of the type used in the experiment on the rocking ratchet.

Optical lattices are periodic potentials for atoms created by the interference of two or more laser fields [12]. In near resonant optical lattices the laser fields produce at once the periodic potential for the atoms and a cooling mechanism, named Sisyphus cooling. We will discuss here the simple case of a

12 Symmetry and Transport in a Rocking Ratchet for Cold Atoms 209

one-dimensional configuration and a $J_\text{g} = 1/2 \rightarrow J_\text{e} = 3/2$ atomic transition. This is the simplest configuration in which Sisyphus cooling takes place.

We consider two counterpropagating laser fields, with orthogonal linear polarizations. They have the same wavelength λ, and they are detuned below atomic resonance. The interference between the two laser fields results into a spatial gradient of polarization ellipticity of period $\lambda/2$. This in turn produces a periodic potential for the atom. More precisely, each atomic ground state $|g, \pm\rangle = |J_\text{g} = 1/2, M = \pm 1/2\rangle$ experiences a periodic potential $U_\pm(z)$ of the form

$$U_\pm(z) = \frac{U_0}{2}\left[-2 \pm \cos kz\right]. \quad (12.13)$$

Here U_0 is the depth of the potential wells. It scales as I_L/Δ, with I_L the total laser intensity and Δ the detuning from atomic resonance.

As the laser fields are near to resonance with the atomic transition, the interaction with the light fields also leads to stochastic transitions between atomic ground states. The rate of this transitions can be quantified by the scattering rate Γ' which scales as I_L/Δ^2. It is therefore possible to vary independently the optical lattice depth U_0 and Γ' by changing simultaneously I_L and Δ.

The stochastic transitions between ground states lead to damping and fluctations. The damping mechanism, named Sisyphus cooling, originates from the combined action of the light shifts and of optical pumping which transfers, through cycles of absorption/spontaneous emission, atoms from one ground state sublevel to the other one. It turns out that in this process the atomic kinetic energy is transformed into potential energy, which is then carried away by a spontaneously emitted photon. In this way, the atoms are slowed down until their kinetic energy is small enough to be trapped in the wells of the optical potential. We notice that the stochastic transitions between the two potentials $U_\pm(z)$ also generates fluctuations of the instantaneous force experienced by the atom. The equilibrium between these cooling and heating mechanisms determine the final kinetic energy of the atoms.

12.4 Rocking Ratchet for Cold Atoms

In a rocking ratchet, Brownian particles experience a periodic potential and an ac force $F(t)$ of zero average. Whenever the relevant symmetries of the system are broken, the particles are set into directed motion.

Optical lattices allows the implementation of rocking ratchets for cold atoms, and we will review here recent experiments [13–16]. As discussed in the previos Section, in near resonant optical lattices the laser fields produce at once a periodic potential for the atoms and a damping mechanism for the atomic motion. The only missing element for a rocking ratchet set-up is the rocking force. We now discuss how to apply an homogeneous ac force $F(t)$ to the atoms in the optical lattice.

In order to generate a time-dependent homogeneous force, one of the lattice beams is phase modulated, and we will indicate by $\alpha(t)$ the time-dependent phase. In the laboratory reference frame the phase modulation of one of the lattice beams results into the generation of a moving optical lattice $U_\pm(z - \alpha(t)/2\,k)$. Consider now the dynamics in the moving reference frame defined by $z' = z - \alpha(t)/2\,k$. In this accelerated reference frame the optical potential is stationery. In addition to the optical periodic potential the atom, of mass m, experiences also an inertial force F in the z direction proportional to the acceleration a of the moving frame

$$F = -ma = \frac{m}{2\,k}\ddot{\alpha}(t). \tag{12.14}$$

In this way, in the accelerated frame of the optical potential the atoms experience an homogeneous force which can be controlled by varying the phase $\alpha(t)$ of one of the lattice beams.

The above described rocking ratchet set-up was recently used to investigate experimentally the relationship between symmetry and transport. We treat here separately the two different cases: a biharmonic driving including two harmonics at frequencies ω and $2\,\omega$, and a multifrequency driving obtained by combining signals at three different frequencies.

12.4.1 Biharmonic Driving

We consider a biharmonic driving of the form of Eq. (12.6). As discussed above, the symmetry analysis predicts a dependence of the current on the phase ϕ of the form $I \sim \sin\phi$. Dissipation introduces a phase lag ϕ_0, and the expected current shows now the dependence $I \sim \sin(\phi - \phi_0)$.

The experimental work of Refs. [13,14] precisely invesigated the relationship between symmetry and transport in the case of biharmonic driving, and examined the effect of dissipation. In that work the depth U_0 of the optical potential, which scales as I_L/Δ, was kept constant while varying the scattering rate Γ', which scales as I_L/Δ^2. This was done by varying simultaneously I_L and Δ, keeping constant their ratio. We notice that as I_L and Δ can be varied only within a finite range, it was not possible to completely suppress dissipation, i.e., obtain $\Gamma' = 0$. However, as it will be discussed in the following, for the driving strength considered in the experiment, the smallest accessible scattering rate results into a phase shift which is zero within the experimental error; i.e., this choice of parameters well approximates the dissipationless case. By then increasing Γ' it is possible to investigate the effects of dissipation.

The experimental results of Refs. [13,14], reported in Figs. 12.1, 12.2 clearly demonstrated the relationship between in symmetry and transport, as predicted by the symmetry analysis. In fact, the measured current of atoms was found to be well approximated by $A\sin(\phi - \phi_0)$. Therefore, by fitting data as those of Fig. 12.1, with the funcion $v/v_{\rm r} = A\sin(\phi - \phi_0)$, the phase shift

12 Symmetry and Transport in a Rocking Ratchet for Cold Atoms   211

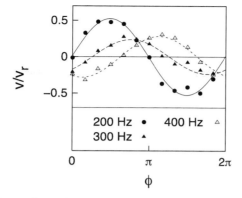

Fig. 12.1. Experimental results for the average atomic velocity, in units of the recoil velocity, as a function of the phase ϕ. The lines are the best fit of the data with the function $v/v_{\rm r} = A\sin(\phi - \phi_0)$. Different data sets corresponds to different scattering rates for a given optical potential depth. The data are labelled by the quantity $\Gamma_{\rm s} = [\omega_{\rm v}/(2\pi)]^2/\Delta$, where $\omega_{\rm v}$ is the vibrational frequency at the bottom of the wells, which is proportional to the scattering rate. Figure from Ref. [14]

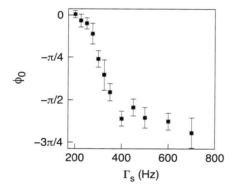

Fig. 12.2. Experimental results for the phase shift ϕ_0 as a function of $\Gamma_{\rm s}$. Figure from Ref. [14]

ϕ_0 was determined as a function of Γ', as reported in Fig. 12.2. The measured phase shift ϕ_0 is zero, within the experimental error, for the smallest scattering rate examined in the experiment. In this case, no current is generated for $\phi = n\pi$, with n integer, as for this value of the phase the system is invariant under time-reversal. The magnitude of the phase shift ϕ_0 increases for increasing scatering rate, and differs significantly from zero. The nonzero phase shift corresponds to current generation for $\phi = n\pi$, i.e., when the system Hamiltonian is invariant under time-reversal transformation. Therefore, the experimental results of Figs. 12.1, 12.2 demonstrate the breaking of the system symmetry by dissipation.

12.4.2 Multifrequency Driving

Recent experiments with multifrequency driving [15, 16] aimed to investigate the transition from periodic to quasiperiodic driving, and to examine how the symmetry analysis is modified in this transition. The multifrequency driving was obtained by combining signals at three different frequencies: ω_1, $2\omega_1$ and ω_2. For ω_2/ω_1 irrational the driving is quasiperiodic. Clearly, in a real experiment ω_2/ω_1 is always a rational number, which can be written as $\omega_2/\omega_1 = p/q$, with p, q two coprime positive integers. However, as the duration of the experiment is finite, by choosing p and q sufficiently large it is possible to obtain a driving which is effectively quasiperiodic on the time scale of the experiment.

Different forms of multifrequency driving allow one to investigate different symmetries. We consider here the driving consisting of the sum of three harmonics:

$$F(t) = A\cos(\omega_1 t) + B\cos(2\omega_1 t + \phi) + C\cos(\omega_2 t + \delta). \tag{12.15}$$

In our treatment, we will neglect the effects of dissipation as we know that it results in an additional phase shift. Consider first the case of periodic driving, with ω_2/ω_1 rational. We already know that for biharmonic driving – $C = 0$ in Eq. (12.15) – the shift symmetry is broken for any value of ϕ, while the time-reversal symmetry is preserved for $\phi = n\pi$, with n integer. A current of the form $I \sim \sin\phi$ is obtained as a result. We now include the third harmonic – $C \neq 0$ in Eq. 12.15. For $\delta = 0$ this additional driving is invariant under time reversal, and therefore the total driving is still invariant under time-reversal for $\phi = n\pi$. Instead, for $\delta \neq 0$ the symmetry under time-reversal is broken and directed transport is allowed also for $\phi = n\pi$. In other words, for $\delta \neq 0$ the third driving leads to an additional phase shift of the current as a function of ϕ. The magnitude of such a shift depends on the phase δ. Taking dissipation also into account, we can conclude that the current will show the dependence $I \sim \sin(\phi - \phi_0)$ where ϕ_0 includes the phase shift produced by dissipation and the phase shift produced by the harmonic at frequency ω_2.

We now turn to the case of a quasiperiodic driving, as obtained in the case of irrational ω_2/ω_1. As discussed in Sec. 12.2 the symmetry analysis for the periodic driving can be generalized to the quasiperiodic case by treating the phases $\Psi_1 = \omega_1 t$ and $\Psi_2 = \omega_2 t$ as independent variables. We notice that the driving considered here, Eq. 12.15, is invariant under the transformation $\Psi_2 \to -\Psi_2 + \chi_2$ for any δ, as δ can be reabsorbed in χ_2. Therefore the invariance under the transformation $\tilde{S}_b$ is entirely determined by the invariance of F under the transformation $\Psi_1 \to -\Psi_1 + \chi_1$, i.e., we recover the results for biharmonic driving: $\tilde{S}_b$ is a symmetry, and therefore directed motion is forbidden, for $\phi = n\pi$. Hence, in the quasiperiodic limit, the third harmonic at frequency ω_2 is not relevant for the symmetry of the system, which is entirely determined by the biharmonic term at frequency ω_1, $2\omega_1$.

12 Symmetry and Transport in a Rocking Ratchet for Cold Atoms

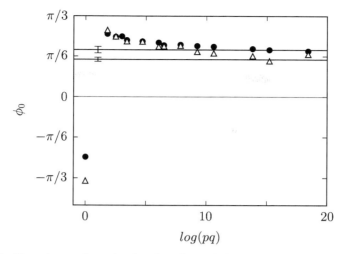

Fig. 12.3. Experimental results for the phase shift ϕ_0 as a function of pq which characterize the degree of periodicity of the driving. The two data sets correspond to different amplitudes of the driving. The two horizontal lines indicate the phase shift ϕ_0 for biharmonic drive, i.e., in the absence of the driving at frequency ω_2. Figure from Ref. [15]

In the experiment, the transition to quasiperiodicity can be investigated by studying the atomic current as a function of ϕ for $\omega_2/\omega_1 = p/q$ with p and q coprimes. By increasing p and q the driving can be made more and more quasiperiodic on the finite duration of the experiment, with the quantity pq a possible measure of the degree of quasiperiodicity. To verify the predictions of the symmetry analysis, the average atomic current was measured as a function of ϕ, for different choices of p and q. The data were fitted with the function $v = v_{\max} \sin(\phi + \phi_0)$. The resulting value for the phase shift ϕ_0 is plotted in Fig. 12.3 as a function of pq.

For small values of the product pq, i.e., for periodic driving, the harmonic at frequency ω_2 leads to a shift which strongly depends on the actual value of pq. For larger values of pq, i.e., approaching quasiperiodicity, the phase shift ϕ_0 tends to a constant value. Such a value was found to be independent of δ, and coincides with the phase shift ϕ_0 measured in the case of pure biharmonic driving (horizontal lines in Fig. 12.3), which is determined by the finite damping of the atomic motion. The experimental results of Fig. 12.3 prove that in the quasiperiodic limit the only relevant symmetries are those determined by the periodic biharmonic driving and by dissipation. For a driving of a form Eq. (12.15), quasiperiodicity therefore restores the symmetries which hold in the absence of the additional driving which produced quasiperiodicity.

12.5 Summary

Rocking ratchets for cold atoms have been experimentally demonstrated using driven optical lattices. The laser fields create both a periodic potential for the atoms, and friction and fluctuations for the atomic dynamics. An homogenous, time-dependent, driving force can be applied by modulating one of the lattice beams. These ratchets, which rely on harmonic mixing between the different components of the driving [17] allowed to experimentally investigate the relationship between symmetry and transport and validated recent theoretical analysis [7, 8, 10]. It was indeed shown that the properties of directed transport are determined by the symmetry of the system.

Further work may include the implementation of ratchets which do not rely on harmonic mixing, such as the *gating* ratchet [4–6], and the study of quantum ratchets [18], in which the transport is determined by the interplay between quantum tunneling and thermal hopping over the barriers.

References

1. R.D. Astumian and P. Hänggi, Phys. Today **55**, 33 (2002)
2. P. Reimann, Phys. Rep. **361**, 57 (2002)
3. P. Hänggi, F. Marchesoni, and F. Nori, Ann. Phys. **14**, 51 (2005)
4. S. Savel'ev, F. Marchesoni, P. Hänggi, and F. Nori, Europhys. Lett. **67**, 179 (2004)
5. S. Savel'ev, F. Marchesoni, P. Hänggi, and F. Nori, Eur. Phys. J. B **40**, 403 (2004)
6. M. Borromeo and F. Marchesoni, Chaos **15**, 026110 (2005)
7. S. Flach, O. Yevtushenko, and Y. Zolotaryuk, Phys. Rev. Lett. **84**, 2358 (2000)
8. O. Yevtushenko, S. Flach, Y. Zolotaryuk, and A.A. Ovchinnikov, Europhys. Lett. **54**, 141 (2001)
9. P. Reimann, Phys. Rev. Lett. **86**, 4992 (2001)
10. S. Flach and S. Denisov, Acta Phys. Pol. **B35**, 1437 (2004)
11. E. Neumann and A. Pikovsky, Eur. Phys. J. B **26**, 219 (2002)
12. G. Grynberg and C. Mennerat-Robillard, Phys. Rep. **255**, 335 (2001)
13. M. Schiavoni, L. Sanchez-Palencia, F. Renzoni, and G. Grynberg, Phys. Rev. Lett. **90**, 094101 (2003)
14. R. Gommers, S. Bergamini, and F. Renzoni, Phys. Rev. Lett. **95**, 073003 (2005)
15. R. Gommers, S. Denisov, and F. Renzoni, Phys. Rev. Lett. **96**, 240604 (2006)
16. R. Gommers, M. Brown, and F. Renzoni, Phys. Rev. A **75**, 053406 (2007)
17. F. Marchesoni, Phys. Lett. A **119**, 221 (1986)
18. P. Reimann, M. Grifoni, and P. Hänggi, Phys. Rev. Lett. **79**, 10 (1997)

Part IV

Metamaterials: From Linear to Nonlinear Features

13

Optical Metamaterials: Invisibility in Visible and Nonlinearities in Reverse

Natalia M. Litchinitser[1] and Vladimir M. Shalaev[2]

[1] The State University of New York at Buffalo, Department of Electrical Engineering, 309 Bonner Hall, Buffalo, New York 14260, USA
natashal@buffalo.edu
[2] School of Electrical and Computer Engineering and Birck Nanotechnology Center, Purdue University, West Lafayette, Indiana 47907, USA
shalaev@purdue.edu

13.1 Introduction

Recent experimental demonstrations of optical metamaterials opened up an entirely new branch of modern optics that can be described as "refractive index engineering" [1–20]. The refractive index of a material is the factor by which an electromagnetic wave is slowed down, compared with a vacuum, when it propagates inside the material. The material properties of conventional materials are largely controlled by the properties of their constituent components, viz., atoms and molecules. Their refractive indices can be modified to some degree by altering material chemical composition, using thermal or electrical tuning, or through nonlinear optical effects. Nevertheless, a majority of existing materials possesses positive, and typically greater than one, index of refraction. In contrast, metamaterials provide almost unlimited opportunities for designing the refractive index through a careful engineering of their constituent components, or meta-atoms. Several examples of engineered optical structures, including magnetic metamaterial and negative index metamaterials (NIMs), are shown in Fig. 13.1. Moreover, metamaterial properties can be tuned [21, 22] and even controlled on a level of a single meta-atom [23]. Basic properties of optical metamaterials will be reviewed in Section 13.1.

Additional design flexibility provided by metamaterials (discussed in Section 13.2) gives rise to new linear and nonlinear optical properties, functionalities, and applications unattainable with conventional materials. In this chapter, we discuss two examples of refractive index engineering in metamaterials that results in truly fascinating phenomena.

One unique potential application, enabled by metamaterials is the possibility of designing a cloak of invisibility. For a long time, the concept of cloaking was primarily associated with myths and science fiction stories. However,

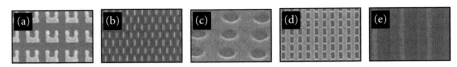

Fig. 13.1. Optical metamaterials. (**a**) Arrays of gold split rings enabling magnetic resonances at visible and telecommunication frequencies, and negative magnetic permeability near 1.5 μm [7], (**b**) paired gold nanorod array providing negative refractive index at 1.5 μm [9], (**c**) array of holes in paired gold films separated by a dielectric (Al_2O_3) layer producing negative refractive index around 2 μm [10], (**d**) the fish-net structure providing negative refractive index at 1.4 μm and the best performance in terms of loss minimization [12], (**e**) coupled gold nanostrips based magnetic metamaterial enabling magnetic response across the entire visible range [20]

the situation radically changed with the emergence of metamaterials. The first metamaterial-based cloak operating in microwave frequency range has recently been proposed [24–26] and demonstrated experimentally [27]. In addition, cloaking devices at very low (near-zero) frequencies [28] and an acoustic cloak [29] were proposed theoretically.

Many civil and military applications would benefit from having cloaks of invisibility operating at optical frequencies; therefore, significant efforts are being devoted to the development of an optical cloak. Recently, a first design and a practical recipe for the realization of a cloaking device operating in an optical spectral range have been proposed [30, 31]. In Section 13.3, we discuss basic theory and proposed metamaterial designs, as well as numerical and experimental results demonstrating cloaking at microwave and optical frequencies.

Another remarkable class of metamaterials includes negative index metamaterials or left-handed materials (LHMs). Although unique properties of NIMs have been predicted theoretically by Veselago about forty years ago [32], their practical realization has become possible only with the appearance of metamaterials [1–20,33–38]. The state-of-the-art experimental results and unusual linear optical properties of metamaterials, including negative refraction [32,34,35] and super-resolution [39–49], have been summarized in recent review articles [1–5].

Recently, it became obvious that besides having very unusual linear properties, NIMs may trigger fundamentally new manifestations of many well-known nonlinear optical phenomena [50–79]. In section 13.4, we discuss new regimes of second-harmonic generation (SHG) and optical parametric amplification (OPA), optical bistability, and novel kinds of solitons predicted in nonlinear NIMs structures.

13.2 Optical Metamaterials: New Degrees of Freedom

The original driving force in the optical metamaterial research was the realization of magnetic and negative index materials in the optical range. However, it was recently realized that the same basic design principles may be applied for enabling a wide range of other unusual phenomena and functionalities.

While the refractive index is one of the basic characteristics of light propagation in continuous media, it does not directly enter into the Maxwell's equations. However, it is closely related to two fundamental physical parameters, specifically, relative dielectric permittivity ϵ and relative magnetic permeability μ that describe material properties through $n = \pm\sqrt{\epsilon\mu}$. Despite an apparently large variety of naturally occurring materials, their properties utilize a rather limited part of the entire (ϵ, μ)-parameter space. Indeed, a majority of naturally existing transparent optical materials possess $\epsilon > 1$ and $\mu \approx 1$ and, therefore, $n > 1$.

Metamaterials, being artificially created micro- or nano-structures, allow a significant expansion of attainable values of material parameters and refractive indices, including ultra-low refractive index materials [80], magnetic metamaterials ($\mu \approx 1$) [6–8, 15, 19, 20], structures with gradually changing refractive index, enabling cloaking and wave concentrator applications [24–31, 81, 82], and negative index materials [9–14, 16–19, 32–38]. Many of these unusual material properties can be achieved near ϵ or μ resonances that can be carefully engineered by adjusting the dimensions, spatial arrangement, and other properties of their constituent components (meta-atoms).

An example of metamaterial structure with resonant electric and magnetic responses [20] is shown in Fig. 13.2. The structure consists of coupled gold nanostrips with varying dimensions. A cross-section of coupled strips pairs is

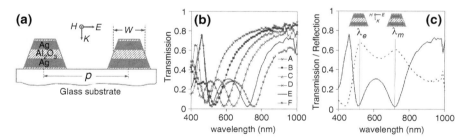

Fig. 13.2. Coupled gold nanostrips based magnetic metamaterial enabling resonant material response across the entire visible range [20]. (a) A cross-section of coupled strips pairs, (b) the transmission spectra for the TM polarization measured in six double-strip metamaterial samples with varying average width parameter w, (c) the transmission (solid line) and reflection (dashed line) spectra with two characteristic wavelengths corresponding to electric and magnetic resonances marked. The inset shows the schematics of the current modes at electric and magnetic resonances, respectively

shown in Fig. 13.2a. The corresponding FE-SEM image of an experimental sample is shown in Fig. 13.1e. Such structures exhibit both magnetic and electric resonances under TM illumination with the magnetic field polarized along the strips, while there are no resonant effects for the TE polarization with the electric field aligned with the strips. Figure 13.2b shows the transmission spectra for the TM polarization measured in six double-strip metamaterial samples with varying average width parameter w (and consequently, varying period of the structure p). These spectra show strong resonant features that shift in wavelength, depending on the strip width, and extend over the entire visible spectrum. In particular, the electric resonance near λ_e, shown in Fig. 13.2c for one of the samples (E), originates from a symmetric current mode, while the magnetic resonance around λ_m results from a circular current, formed by the anti-symmetric current flows in the upper and lower strips.

A possibility of the realization of magnetic resonances at optical frequencies was also shown to have an important impact on nonlinear optical properties of metamaterials [51,59,61]. Several orders of magnitude enhancements of the second- and third-harmonic generation from metamaterial thin film composed of gold split-ring resonators excited at 1.5 µm were experimentally observed when magnetic-dipole resonances were excited, as compared to the case of just electric-dipole resonances [61]. Although optical metamaterials are currently only available in a form of sub-wavelength thin films, these results clearly indicate a pronounced effect of the metamaterial nanostructure on their basic nonlinear properties. While no comprehensive microscopic theory describing the nonlinear interaction of light with metamaterial structures is available so far, in Section 13.4, we discuss several unusual regimes of nonlinear wave interactions in bulk metamaterials assuming effective nonlinear coefficients $\chi^{(2)}$ or $\chi^{(3)}$ as well as touch upon the first steps toward understanding the effects of a material's nanostructure.

Numerous advantages of using engineered resonant structures often come at a price of limited useful bandwidth for a particular application, increased absorption (losses), and non-negligible reflection owing to impedance mismatch [1,3,12]. Finding solutions to these problems, acceptable trade-offs, and optimized or alternative designs are some of the directions of current research in the field of optical metamaterials.

13.3 A Route to Invisibility

Owing to unique and controllable properties of metamaterials, making objects invisible to the naked eye or the radar is likely to become a widespread reality. Two general metamaterial-based approaches rely on either cancellation [83–90] of the incoming radiation or redirecting waves around the object [24–31].

One form of cancellation-based cloaking, theoretically proposed by Alu and Engheta [83–86], utilizes a plasmonic or metamaterial spherical shell with low-positive or negative permittivity. If the dielectric permittivity ϵ of the cover material is less than that of the background material ϵ_0, the local electric polarization vector $P = (\epsilon - \epsilon_0)E$ changes its sign, resulting in the overall dipole moment cancellation. This is a simple and intuitive technique, not requiring complex metamaterial designs; however, it can only be used for the cloaking of sub-wavelength-size objects and also is "object-specific", that is, the cloak has to be tailored for a particular object.

Also, somewhat similar idea was discussed long before the emergence of metamaterials in a context of coated compound ellipsoids composed of an inner ellipsoidal core and an outer confocal ellipsoidal shell with different dielectric constants [87]. Under a plane wave illumination and for certain combinations of dielectric permittivities, the scattering vanishes, thus making the ellipsoids invisible.

Another "cancellation-based" approach was proposed by Milton and Nicorovici [88–90], who considered a polarizable line dipole placed in the vicinity of a coated cylinder with a core dielectric constant $\epsilon_c = \epsilon_0 = -\epsilon_s$ where ϵ_s is the permittivity of the coating material. Under the action of an external quasistatic transverse magnetic field, both the coated cylinder and the polarizable line dipole become invisible. This effect can be explained by the fact that the resonant field generated by a polarizable line dipole acts back on the line dipole and effectively cancels the fields from outside sources. This approach was also extended to the plane-parallel slab of metamaterial of thickness d, known as a superlens. When a polarizable line dipole is located less than a distance $d/2$ from the lens, it is cloaked, owing to the presence of a resonant field in front of the lens. Nevertheless, the main limitation of this approach is that cloaking is limited to sub-wavelength size objects.

Before proceeding to a fundamentally different cloaking approach, it may be important to formulate the requirements for an ideal cloaking device. First, such a device should be object-independent, and macroscopic, that is, suitable for concealing large objects. Second, it should not reflect, scatter, or absorb any light, introduce any phase shifts, or produce a shadow. Third, it should operate for non-polarized light and in a wide range of frequencies simultaneously.

The redirection-based approach appears to satisfy many, although not to all the above requirements and, therefore, appears to be a more general technique in comparison to the previous two. The main goal is to design a cover (i.e., a cloak) that would guide electromagnetic waves around the object. This approach has been proposed independently by Leonhardt [25, 26] within the geometrical optics approximation and by Pendry et al. [24] within the frames of full Maxwell's equations. It is based on the form invariance of Maxwell's equations that allows us to map the coordinate transformation to a set of material parameters ϵ and μ. Basic design equations for the redirection approach will be summarized in section 13.3.1. Metamaterial technology allows precise control over material parameters, and therefore is essential for such mapping. Recently, first designs for microwave and optical cloaks based on this technique were reported as discussed in section 13.3.2. Moreover, cloaking at microwave frequencies was demonstrated experimentally [27].

13.3.1 Transformation Approach

In order to redirect waves around the object, either the space around the object should be deformed, assuming that material properties stay the same, or the material properties should be modified around the object. The former approach is referred to as topological interpretation, while the latter is called a material interpretation. Thanks to the form invariance of Maxwell's equations, these two interpretations are equivalent. More specifically, under a coordinate transformation, the form of Maxwell equations should remain invariant while new ϵ and μ would contain the information regarding the coordinate transformation and the original material parameters. This important property of Maxwell's equations that was first studied and utilized in a totally different context of computational studies of complex systems involving several length scales [91], forms the basis of the transformation method [30,31,81,82,92–94]. The main design tools of this approach are given by

$$\epsilon^{i'j'} = \left|\det\left(\Lambda_i^{i'}\right)^{-1}\right| \Lambda_i^{i'} \Lambda_j^{j'} \epsilon,$$

$$\mu^{i'j'} = \left|\det\left(\Lambda_i^{i'}\right)^{-1}\right| \Lambda_i^{i'} \Lambda_j^{j'} \mu \qquad i,j = 1,2,3, \qquad (13.1)$$

where it was assumed that the original space is isotopic and transformations are time invariant, and $\Lambda_\alpha^{\alpha'} = \partial x^{\alpha'}/\partial x^\alpha$ are the elements of the Jacobian transformation matrix.

The general design strategy using the transformation approach includes two main steps. In the first step, a coordinate transformation of the space with desired property is built. In the next step, a set of material properties that would realize this property of the transformed space in the original space using Eqs. (13.1) is calculated. Certainly, an important question would be whether a particular set of material parameters predicted by Eqs. (13.1) is physically realizable. In many cases, metamaterials bring about a positive answer.

Finally, the transformation approach is rather general and is not limited to cloaking applications. Figure 13.3 illustrates several examples, including (a) square cloak, (b) wave concentrator, and (c) wave rotator, all designed using this technique.

13.3.2 Cloaking Device: From Microwaves to Optics

The field of metamaterial-based cloaking is very new and rapidly developing. The first experimental demonstration of a cloaking device at microwave

13 Optical Metamaterials: Invisibility and Nonlinearities

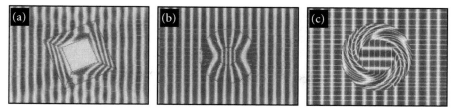

Fig. 13.3. The examples of the structures that can be built using the transformation approach. (**a**) Square cloak [81], (**b**) Wave concentrator [81], (**c**) Wave rotator [82]

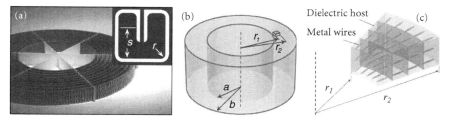

Fig. 13.4. (**a**) The cylindrical cloak used in the microwave experiments [27], the inset shows the SRR with two adjustable parameters, (**b**) a schematic of the optical cloak, (**c**) a fraction of the cylindrical cloak built with metal wires embedded in a dielectric material [30]

frequencies has been reported only a few months after the original theoretical design was published [24–27]. In this pioneering experiment, the object, a copper cylinder, was concealed by the cylindrical cloak built using split-ring resonators (SRR) positioned with their axes along the radial direction as shown in Fig. 13.4a.

The coordinate transformation that compresses the cylindrical region $0 < r < b$ in space into the shell $a < r' < b$ is given by [27]

$$r' = \frac{b-a}{b} r + a, \quad \theta' = \theta, \quad z' = z. \tag{13.2}$$

The design equations (13.1) give the following material parameters [27,94]

$$\epsilon_r = \mu_r = \frac{r-a}{r}, \quad \epsilon_\theta = \mu_\theta = \frac{r}{r-a}, \quad \epsilon_z = \mu_z = \left(\frac{b}{b-a}\right)^2 \frac{r-a}{r}, \tag{13.3}$$

where primes were omitted to emphasize that these parameters are transformed material parameters in an original (untransformed) coordinate space.

First experiments were designed and performed for the TE polarization with the electric field polarized along the cylinder axis. In this case the original parameters given by Eqs. (13.3) can be significantly simplified. One such possibility is a reduced parameter set that was used in the first experiments, given by [27,94]

$$\epsilon_z = \left(\frac{b}{b-a}\right)^2, \quad \mu_r = \left(\frac{r-a}{r}\right)^2, \quad \mu_\theta = 1. \tag{13.4}$$

This set of parameters contains only one spatially inhomogeneous component μ_r and also eliminates infinite values of material parameters components that follow from Eqs. (13.3). However, while the ideal cloaking parameters described by Eqs. (13.3) assure power-flow bending and retaining the phase front and guarantee no reflection at interface with free space, a set of reduced design parameters given by Eqs. (13.4) is not reflectionless. Indeed, for normal incidence, the impedance of the simplified parameter metamaterial is

$$Z = \sqrt{\frac{\mu_\theta}{\epsilon_z}} = \frac{b-a}{b}. \tag{13.5}$$

Then, the power reflection at normal incidence can be estimated as $R = [a/(2b-a)]^2$ and would be negligible only when $b \gg a$.

The material parameters prescribed by the design equations (13.4) were realized in the experiment using the SRR with two main adjustable parameters: the length of the split s and the radius r of the corners shown in the inset in Fig. 13.4a. These parameters can be used to tune the magnetic and electric resonance, respectively. Although the invisibility demonstrated in these experiments was not perfect, the principal feasibility of the electromagnetic cloaking mechanism on a basis of metamaterial design was clearly demonstrated.

The next important step was to examine the possibility of cloaking in the optical range. Very recently, the first theoretical design of a non-magnetic cloaking device operating at optical frequencies was proposed by the Purdue team [30]. While magnetism at optical frequencies has previously been demonstrated, it is still considered to be a challenging task that can only be realized in resonant structures with relatively high loss. However, it was realized that an optical cloak for the TM polarization can be built without any magnetism. In this case, Eqs. (13.3) are replaced with the following set of reduced parameters [30]

$$\mu_z = 1, \quad \epsilon_\theta = \left(\frac{b}{b-a}\right)^2, \quad \epsilon_r = \left(\frac{b}{b-a}\right)^2 \left(\frac{r-a}{r}\right)^2. \tag{13.6}$$

Constant, greater than one azimuthal dielectric permittivity component can easily be achieved in conventional dielectrics. A crucial part of the design is the realization of required radial distribution of dielectric permittivity varying from zero to one. This can be achieved using subwavelength metal wires aligned along the radii of the annular cloak shell as shown in Figs. 13.4b and 13.4c. A detailed effective-medium theory-based recipe for determining the physical parameters of the cloak, including the shape factor of the device $R_{ab} = a/b$ and metal filling fraction distribution along the radial direction, can be found in Ref. [30]. Other potential implementations of design parameters (13.6) include chains of metal nanoparticles and thin continuous and semi-continuous strips.

13 Optical Metamaterials: Invisibility and Nonlinearities   225

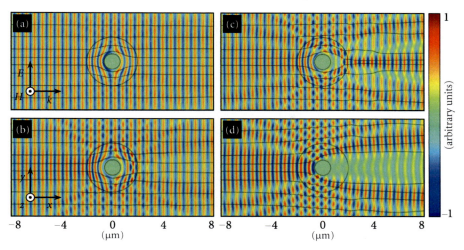

Fig. 13.5. Numerical simulations of the magnetic field mapping around the cloaked cylinder [30]. (**a**) An ideal cloak with parameters determined by Eqs. (13.3), (**b**) the non-magnetic cloak with parameters given by Eqs. (13.6), (**c**) the designed metal wire composite cloak, (**d**) wave propagation without the cloak

Figures 13.5a and 13.5b show the results of full-wave numerical simulation illustrating the performances of an ideal cloaking device with material parameters given by Eqs. (13.3) and that of the cloak, based on reduced parameters of Eq. (13.6). The latter one shows small but non-negligible amount of scattering owing to the impedance mismatch at the outer boundary of the cloak designed using Eqs. (13.6). These results should be compared to the performance of the non-magnetic cloaking device composed of prolate spheroidal silver nanoparticles embedded in a silica tube, as shown in Fig. 13.5c. While the fields are slightly perturbed in this case, the overall performance closely resembles that of the previous two cases. Finally, Fig. 13.5d illustrates the no cloak case, clearly showing the distortions of the fields around the object (metallic cylinder). As expected, a significant shadow is formed behind the object.

We note that both originally proposed microwave and optical cloaking devices perfectly satisfy the first criterion for the ideal cloak formulated in the beginning of Section 13.3, that is, the cloak is object-independent and does not impose any limitations of the size of the object. The second criterion, related to reflection, scattering, absorption, phase shifts, and shadows, is not fully satisfied due to the impedance mismatch related to reduced design parameters (13.6) and small but non-negligible material absorption.

Additionally, these first cloaks were specifically designed for either the TE or the TM polarizations, allowing significant simplifications of the design conditions (13.3). Finally, the last criterion is the most challenging; how to make a cloak operation broadband remains an open question.

The cloaking devices, in their current implementation are inherently narrowband. Indeed, since the refractive index of the cloaking shell varies from zero to one, a phase velocity of light inside the shell is greater than the velocity of light in a vacuum. While this condition itself does not contradict any law of physics, it implies that the material parameters must be frequency-dependent or dispersive. However, even a narrowband cloak, once realized experimentally, may have a number of practical applications, including night-imaging systems or hiding objects from light designators. Nevertheless, designing a broadband cloaking device is of interest from fundamental as well as practical standpoints.

Another feature of the cloaking device that is not often mentioned, but may be of rather general character, is that the invisibility from outside is accompanied by the darkness inside the cloak and inability to see from within the cloak. This property may or may not be desirable in practical applications.

It is noteworthy that first designs of the cloaking devices assumed inactive objects that do not generate electromagnetic waves. However, cloaking of active devices, including those emitting or absorbing radiation are certainly of practical interest and was recently addressed in a theoretical paper by Greenleaf et al [95].

Finally, until recently all cloaking devices designed using the transformation method relied on a linear transformation such as Eqs. (13.2) and the sets of reduced parameters (13.4) and (13.6). Reduced parameters are easier to implement than exact material parameters described by Eqs. (13.3), but a shortcoming of using these parameters is that the cloak is not reflectionless. However, recently it was pointed out [31] that at least from mathematical viewpoint, there are numerous ways of compressing a cylindrical or spherical region $0 < r < b$ into an annular region $a < r' < b$ and the transformation function does not have to be linear. As a result, a novel approach utilizing a high-order coordinate transformation that eliminate undesired reflections at the outer boundary of the non-magnetic optical cloak was proposed [31]. Figure 13.6 compares the results of full-wave field mapping simulations for a

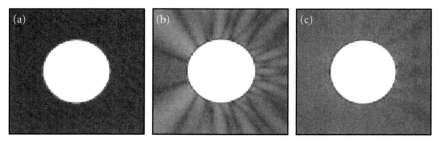

Fig. 13.6. Full-wave simulations of the performance of (**a**) an ideal linear cloak, (**b**) the linear non-magnetic cloak with $p = 0$, and (**c**) the quadratic cloak with $p = a/b^2$ [31]

metal cylinder inside (a) an ideal cloak, (b) the non-magnetic cloak designed using a linear transformation, and (c) the non-magnetic cloak built using a quadratic transformation given by

$$r' = f(r) = \left[\frac{b-a}{b} + p(r-b)\right] r + a, \quad (13.7)$$

where p is a flexible parameter that can be chosen to facilitate $Z|_{r'=b} = 1$. In the example shown in Fig. 13.6c $p = a/b^2$. A linear transformation used in the original design follows from Eq. (13.7) if $p = 0$. Figure 13.6c confirms that the cloaking device designed using the quadratic transformation results in negligible reflectance.

13.4 Nonlinear Optics with Backward Waves in Negative Index Materials

An important class of metamaterials that originally inspired the development of the entire field of metamaterials research is negative index metamaterials. Recently, there has been a significant interest in nonlinear optical processes in NIMs. First of all, why would the nonlinear optics be different in NIMs as compared to conventional materials? To answer this question, we will first outline some of the remarkable linear properties of NIMs.

As their name suggests, these materials have a negative index of refraction. As a result, light refracts "negatively" as illustrated in Fig. 13.7b in contrast to conventional, or "positive" refraction as shown in Fig. 13.7a [4]. Figure 13.7c shows the experimental results demonstrating negative refraction at microwave frequencies [35]. Another fundamental characteristic of NIMs is their inherent frequency dependence; that is, both ϵ and μ are functions of frequency. In fact, the refractive index is negative only in a limited range of frequencies – a direct result of their resonant nature. Consequently, the same

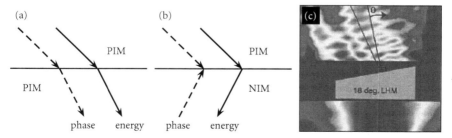

Fig. 13.7. (a) Conventional (or positive) refraction, (b) negative refraction, (c) results of a microwave scattering experiment demonstrating negative refraction in left-handed metamaterial prism [35]

material may act as a NIM in one range of frequencies, and as a PIM at other frequencies.

Moreover, Fig. 13.7b shows one of the most important properties of NIMs – opposite directionality of the phase velocity (or wave vector) and the energy velocity (or Poynting vector). Waves in NIMs with the phase velocity pointing toward the interface are often referred to as backward waves. It should be mentioned that backward waves were discussed long before Veselago's work and metamaterial demonstration, in contexts ranging from mechanical to optical systems [96–101].

It turned out that backward waves in combination with the strong frequency dependence of the material parameters play an especially important role in nonlinear optics of NIMs. In particular, they facilitate a fundamentally new regime of phase matching: backward phase matching. Since many phase-sensitive nonlinear processes, including second-harmonic generation, sum- and difference-frequency generation, parametric amplification and oscillation, as well as four-wave mixing, strongly rely on phase matching, their manifestation in NIMs is expected to differ as compared to PIMs. Fundamentally new regimes of SHG [52–62], three- and four-wave mixing [57,58,63–65] enabled by unusual phase-matching conditions in NIMs have been proposed. Realization of SHG in NIMs was predicted to enable nonlinear-optical mirrors, converting 100% of the incident radiation into a reflected second harmonic [55–57] and a novel type of nonlinear lens that can provide a sub-wavelength image of the source at the second-harmonic frequency, while it is opaque at the fundamental frequency [43]. Such lenses might provide a practical solution for loss mitigation at the fundamental frequency. Two of these important nonlinear phenomena, namely second-harmonic generation and optical parametric amplification in NIMs will be considered in more detail in sections 13.4.1 and 13.4.2, respectively.

Backward waves can also facilitate an effective feedback mechanism in wave-guiding structures. This novel mechanism will be exemplified by a nonlinear coupler with one channel filled with NIM. It turns out that introducing a NIM in one of the channels dramatically changes both linear and nonlinear transmission characteristics of such a coupler [70]. Optical bistability and gap solitons in nonlinear optical couplers will be discussed in section 13.4.3.

Another important manifestation of negative refraction is a negative phase shift (phase advance) that was actually measured in thin films of NIMs in the first experiments [9, 10]. Recently, several novel device applications that rely on such phase shifts introduced by NIMs have been proposed. These include miniaturized optical waveguides, resonators and laser cavities, phase compensators/conjugators, and nonreciprocal (diode-like) applications [66–69, 102–105]. Novel regimes of optical bistability and potentially useful nonlinear transmission properties of layered structures containing thin films of NIMs will be discussed in section 13.4.4. Finally, since NIMs are artificial nanostructured materials, a natural question is how the underlying nanostructure influences the nonlinear wave propagation in NIMs. While this fundamental issue is not

13 Optical Metamaterials: Invisibility and Nonlinearities   229

fully explored and understood, some initial theoretical considerations will be outlined in section 13.4.5.

13.4.1 Second-harmonic Generation

Discovery of the SHG at optical frequencies is considered as the first milestone of nonlinear optics [106]. Naturally, it was one of the first nonlinear processes examined in a context of NIMs. As mentioned in the introduction, backward phase-matching enables principally new regimes of SHG in NIMs.

The basic idea of backward phase-matching is illustrated in Fig. 13.8a. Let us consider the material that is a NIM at the fundamental frequency ω and that is a PIM at the second-harmonic frequency 2ω. If the energy flow of the fundamental frequency travels from left to right, the phase of the wave at the same frequency should move in the opposite direction, that is, from right to left. The phase-matching requirement $k_{2\omega} = 2 k_\omega$ can be satisfied if the phase of the second harmonic also travels from right to left. Since the second harmonic propagates in the PIM, its energy flow is co-directed with the phase velocity and, therefore, the energy propagates from right to left as well, as shown in Fig. 13.8a. Figure 13.8b illustrates the conventional PIM case.

The equations describing the SHG process in NIMs can be written in the following form:

$$\frac{\partial A_\omega}{\partial z} = -\mathrm{i}\frac{2 K\omega^2 \mu_\omega}{c^2 k_\omega} A_{2\omega} A_\omega^* \exp(-\mathrm{i}\Delta k z) \tag{13.8}$$

$$\frac{\partial A_{2\omega}}{\partial z} = \mathrm{i}\frac{4 K\omega^2 \mu_{2\omega}}{c^2 k_{2\omega}} A_\omega^2 \exp(-\mathrm{i}\Delta k z) \tag{13.9}$$

where A_ω and $A_{2\omega}$ are slowly varying amplitudes of the fundamental and second-harmonic waves, respectively, $\Delta k = 2 k_\omega - k_{2\omega}$, $K = 2\pi c^{-2}\chi^{(2)}(2\omega) = \pi c^{-2}\chi^{(2)}(\omega)$. Note that the SHG process is studied here for the case of

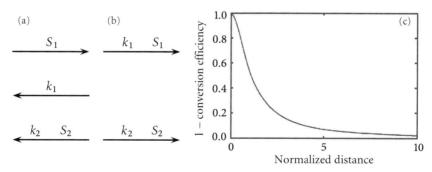

Fig. 13.8. (a) A schematic of SHG in NIM, (b) a schematic of SHG in PIM, (c) the deviation of the conversion efficiency from unity [56]

nonlinear polarization instead of considering nonlinear magnetization as in Refs. [56, 57].

Assuming that the phase-matching condition is satisfied, the spatially invariant Manley-Rowe relations take the form

$$|A_\omega|^2 - |A_{2\omega}|^2 = C \qquad (13.10)$$

Equation (13.10) reflects one of the important differences between the NIM and PIM cases, since in a conventional PIM case, the Manley-Rowe relations require that the sum of the squared amplitudes is constant. This unusual form of Manley-Rowe relations in NIMs results from the fact that the Poynting vectors for the fundamental and the second harmonic are antiparallel, while their wavevectors are parallel.

It is noteworthy that the boundary conditions for the fundamental and second-harmonic waves in the NIM case are specified at opposite interfaces of the slab of a finite length L in contrast to the PIM case, where both conditions are specified at the front interface. Owing to such boundary conditions, the conversion at any point within the NIM slab depends on the total thickness of the slab. In the limit of semi-infinite NIM, $C = 0$ as both waves disappear at infinity and Eq. (13.10) predicts 100% conversion efficiency of the incoming wave at the fundamental frequency to the second harmonic frequency propagating in the opposite direction. As a result, the NIM slab acts as a nonlinear mirror as shown in Fig. 13.8c [55–57].

13.4.2 Optical Parametric Amplification: Loss Compensation in NIMs

Losses are one of the major obstacles that may delay many practical applications of optical NIMs. It was shown that due to causality requirements, low-loss resonant NIMs cannot be realized without the incorporation of some active components [64]. It was also argued that losses cannot be completely compensated or even significantly reduced in NIMs without losing the negative refractive properties [107]. However, the latter statement that relies on linear Kramers-Kronig (KK) relations, applies to a purely linear system, while in a general nonlinear case, the KK relations are either not applicable or should be modified [108]. Therefore, nonlinear optical effects, such as optical parametric amplification (OPA), have a strong potential for loss compensation.

The basic idea of loss compensation using the OPA is to use the electromagnetic waves with the frequencies outside the negative index frequency range to provide the loss-balancing OPA at frequencies corresponding to a negative index of refraction. Indeed, recently, parametric amplification has been demonstrated experimentally, although not in optical systems, but in negative index nonlinear transmission line media [64].

Two basic approaches for loss compensation using the OPA have been investigated theoretically. One possibility relies on quadratic nonlinearity of the NIM [57, 58]. In this case, a strong pump field at frequency ω_3 interacts with the signal at a frequency ω_1. As a result, the signal is amplified, and a new wave, an idler, at a frequency $\omega_2 = \omega_3 - \omega_1$ is generated. Quadratic nonlinearity may be introduced by inserting nonlinear elements into the NIM's meta-atoms. For example, it was proposed that at microwave frequencies it can be realized using diodes as nonlinear insertions into NIM's meta-atoms such as SRR.

An alternative approach does not require strong nonlinear response of the building blocks of the NIM. Instead, it employs embedded four-level centers that can be controlled independently from the NIM parameters, resulting in a possibility of realization of frequency-tunable transparency windows in NIMs [65]. This technique relies on a four-wave interaction process in a medium with cubic nonlinearity. In this case, two control (pump) fields at frequencies ω_3 and ω_4 and a signal field at ω_1 combine to generate an idler at $\omega_2 = \omega_3 + \omega_4 - \omega_1$, which is amplified and contributes back to the signal field through the four-wave mixing process which results in strongly enhanced OPA. Importantly, using realistic material parameters it was shown that the transparency and gain in such system can be achieved without noticeable changes in linear negative refractive index [65].

Both quadratic and cubic parametric amplification processes strongly rely on phase matching between the interacting waves. As a result of opposite directionality of the phase and energy velocities, backward phase-matching takes place in both quadratic and cubic nonlinearity cases, shown schematically in Figs. 13.9a and 13.9b. In particular, in the quadratic nonlinearity case, the amplification factor for the signal wave and the conversion efficiency for the idle wave show the oscillatory behavior even at $\Delta k = 0$, as shown in Fig. 13.9c. This is in sharp contrast to the conversion efficiency behavior in PIMs, which is shown schematically in Fig. 13.9d. In addition, the amplification threshold

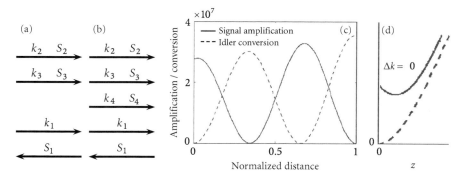

Fig. 13.9. Optical parameteric amplification in NIMs with (**a**) quadratic [58] and (**b**) cubic [65] nonlinearities. The phase-matched amplification factor for the signal wave (solid curve), and the conversion factor for the idler wave (dashed curve) in the (**c**) NIM and (**d**) PIM with quadratic nonlinearity and absorption [58]

in NIMs appears for the product gL rather than for g as in PIMs, where g is the factor proportional to the product of quadratic nonlinear susceptibility and the intensity of the pump field. The important advantage of the backward OPA in NIMs is effective distributed feedback, which enables oscillations without a cavity. In the NIM case, each spatial point serves as a source for the generated wave in the reflected direction, whereas the phase velocities of all interacting waves are co-directed. This is also true in the SHG case.

13.4.3 Bistability in Couplers

The same mechanism that enables backward phase-matching and unusual manifestation of SHG and OPA processes, gives rise to other remarkable effects in a context of nonlinear optical couplers. Nonlinear couplers have attracted significant attention owing to their strong potential for all-optical processing applications. Transmission properties of a nonlinear coherent directional coupler were originally studied by Jensen [109], who concluded that a coupler consisting of two channels made of conventional homogeneous nonlinear materials is not a bistable device.

Bi- (or multi-)stability is a phenomenon in which the system exhibits two (or more) steady transmission states for the same input intensity [110]. Optical bistability has been predicted and experimentally realized in various settings including a Fabry-Perot resonator filled with a nonlinear material, layered periodic structures and nonlinear couplers with external feedback mechanisms [108, 110, 111].

Recently, it was shown that a nonlinear optical coupler with one of the channels filled with NIM can be bistable [70]. Moreover, the entirely uniform PIM-NIM coupler structure supports gap solitons – a feature commonly associated with periodic structures. As shown in Fig. 13.10a, in PIM-NIM couplers, wavevectors in both channels are co-directed, while the Poynting

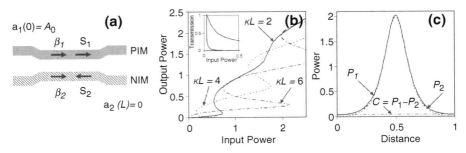

Fig. 13.10. (a) A schematic of a nonlinear PIM-NIM coupler, (b) output power $P_1(L)$ as a function of input power $P_1(0)$ for three values of coupling coefficient; inset shows the transmission coefficient, (c) spatial distributions of P_1 (solid line), of P_2 (dashed line), and the constant of motion C (dot-dashed line) versus distance at transmission resonance shown in the inset in (b) [70]

vectors are counter-directed, producing an "effective" feedback mechanism in these structures.

Continuous wave propagation in a nonlinear PIM-NIM coupler can be described by the following system of equations

$$i\frac{\partial a_1}{\partial z} + \kappa_{12} a_2 \exp(-i\delta z) + \gamma_1 |a_1|^2 a_1 = 0,$$
$$-i\frac{\partial a_2}{\partial z} + \kappa_{21} a_1 \exp(i\delta z) + \gamma_2 |a_2|^2 a_2 = 0, \quad (13.11)$$

where a_1 and a_2 are the complex normalized amplitudes of the modes in the PIM and NIM channels respectively, κ_{12} and κ_{21} are the coupling coefficients, $\delta = \beta_1 - \beta_2$ is the mismatch between the propagation constants in the individual channels. In a the simplest case of $\kappa_{12} = \kappa_{21} \equiv \kappa$ and $\gamma_1 = \gamma_2 \equiv \gamma$, one of the constants of motion is given by

$$C = P_1 - P_2 \quad (13.12)$$

where $P_1 = A_1^2$, $P_2 = A_2^2$ with $a_1 = A_1 \exp(i\phi_1)$ and $a_2 = A_2 \exp(i\phi_2)$, assuming A_1, A_2, ϕ_1 and ϕ_2 are real functions of z. The expression in Eq. (13.12) should be compared to that of conventional PIM-PIM nonlinear coupler, in which case $C = P_1 + P_2$.

In the case of $\delta = 0$, the solutions for P_1 and P_2 are found in the form

$$P_1(z) = C \frac{dn(2\kappa(z-L)/m, m) + 1}{2\, dn(2\kappa(z-L)/m, m)},$$
$$P_2(z) = C \frac{1 - dn(2\kappa(z-L)/m, m)}{2\, dn(2\kappa(z-L)/m, m)}, \quad (13.13)$$

where $m = k/\sqrt{1+k^2} = 1/\sqrt{1+(\gamma C/4\kappa)^2}$, $dn(z', k')$ is the Jacobi elliptic function [112]. The parameter C can be found using the transcendental equation

$$A_0^2 = C \frac{dn(2\kappa L/m, m) + 1}{2\, dn(2\kappa L/m, m)}. \quad (13.14)$$

Figure 13.10b shows the transmission coefficient as a function of input power $P_1(0)$ for three values of coupling coefficient. As the coupling increases, the effective feedback mechanism establishes, and the PIM-NIM nonlinear coupler becomes bistable or, more generally, multi-stable as illustrated Fig. 13.10b. Its transmission characteristics are very similar to those of distributed feedback (DFB) structures.

The phenomenon of bistability in DFBs is closely related to the notion of gap solitons [113–115]. As shown in the inset in Fig. 13.10b, the transmission coefficient approaches $\Im = 1$ at a certain input power, suggesting the existence of transmission resonance. At this resonance, spatial power distributions $P_1(z)$ (solid line) and $P_2(z)$ (dashed line) peak in the middle of the structure as shown in Fig. 13.10c. The dot-dashed line in Fig. 13.10c shows the constant of the motion $C = P_1 - P_2$. These results indicate that at the transmission resonance incident light is coupled to a soliton-like static entity that has its maximum in the middle of the structure and is known as a gap soliton. While optical bistability and gap solitons are not commonly observed in homogeneous couplers without the external feedback, in PIM-NIM coupler these phenomena result from the inherent property of NIMs, the opposite directionality of the phase and energy velocities, which provides an "effective" feedback mechanism.

13.4.4 Bistability in Layered Structures

As discussed in Section 13.2, the state-of-the-art optical NIMs were realized only in a form of sub-wavelength thin films. While no propagation effects or negative refraction, as such, can be observed in these films, the negative index of refraction reveals itself in a phase advancement (negative phase shift), which is in contrast to the phase retardation (positive phase shift) in conventional PIMs. The simplest way of introducing nonlinearities to existing NIM films is to place an overlay of nonlinear material such as nematic liquid crystals, having very high nonlinear $\chi^{(3)}$ coefficients of $\sim 10^{-9}\,\mathrm{m^2/W}$ on top of the NIM thin film as shown in Fig. 13.11a [69].

In order to understand the effect of the NIM thin film on the transmission properties of such a bilayered structure, three configurations can be compared: (a) a single nonlinear slab, (b) a nonlinear slab combined with the PIM layer, and (c) a nonlinear slab combined with the NIM layer. A nonlinear film surrounded by a linear dielectric with a high refractive index can be considered as a resonator and is known to exhibit bistability and, more generally, multistability, when illuminated at an angle θ_{in}, such that $\theta_{\mathrm{res}} < \theta_{\mathrm{in}} < \theta_{\mathrm{TIR}}$, where θ_{res} is the angle corresponding to the resonant peak nearest to the angle of total internal reflection (TIR) θ_{TIR} in the linear transmission curve. In this configuration, transmission in the linear regime is low as shown in Fig. 13.11b.

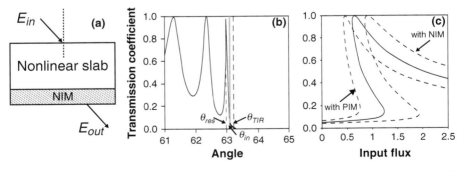

Fig. 13.11. (a) A schematic of a bilayer structure comprising a nonlinear slab and a linear NIM thin film, (b) linear transmission spectra versus angle, (c) nonlinear transmission spectra versus input flux

However, as the incident intensity increases, in the case of self-focusing Kerr nonlinearity, the nonlinear refractive index increases, resulting in a shift of both θ_{TIR} and θ_{res} to larger values. Simultaneously, the transmission coefficient becomes a multi-valued function of the input flux, leading to formation of a hysteresis loop as shown in Fig. 13.11c. Moreover, numerical simulations reveal strong sensitivity of the width and depth of the hysteresis to the changes of the material parameters of the NIM layer. Although the NIM film is very thin, the effect of this phase shift on the nonlinear optical response of the entire structure turns out to be very significant. As discussed above, negative refraction reveals itself in a phase advance or a negative phase shift. Therefore, in the case of a NIM thin layer, the "resonator" length decreases, implying that the intensity dependent nonlinear index change required for switching the transmission to the high-transmission state should increase, which is in contrast to the case of PIM thin film. Figure 13.11c confirms this prediction. Owing to high sensitivity of the nonlinear response to the NIM's parameters, these results may be particularly useful for characterization of NIMs.

13.4.5 Solitons in Resonant Plasmonic Nanostructures

Previous sections mostly dealt with continuous wave effects in optical NIMs. Several aspects of temporal and spatial dynamics and soliton propagation in NIMs have been recently addressed by Scalora et al. [116, 117], who have derived a generalized nonlinear Schrödinger equation taking into account frequency dependence of both dielectric permittivity and magnetic permeability, and describing the propagation of ultrashort pulses in a wide class of magnetically active metamaterials. Novel types of gap solitons in NIMs have been predicted in periodic NIM-PIM structures as well as in a single slab of NIM [71, 78]. However, in most studies of nonlinear propagation effects in NIMs, the origin of the nonlinear response has either not been specified, or the model developed for a particular case of nonlinear response of the SRR-based microwave NIMs [51] was utilized.

The first step toward understanding the effects of material nanostructure was made in Ref. [76]. Short pulse propagation was studied in the simplest case of a nanostructure consisting of metallic nanoparticles embedded in a glass host, such that the resonance frequencies of the host medium and of the nanoparticles are well separated. Such structures with, for example, plasmonic resonance frequency in the visible and the resonance of the host material in the ultraviolet can be realized using silver or gold spherical or spheroidal nanoparticles embedded in SiO_2.

The origin and magnitude of third-order nonlinearity in such nanostructures are well characterized theoretically and experimentally [118–120]. In particular, it was shown that quantum effects in metal nanoparticles driven by a resonant optical field play an important role in inducing a strong nonlinear response.

Light interaction with metal nanoparticles can be described by a system of equations consisting of Maxwell's equation for the electric field and an oscillator equation describing the plasmonic oscillations, a so-called Maxwell-Duffing model. Four-parameter solitary wave solutions of this Maxwell-Duffing system of equations revealed many unique properties related to propagation and collisional effects. Their collision dynamics was found to be dependent on initial solitary wave parameters, and particularly on phase, leading to principally different regimes of interactions: in one regime, the collisions are quasi-elastic; in another regime, the collisions are inelastic. A strong sensitivity to the initial phase of the soliton makes this system potentially useful for a number of phase-controlled applications.

Although a simple nanoparticle-based model may seem to be not directly related to double-resonant metamaterials of interest, at least in a particular case of a metamaterials with a nonlinear electric response and a linear magnetic response, the governing nonlinear equations for the electric and magnetic fields decouple. As a result, a system of equations describing light interaction with double-resonant NIMs is very similar to that derived for the simplest nanostructures composed of metallic nanoparticles when $\mu = 1$. Therefore, many of the effects predicted for the simple structure may be relevant for understanding the properties of NIMs.

Finally, since it was shown in context of the SRR-based microwave NIMs that magnetic nonlinearities' contribution may be even more pronounced than that of electric nonlinearities [51, 59, 61] taking into account the effects of nonlinear magnetization and examining its effect in optical nanostructures may bring about new and exciting phenomena.

13.5 Summary

In this chapter, we reviewed the results of truly fascinating opportunities for light manipulation enabled by metamaterials. Although a tremendous progress in fabrication, characterization, and basic theory of light interaction with metamaterials was made over the last few years, the field of optical metamaterials is still in its early stage of development. Many fundamental and practical challenges need to be resolved in order to exploit the full potential of these unique structures.

To date, a majority of cloaking devices and negative index material designs relied on passive materials with resonant material properties. The main limitations of this approach are significant losses and limited bandwidth. At least one or even both problems may potentially be addressed by using active structures. From this perspective, nonlinear optical effects may be of foremost importance.

Finally, cloaking and NIMs are the first, but most likely not the only examples of the unique capabilities of metamaterials to make the seemingly impossible a reality.

Acknowledgements

The authors gratefully acknowledge the support of the Army Research Office through Grants W911NF-07-1-0343 and 50342-PH-MUR.

References

1. V.M. Shalaev, Nature Photonics **1**, 41 (2007)
2. C.M. Soukoulis, S. Linden, and M. Wegener, Science **315**, 47 (2007)
3. T.A. Klar, A.V. Kildishev, V.P. Drachev, and V.M. Shalaev, IEEE J. of Selected Topics in Quantum Electronics **12**, 1106 (2006)
4. V.G. Veselago, L. Braginsky, V. Shklover, and C. Hafner, J. of Computational and Theoretical Nanoscience **3**, 189 (2006)
5. V.G. Veselago and E.E. Narimanov, Nature Materials **5**, 759 (2006)
6. G. Dolling, C. Enrich, M. Wegener, J.F. Zhou, C.M. Soukoulis, and S. Linden, Opt. Lett. **30**, 3198 (2005)
7. C. Enkrich, M. Wegener, S. Linden, S. Burger, L. Zschiedrich, F. Schmidt, J. Zhou, T. Koschny, and C.M. Soukoulis, Phys. Rev. Lett. **95**, 203901 (2005)
8. S. Zhang, W. Fan, B.K. Minhas, A. Frauenglass, K.J. Malloy, and S.R.J. Brueck, Phys. Rev. Lett. **94**, 037402 (2005)
9. V.M. Shalaev, W. Cai, U.K. Chettiar, H. Yuan, A.K. Sarychev, V.P. Drachev, and A.V. Kildishev, Opt. Lett. **30**, 3356 (2005)
10. S. Zhang, W. Fan, N.C. Panoiu, K.J. Malloy, R.M. Osgood, and S.R.J. Brueck, Phys. Rev. Lett. **95**, 137404 (2005)
11. U.K. Chettiar, A.V. Kildishev, T.A. Klar, V.M. Shalaev, Opt. Express **14**, 7872 (2006)
12. G. Dolling, C. Enkrich, M. Wegener, C.M. Soukoulis, and S. Linden, Opt. Lett. **31**, 1800 (2006)
13. V.P. Drachev, W. Cai, U.K. Chettiar, H.-K. Yuan, A.K. Sarychev, A.V. Kildishev, G. Klimeck, and V.M. Shalaev, Laser Phys. Lett. **3**, 49 (2006)
14. A.V. Kildishev, W. Cai, U.K. Chettiar, H.-K. Yuan, A.K. Sarychev, V.P. Drachev, and V.M. Shalaev, J. Opt. Soc. Am. B **23**, 423 (2006)
15. A.K. Sarychev, G. Shvets, and V.M. Shalaev, Phys. Rev. E **73**, 036609 (2006)
16. S. Zhang, W. Fan, N.C. Panoiu, K.J. Malloy, R.M. Osgood, and S.R.J. Brueck, Opt. Express **14**, 6778 (2006)
17. G. Dolling, M. Wegener, C.M. Soukoulis, and S. Linden, Opt. Lett. **32**, 53 (2007)
18. G. Dolling, M. Wegener, and S. Linden, Opt. Lett. **32**, 551 (2007)
19. U.K. Chettiar, A.V. Kildishev, H.-K. Yuan, W. Cai, S. Xiao, V.P. Drachev, and V.M. Shalaev, V. M., Opt. Lett. **32**, 1671 (2007)
20. W. Cai, U.K. Chettiar, H.-K. Yuan, V.C. de Silva, A.V. Kildishev, V.P. Drachev, and V.M. Shalaev, Opt. Express **15**, 3333 (2007)
21. I.V. Shadrivov, S.K. Morrison, and Y.S. Kivshar, Opt. Express **14**, 9344 (2006)
22. D.H. Werner, D.-H. Kwon, I.-C. Khoo, A.V. Kildishev, and V.M. Shalaev, Opt. Express **15**, 3342 (2007)
23. A. Degiron, J.J. Mock, and D.R. Smith, Opt. Express **15**, 1115 (2007)
24. J.B. Pendry, D. Schurig, and D.R. Smith, Science **312**, 1780 (2006)
25. U. Leonhardt, Science **312**, 1777 (2006)

26. U. Leonhardt and T.G. Philbin, New J. Phys. **8**, 247 (2006)
27. D. Schurig, J.J. Mock, B.J. Justice, S.A. Cummer, J.B. Pendry, A.F. Starr, and D.R. Smith, Science Express Manuscript Number 113362 (2006)
28. B. Wood and J.B. Pendry, J. Phys. **19**, 076208 (2007)
29. S.A. Cummer and D. Schurig, New J. Phys. **9**, 45 (2007)
30. W. Cai, U.K. Chettiar, A.V. Kildishev, and V.M. Shalaev, Nature Photonics **1**, 224 (2007)
31. W. Cai, U.K. Chettiar, A.V. Kildishev, G.W. Milton, and V.M. Shalaev, Appl. Phys. Lett. **91**, 111105 (2007)
32. V.G. Veselago, Sov. Phys. Usp. **10**, 509 (1968)
33. D.R. Smith, W.J. Padilla, D.C. Vier, S.C. Nemat-Nasser, and S. Schultz, Phys. Rev. Lett. **84**, 4184 (2000)
34. R.A. Shelby, D.R. Smith, and S. Schultz, Science **292**, 77 (2001)
35. A.A. Houck, J.B. Brock, and I.L. Chuang, Phys. Rev. Lett. **90**, 137401 (2003)
36. S. Linden, C. Enkrich, M. Wegener, J. Zhou, T. Koschny, and C.M. Soukoulis, Science **306**, 1351 (2004)
37. T.J. Yen, W.J. Padilla, N. Fang, D.C. Vier, D.R. Smith, J.B. Pendry, D.N. Basov, and X. Zhang, Science **303**, 1494 (2004)
38. T.F. Gundogdu, I. Tsiapa, A. Kostopoulos, G. Konstantinidis, N. Katsarakis, R.S. Penciu, M. Kafesaki, E.N. Economou, T. Koschny, and C.M. Soukoulis, Appl. Phys. Lett. **89**, 084103 (2006)
39. J.B. Pendry, Phys. Rev. Lett. **85**, 3966 (2000)
40. R. Merlin, Appl. Phys. Lett. **84**,1290 (2004)
41. N. Fang, H. Lee, C. Sun, and X. Zhang, Science **308**, 534 (2005)
42. V.A. Podolskiy and E.E. Narimanov, Opt. Lett. **30**, 75 (2005)
43. A.A. Zharov, N.A. Zharova, I.V. Shadrivov, and Y.S. Kivshar, Appl. Phys. Lett. **87**, 091104 (2005)
44. R.J. Blaikie, D.O.S. Melville, and M.M. Alkalsi, Microelectronic Engineering **83**, 723 (2006)
45. Z. Jacob, L.V. Alekseyev, and E. Narimanov, Opt. Express **14**, 8247 (2006)
46. T. Taubner, D. Korobkin, Y. Urzhumov, G. Shvets, and R. Hillenbrand, Science **313**, 1595 (2006)
47. Z. Liu, S. Durant, H. Lee, Y. Pikus, Y. Xiong, C. Sun, and X. Zhang, Opt. Express **15**, 6947 (2007)
48. Z. Liu, H. Lee, Y. Xiong, C. Sun, and X. Zhang, Science **315**, 1686 (2007)
49. I.I. Smolyaninov, Y.-J. Hung, and C.C. Davis, Science **315**, 1699 (2007)
50. J.B. Pendry, A.J. Holden, D.J. Robbins, and W.J. Stewart, IEEE Trans. Microwave Theory Tech. **47**, 2075 (1999)
51. A.A. Zharov, I.V. Shadrivov, and Y.S. Kivshar, Phys. Rev. Lett. **91**, 037401 (2003)
52. V.M. Agranovich, Y.R. Shen, R.H. Baughman, and A.A. Zakhidov, Phys. Rev. B **69**, 165112 (2004)
53. N. Mattiucci, G. D'Aguanno, M.J. Bloemer, and M. Scalora, Phys. Rev. E **72**, 066612 (2005)
54. G. D'Aguanno, N. Mattiucci, M.J. Bloemer, and M. Scalora, Phys. Rev. E **73**, 036603 (2006)
55. V. Shadrivov, A.A Zharov, and Y.S. Kivshar, J. Opt. Soc. Am. B **23**, 529 (2006)
56. A.K. Popov, V.V. Slabko, and V.M. Shalaev, Laser Phys. Lett. **3**, 293 (2006)

57. A.K. Popov and V.M. Shalaev, Appl. Phys. B **84**, 131 (2006)
58. A.K. Popov and V.M. Shalaev, Opt. Lett. **31**, 2169 (2006)
59. M.W. Klein, C. Enkrich, M. Wegener, and S. Linden, Science **313**, 502 (2006)
60. M. Scalora, G. D'Aguanno, M.J. Bloemer, M. Centini, D. de Ceglia, N. Mattiucci, and Y.S. Kivshar, Opt. Express **14**, 4746 (2006)
61. M.W. Klein, M. Wegener, N. Feth, and S. Linden, Opt. Express **15**, 5238 (2007)
62. D. de Ceglia, A. D'Orazio, M. de Sario, V. Petruzzelli, F. Prudenzano, M. Centini, M.G. Cappeddu, M.J. Bloemer, and M. Scalora, Opt. Lett. **32**, 265 (2007)
63. M.V. Gorkunov, I.V. Shadrivov, and Y.S. Kivshar, Appl. Phys. Lett. **88**, 071912 (2006)
64. A.B. Kozyrev, H. Kim, and D.W. van der Weide, Appl. Phys. Lett. **88**, 264101 (2006)
65. A.K. Popov, S.A. Myslivets, T.F. George, and V.M. Shalaev, Opt. Lett. **32**, 3044 (2007)
66. M.W. Feise, I.V. Shadrivov, and Y.S. Kivshar, Appl. Phys. Lett. **85**, 1451 (2004)
67. M.W. Feise, I.V. Shadrivov, and Y.S. Kivshar, Phys. Rev. E **71**, 037602 (2005)
68. R.S. Hegde and H. Winful, Microwave and Opt. Technol. Lett. **46**, 528 (2005)
69. N.M. Litchinitser, I.R. Gabitov, A.I. Maimistov, and V.M. Shalaev, Opt. Lett. **32**, 151 (2007)
70. N.M. Litchinitser, I.R. Gabitov, and A.I. Maimistov, Phys. Rev. Lett. **99**, 113902 (2007)
71. G. D'Aguanno, N. Mattiucci, M. Scalora, and M.J. Bloemer, Phys. Rev. Lett. **93**, 213902 (2004)
72. I.V. Shadrivov, N. Zharova, A. Zharov, and Y.S. Kivshar, Opt. Express **13**, 1291 (2005)
73. I.V. Shadrivov and Y.S. Kivshar, J. Opt. A **7**, S68 (2005)
74. A.D. Boardman, P. Egan, L. Velasco, and N. King, J. Opt. A **7**, S57 (2005)
75. A.D. Boardman, L. Velasco, N. King, and Y. Rapoport, J. Opt. Soc. Am. B **22**, 1443 (2005)
76. I.R. Gabitov, R.A. Indik, N.M. Litchinitser, A.I. Maimistov, V.M. Shalaev, and J.E. Soneson, J. Opt. Soc. Am. B **23**, 535 (2006)
77. A.I. Maimistov, I.R. Gabitov, and E.V. Kazantseva, Optics and Spectroscopy **102**, 90 (2007)
78. R.S. Hegde and H. Winful, Opt. Lett. **30**, 1852 (2005)
79. M. Marklund, P.K. Shukla, and L. Stenflo, Phys. Rev. E **73**, 0376011 (2006)
80. B.T. Schwartz and R. Piestun, J. Opt. Soc. Am. B **20**, 2448 (2003)
81. M. Rahm, D. Schurig, D.A. Roberts, S.A. Cummer, D.R. Smith, and J.B. Pendry, Photonics Nanostruct. Fundam. Appl. **6**, 87 (2008)
82. H. Chen and C.T. Chan, Appl. Phys. Lett. **90**, 241105 (2007)
83. A. Alù and N. Engheta, Phys. Rev. E **72**, 016623 (2005)
84. A. Alù and N. Engheta, Opt. Express **15**, 3318 (2007)
85. M.G. Silveirinha, A. Alù, and N. Engheta, Phys. Rev. E **75**, 036603 (2007)
86. A. Alù and N. Engheta, Opt. Express **15**, 7578 (2007)
87. M. Kerker, J. Opt. Soc. Am. **65**, 375 (1975)
88. N.A. Nicorovici, R.C. McPhedran, and G.W. Milton, Phys. Rev. B **49**, 8479 (1994)
89. G.W. Milton, and N.A. Nicorovici, Proc. R. Soc. Lond. A **462**, 3027 (2006)

90. N.A. Nicorovici, G.W. Milton, R.C. McPhedran, and L.C. Botten, Opt. Express **15**, 6314 (2007)
91. A.J. Ward, and J.B. Pendry, J. Modern Opt. **43**, 773 (1996)
92. E.J. Post, *Formal Structure of Electromagnetics*, Wiley, New York (1962)
93. D. Schurig, J.B. Pendry, and D.R. Smith, Opt. Express **14**, 9794 (2006)
94. S.A. Cummer, B.-I. Popa, D. Schurig, D.R. Smith, and J.B. Pendry, Phys. Rev. E **74**, 036621 (2006)
95. A. Greenleaf, Y. Kurylev, M. Lassas, and G. Uhlmann, arXiv:math/0611185v3 (2007)
96. H. Lamb, Proc. Lond. Math. Soc. **1**, 473 (1904)
97. A. Schuster, *An Introduction to the Theory of Optics*, Edward Arnold, London (1904)
98. M. von Laue, Ann. Phys. **18**, 523 (1905)
99. H.C. Pocklington, Nature **71**, 607 (1905)
100. D.V. Sivukhin, Opt. Spektrosk. **3**, 308 (1957)
101. V.M. Agranovich and V.L. Ginzburg, *Crystal Optics with Spatial Dispersion, and Excitons*, Springer, Berlin (1984)
102. A. Alù and N. Engheta, IEEE Trans. on Antennas and Propagation **51**, 2558 (2003)
103. A. Alù and N. Engheta, IEEE Trans. on Microwave Theory and Techniques **52**, 199 (2004)
104. N. Engheta and R.W. Ziolkowski, IEEE Trans. Microwave Theory and Techniques **53**, 1535 (2005)
105. R.W. Ziolkowski, J. Opt. Soc. Am. B **23**, 451 (2006)
106. P.A. Franken, A.E. Hill, C.W. Peters, and G. Weinreich, Phys. Rev. Lett. **7**, 118 (1961)
107. M.I. Stockman, Phys. Rev. Lett. **98**, 177404 (2007)
108. R.W. Boyd, *Nonlinear optics*, second edition, Elsevier (2003)
109. S.M. Jensen, IEEE J. Quantum Electron. **18**, 1580 (1982)
110. H.M. Gibbs, *Optical Bistability*, Academic Press, Orlando (1985)
111. H.G. Winful, J.H. Marburger, and E. Garmire, Appl. Phys. Lett. **35**, 379 (1979)
112. M. Abramowitz and I.A. Stegun (Eds.), *Handbook of Mathematical Functions with Formulas, Graphs, and Mathematical Tables*, 9th printing, Dover, New York (1972)
113. C.M. de Sterke and J.E. Sipe, in *Progress in Optics*, edited by E. Wolf, Vol. 33, pp. 203–260, Elsevier, Amsterdam (1994)
114. W. Chen and D.L. Mills, Phys. Rev. Lett. **58**, 160 (1987)
115. B.J. Eggleton, R.E. Slusher, C.M. de Sterke, P.A. Krug, and J.E. Sipe, Phys. Rev. Lett. **76**, 1627 (1996)
116. M. Scalora, M. Syrchin, N. Akozbek, E.Y. Poliakov, G. D'Aguanno, N. Mattiucci, M.J. Bloemer, and A.M. Zheltikov, Phys. Rev. Lett. **95**, 013902 (2005); Erratum, Phys. Rev. Lett. **95**, 239902 (2005)
117. M. Scalora, G. D'Aguanno, N. Mattiucci, N. Akozbek, M.J. Bloemer, M. Centini, C. Sibilia, and M. Bertolotti, Phys. Rev. E **72**, 066601 (2005)
118. K. Uchida, S. Kaneko, S. Omi, C. Hata, H. Tanji, Y. Asahara, A.J. Ikushima, T. Tokizaki, A. Nakamura, J. Opt. Soc. Am. B **11**, 1236 (1994)
119. S.G. Rautian, JETP **85**, 451 (1997)
120. V.P. Drachev, A.K. Buin, H. Nakotte, and V.M. Shalaev, Nano Lett. **4**, 1535 (2004)

14
Nonlinear Metamaterials

Ilya V. Shadrivov

Nonlinear Physics Centre, Research School of Physics and Engineering, Australian National University, Canberra ACT 0200, Australia
ivs124@rsphysse.anu.edu.au

14.1 Introduction

Nature has provided us with a wide range of materials exhibiting various electromagnetic properties. However, theoretical speculations [1] have suggested that having materials with some particular unnatural characteristics would enable us to observe very unusual and potentially useful effects, including negative refraction for interface scattering, inverse light pressure, reverse Doppler and Vavilov-Cherenkov effects, etc. Recently, the way to artificially make materials with desired properties was suggested theoretically [2, 3] and such materials were prepared experimentally [4–6]. Composite materials were made of a mixture of electric and magnetic resonators, so that they can provide simultaneously negative dielectric and magnetic response. In particular, the simplest composite materials of this type are created by a mesh of metallic wires and split-ring resonators (SRRs), and their unique properties are associated with negative real parts of magnetic permeability and dielectric permittivity. Such composite materials are often referred to as *left-handed materials* (LHMs) or *materials with negative refractive index*. Further developments in the area of complex resonant metamaterials have shown their potential for creating of an electromagnetic cloak [7].

In this Chapter we will present our theoretical and experimental studies of some of the properties of *nonlinear metamaterials*. The nonlinearity of the resonant composite materials is nontrivial since they have strong frequency dispersion, which is basically determined by the geometry of the structure. The possibility to control the effective parameters of the metamaterial using nonlinearity was first suggested in Refs. [8,9]. The nonlinear response of metamaterials is significantly enhanced compared to conventional bulk nonlinear dielectrics, because the microscopic electric field in the vicinity of the metallic particles forming left-handed structure can be much higher than the macroscopic electric field carried by the propagating wave. We believe our findings may stimulate further experiments in this field, as well as the studies of nonlinear effects for plasmonic applications, and in photonic crystals, where the phenomenon of negative refraction is analyzed now very intensively [10,11].

We demonstrate that the dominant nonlinear properties of metamaterials arise from the hysteresis-type dependence of the magnetic permeability on the magnetic field intensity in the electromagnetic wave propagating through the material. It allows changing the material properties from left- to right-handed and back. Using the finite-difference time-domain simulations, we study the wave scattering from a slab of a nonlinear left-handed material and discuss a possibility of generation and propagation of spatiotemporal solitons in such materials. We demonstrate also that the nonlinear left-handed metamaterials can support self-trapped localized beams, *spatial electromagnetic solitons*. We also discuss the physical mechanisms and novel effects in the parametric processes such as second-harmonic generation, which can take place in metamaterials. We demonstrate a novel type of the exact spatio-temporal phase matching between the backward propagating wave of the fundamental frequency and the forward propagating wave of the second harmonic.

This Chapter is organized as follows. In Sec. 14.2 we present an overview of nonlinear properties of left-handed metamaterials for the example of a lattice of SRRs and wires embedded in a nonlinear dielectric. In Sec. 14.3 we show first experimental results on observation of the transmission properties of nonlinear microwave metamaterial. Sec. 14.4 is devoted to the numerical studies of the nonlinear metamaterial by means of the finite-difference time-domain (FDTD) method. In Sec. 14.5 we discuss the structure of electromagnetic solitons supported by the nonlinear left-handed materials with hysteresis-type nonlinear response. In Sec. 14.6 we study second-order nonlinear effects in metamaterials, such as second-harmonic generation (SHG). We derive coupled equations for describing the process of SHG for particular model of metamaterial.

14.2 Nonlinear Response of Metamaterials: Theory

We consider a three-dimensional composite structure in the form of a cubic lattice of conducting wires and SRRs. We assume that the unit-cell size d_{cell} of the structure is much smaller than the wavelength of the electromagnetic wave propagating in the material. For simplicity, we choose the single-ring geometry of a lattice of SRRs. The results obtained for this case are qualitatively similar to those obtained in more involved cases of double SRRs near low-frequency resonance, for which the currents in both rings of the double SRR are in-phase.

The negative real part of the effective dielectric permittivity of such a composite structure appears due to the metallic wires whereas a negative magnetic permeability becomes possible due to the SRR lattice. As a result, these materials demonstrate the properties of negative refraction in a finite frequency range, i.e., $\omega_0 < \omega < \min(\omega_{\text{p}}, \omega_{\|\text{m}})$, where ω_0 is the eigenfrequency of the array of SRRs, $\omega_{\|\text{m}}$ is some characteristic frequency, which we call the frequency of the longitudinal magnetic plasmon, ω_p is the effective plasma frequency, and ω is the angular frequency of the propagating electromagnetic waves, $(\mathcal{E}, \mathcal{H}) \sim (\boldsymbol{E}, \boldsymbol{H}) \exp(i\omega t)$. The SRR can be described as an effective LC oscillator (see, e.g., Ref. [12]) with capacitance of the SRR gaps, as well as an effective inductance and resistance.

If we embed the structure in a nonlinear dielectric, we expect it to exhibit quite unusual properties. The nonlinearity here must become complex in the sense that both the dielectric permittivity and the magnetic permeability will change with variation of the external electromagnetic fields. Nonlinear dielectric response is determined by the bulk of the nonlinear dielectric, and frequency-dependent contribution will be provided by wires. Since we assume that the nonlinear dielectric does not possess magnetic properties itself, it seems that the magnetic response of the composite will still be determined by the array of SRRs. However, the response of the SRRs depends on the field pattern in the vicinity of the resonators, which, in turn, depends on the properties of the dielectric. Thus, both dielectric and magnetic responses of the composite are nonlinear.

As we just mentioned above, the nonlinear magnetic response will be determined by the dielectric which is close to the SRRs, and, in particular, in SRR slits where the electric field is the strongest. The dielectric response is due to bulk of the nonlinear dielectric. This suggests the way to *engineer the nonlinear response* of the composite structure. E.g., one can include small amount of nonlinear material to the SRRs, and the whole composite will have nonlinear magnetic properties only. Placing dielectric everywhere in the composite except the very vicinity of the SRRs will produce nonlinear dielectric properties, with linear magnetic response.

14.2.1 Nonlinear Magnetic Permeability

Firstly, we assume that only the slits of the SRRs are filled with nonlinear dielectric with a permittivity that depends on the intensity of the electric field $|E|^2$ in a rather general form, $\varepsilon_D = \varepsilon_D(|\boldsymbol{E}|^2)$. For the calculations presented below, we take the dependence that corresponds to the Kerr-type nonlinear response, $\varepsilon_D = \varepsilon_l + \alpha |E|^2 / E_c^2$, where ε_l is the linear part of the dielectric permittivity, E_c is a characteristic electric field strength, $\alpha = +1$ for focusing nonlinearity and $\alpha = -1$ for defocusing nonlinearity.

The nonlinear magnetic response of the composite material comes from the lattice of resonators since the SRR capacitance (and, therefore, the SRR eigenfrequency) depends on the strength of the local electric field in a narrow slit (we assume here that the the capacitance of the SRR is due to its slit only). The intensity of the local electric field in the SRR gap, E_g, depends on the electromotive force in the resonator loop, which is induced by the magnetic field. Therefore, the effective magnetic permeability μ_{eff} depends on the macroscopic (average) magnetic field $\boldsymbol{H}$, and this dependence can be found [8,13] in the three-dimensional case in the form,

$$\mu_{\text{eff}}(\boldsymbol{H}) = 1 + \frac{F\omega^2}{\omega_{\text{0NL}}^2(\boldsymbol{H}) - \omega^2(1 + F/3) + i\Gamma\omega}, \quad (14.1)$$

where

$$\omega_{\text{0NL}}^2(\boldsymbol{H}) = \left(\frac{c}{a}\right)^2 \frac{d_{\text{g}}}{[2\pi r_{\text{w}}\varepsilon_{\text{D}}(|\boldsymbol{E}_{\text{g}}(\boldsymbol{H})|^2)]} \quad (14.2)$$

is the eigenfrequency of nonlinear oscillations, $\Gamma = c^2/4\pi\sigma ar_{\text{w}}$ is the damping coefficient, $F = \pi^2 a^3/2\, d_{\text{cell}}^3 [\ln(8\,a/r_{\text{w}}) - 7/4]$ is the filling factor, a is the SRR radius, r_{w} is the radius of the SRR wire, σ is the conductivity of the wires, E_{g} is the strength of the electric field in the SRR slit, c is the speed of light. It is important to note that Eq. (14.1) has a simple physical interpretation: the resonant frequency of the artificial magnetic structure depends on the amplitude of the external magnetic field and, in turn, this leads to the intensity-dependent function μ_{eff}.

Figures 14.1 and 14.2 summarize different types of nonlinear composites which are characterized by the dependence of the dimensionless frequency of the external field $\Omega = \omega/\omega_0$, for both *focusing* (Figs. 14.1a, 14.1b and Figs. 14.2a, 14.2b) and *defocusing* (Figs. 14.1c, 14.1d and Figs. 14.2c, 14.2d)

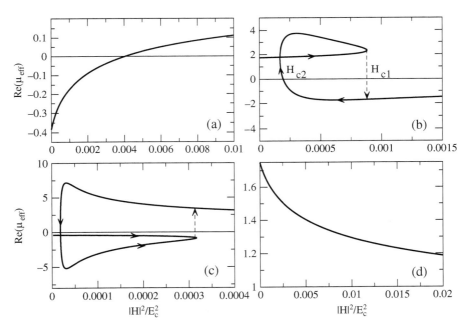

Fig. 14.1. Real part of the effective magnetic permeability vs. normalized intensity of the magnetic field for $\Gamma/\omega_0 = 0.05$: (a) $\Omega > 1$, $\alpha = 1$, (b) $\Omega < 1$, $\alpha = 1$, (c) $\Omega > 1$, $\alpha = -1$, (d) $\Omega < 1$, $\alpha = -1$ [8]

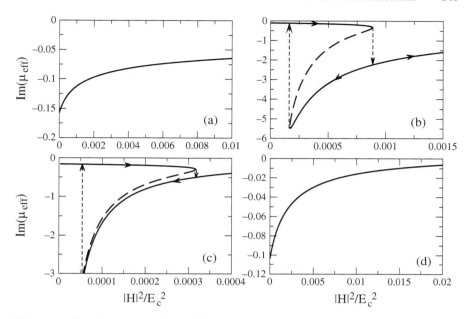

Fig. 14.2. Imaginary part of the effective magnetic permeability vs. intensity of the magnetic field for $\gamma = 0.05$: (a) $\Omega > 1$, $\alpha = 1$, (b) $\Omega < 1$, $\alpha = 1$, (c) $\Omega > 1$, $\alpha = -1$, and (d) $\Omega < 1$, $\alpha = -1$. Dashed curves show the branches of unstable solutions [8]

nonlinearity of the dielectric. The actual values for Ω used in computations are 1.2 and 0.8.

The critical fields for switching between the LH and RH states, shown in Fig. 14.1 can be reduced to a desirable value by choosing the frequency close to the resonant frequency of SRRs. We want to emphasize that strong losses can suppress nonlinear resonance and multistable behaviour. With low enough losses, even for a relatively large difference between the SRR eigenfrequency and the external frequency, as in Fig. 14.1b where $\Omega = 0.8$ (i.e. $\omega = 0.8\omega_0$), the switching amplitude of the magnetic field is $\sim 0.03\,E_c$. The characteristic values of the focusing nonlinearity can be estimated for some materials such as n-InSb for which $E_c = 200\,\text{V/cm}$ [14]. As a result, the strength of the critical magnetic field is found as $H_{c1} \approx 1.6\,\text{A/m}$. Strong defocusing properties for microwave frequencies are found in $\text{Ba}_x\text{Sr}_{1-x}\text{TiO}_3$ (see Ref. [15] and references therein). The critical nonlinear field of a thin film of this material is $E_c = 4 \times 10^4\,\text{V/cm}$, and the corresponding field of the transition from the LH to RH state (see Fig. 14.1c) can be found as $H_{c1} \approx \sqrt{0.003} \times 4/3\,(\text{CGS}) = 5.8\,\text{A/m}$.

14.2.2 Nonlinear Dielectric Permittivity

Now we analyze the dielectric properties of the composite. We suppose that contribution to the dielectric function given by the array of wires is much stronger then that from SRRs. In this case, we can obtain the following expression for the effective nonlinear dielectric permittivity [8]

$$\varepsilon_{\text{eff}}(|E|^2) = \varepsilon_{\text{D}}(|E|^2) - \frac{\omega_{\text{p}}^2}{\omega(\omega - i\gamma_\varepsilon)} , \qquad (14.3)$$

where $\omega_{\text{p}} \approx (c/d)[2\pi/\ln(d/r)]^{1/2}$ is the effective plasma frequency, and $\gamma_\varepsilon = c^2/2\,\sigma S \ln(d/r)$ is an effective wire cross-section. The second term on the right-hand side of Eq. (14.3) is in complete agreement with the earlier result obtained by Pendry and co-authors [2]. One should note that the low losses case, i.e. $\gamma_\varepsilon \ll \omega$, corresponds to the condition $\delta \ll r$, i.e, when the wires are thick with respect to the skin-layer depth.

14.3 Nonlinear Metamaterials: Experiments

We have manufactured a two-dimensional nonlinear metamaterial consisting of periodic arrays of wires and nonlinear split-ring resonators (see Fig. 14.3). Each SRR contained an additional slit with variable capacity diode [16], which provided power-dependent response to the external electromagnetic field. The size of the array is $29 \times 4 \times 1$ resonator with unit cell size of 1 cm.

The experiment was performed in parallel-plate waveguide with the separation between conducting planes of 12 mm. The wires in metamaterial were

Fig. 14.3. Photograph of nonlinear metamaterial. Each SRR in metamaterial contains varactor diode (Skyworks SMV1405)

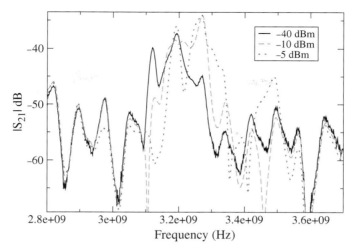

Fig. 14.4. Transmission coefficient from the output of the amplifier to the receiving port of the VNA for different powers of the VNA

extending above and below the metamaterial in order to provide electric contact with the plates of the waveguide. The wave was excited by a wire antenna with dielectric coating which was placed at the surface of the metamaterial slab (see Fig. 14.3). Such position of the antenna was chosen in order to deliver enough power to the metamaterial to observe nonlinear effects. The source antenna was connected to the output of the amplifier (HP83020A) which provided gain of 38 dB in the frequency range of interest to the signal from the vector network analyzer (VNA). The receiving antenna was placed 2 cm behind the metamaterial slab. Transmission coefficient as a function of frequency is shown in Fig. 14.4 for different values of the output power of the VNA. We note, that the power indicated in the Figure is output power of VNA, which is then amplified by 38 dB before reaching the source antenna. However, the energy radiated by antenna cannot be easily estimated, since the antenna is impedance-mismatched with the waveguide, and presence of the nonlinear metamaterial in the vicinity of the antenna also modifies impedance. Shift of the left-handed transmission band with the change of the power is clearly seen, and it confirms the predictions of Section 14.2.1. The response of the metamaterial outside of the resonance band did not depend on the power of the source, indicating linear behaviour of the metamaterial.

14.4 Nonlinearity-controlled Transmission

In order to verify the specific features of the left-handed metamaterials introduced by their nonlinear response, in this section we study the scattering of electromagnetic waves from the nonlinear metamaterial discussed above.

In particular, we perform the FDTD numerical simulations of the plane wave interaction with a slab of LHM of a finite thickness [17].

Following Ref. [17], we study the temporal dynamics of the wave scattering by a finite slab of nonlinear metamaterial. For simplicity, we consider a one-dimensional problem that describes the interaction of the plane wave incident at the normal angle from air on a slab of metamaterial of a finite thickness. We consider *two types of nonlinear effects*: (i) nonlinearity-induced suppression of the wave transmission when initially transparent left-handed material becomes opaque with the growth of the input amplitude, and (ii) nonlinearity-induced transparency when an opaque metamaterial becomes left-handed (and therefore transparent) with the growth of the input amplitude. The first case corresponds to the dependence of the effective magnetic permeability on the external field shown in Figs. 14.1a and 14.1c, when initially negative magnetic permeability (we consider $\varepsilon < 0$ in all frequency range) becomes positive with the growth of the magnetic field intensity. The second case corresponds to the multi-valued dependence shown in Fig. 14.1b.

In all numerical simulations, we use *linearly growing* amplitude of the incident field within the first 50 periods, that becomes constant afterwards. The slab thickness is selected as $1.3\,\lambda_0$ where λ_0 is a free-space wavelength. For the parameters we have chosen, the metamaterial is left-handed in the linear regime for the frequency range from $f_1 = 5.787$ GHz to $f_2 = 6.05$ GHz.

Our simulations show that for the incident wave with the frequency $f_0 = 5.9$ GHz (i.e. inside the left-handed transmission band), electromagnetic field reaches a steady state independently of the sign of the nonlinearity. In the linear regime, the effective parameters of the metamaterial at the frequency f_0 are: $\varepsilon = -1.33 - 0.01\,\mathrm{i}$ and $\mu = -1.27 - 0.3\,\mathrm{i}$; this allows excellent impedance matching with surrounding air. The scattering results in a vanishing reflection coefficient for small incident intensities.

Reflection and transmission coefficients are qualitatively different for two different types of infilling nonlinear dielectric. For the defocusing nonlinearity, the reflection coefficient varies from low to high values when the incident field exceeds some threshold value. Such a sharp transition can be explained in terms of the hysteresis behavior of the magnetic permeability shown in Fig. 14.1c. When the field amplitude in metamaterial becomes higher than the critical amplitude (shown by a dashed arrow in Fig. 14.1c), magnetic permeability changes its sign, and the metamaterial becomes opaque. Our FDTD simulations show that for overcritical amplitudes of the incident field, the opaque region of positive magnetic permeability appears inside the slab [17]. The magnetic permeability experiences an abrupt change at the boundary between the transparent and opaque regions.

For the focusing nonlinearity (see Fig. 14.5), the dependence of the reflection and transmission coefficients on the amplitude of the incident field is smooth. This effect originates, firstly, from a gradual detuning from the impedance matching condition, and, for higher powers, from the appearance

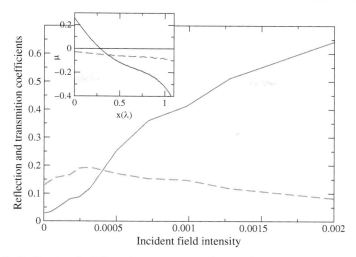

Fig. 14.5. Reflection (solid) and transmission (dashed) coefficients for a nonlinear metamaterial slab vs. the incident field intensity H^2/E_c^2, for the focusing nonlinearity. Inset shows real (solid) and imaginary (dashed) parts of magnetic permeability inside the slab in one of the high-reflectivity regimes [17]

of an opaque layer (see the inset in Fig. 14.5) with a positive value of the magnetic permeability that is a continuous function of the coordinate inside the slab.

Now we consider another interesting case when initially opaque metamaterial becomes transparent with the growth of the incident field amplitude. We take the frequency of the incident field to be $f_0 = 5.67\,\text{GHz}$, so that magnetic permeability is positive in the linear regime and the metamaterial is opaque. In the case of self-focusing nonlinear response ($\alpha = 1$), it is possible to switch the material properties to the regime with negative magnetic permeability (see Fig. 14.1b) making the material slab left-handed and therefore transparent. Moreover, one can expect the formation of self-focused localized states inside the composite, the effect which was previously discussed for the interaction of the intense electromagnetic waves with over-dense plasma [18–20].

For the lower incident power, or in linear regime, we observe total reflection from the metamaterial slab. However, in a strongly nonlinear regime, we observe the effect of the dynamical self-modulation of the reflected electromagnetic wave that results from the periodic generation of the self-localized states inside the metamaterial (see Fig. 14.6). Such localized states resemble *temporal solitons*, which transfer the energy away from the interface. Figure 14.6c

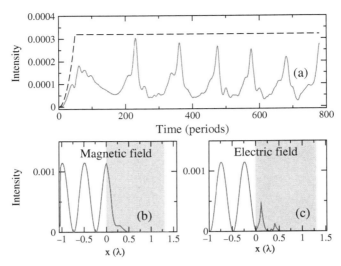

Fig. 14.6. (a) Reflected (solid) and incident (dashed) wave intensity H^2/E_c^2 for the overcritical nonlinear regime, (b), (c) Distribution of the magnetic and electric fields at the end of simulation time; the metamaterial is shaded [17]

shows an example when two localized states enter the metamaterial. These localized states appear on the jumps of the magnetic permeability and, as a result, we observe a change of the sign of the electric field derivative at the maximum of the soliton intensity, and subsequent appearance of transparent regions in the metamaterial. Unlike all previous cases, the field structure in this regime does not reach any steady state for high enough intensities of the incident field.

14.5 Electromagnetic Spatial Solitons in Metamaterials

Similar to other nonlinear media [21], nonlinear left-handed composite materials can support self-trapped electromagnetic waves in the form of *spatial solitons* [22]. Such solitons possess interesting properties because they exist in materials with a hysteresis-type (multi-stable) nonlinear magnetic response. Below, we describe novel and unique types of single- and multi-hump (symmetric, antisymmetric, or even asymmetric) backward-wave spatial electromagnetic solitons supported by the nonlinear magnetic permeability.

Due to the multi-valued function of magnetic permeability, there exist a wide range of soliton families in metamaterials [22]. Apart from fundamental soliton, (see Fig. 14.7a), more complex localized solutions can be found, some examples are shown in Figs. 14.7b–d. Such localized states exist due to induced regions where the metamaterial has different value of magnetic permeability (or even different sign), and these regions act as self-induced waveguides [22].

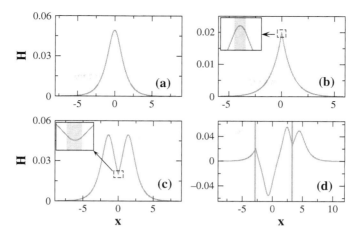

Fig. 14.7. Examples of different types of solitons: (**a**) fundamental soliton, (**b**), (**c**) solitons with one domain of, respectively, negative and positive magnetic permeability (shaded area in the inset), (**d**) soliton with two different domains. Insets in (b), (c) show the magnified regions of the steep change of the magnetic field [22]

For the multi-valued nonlinear magnetic response, the domains with different values of magnetic permeability "excited" by the spatial soliton can be viewed as effective induced left-handed waveguides which make possible the existence of single- and multi-hump solitons. Due to the existence of such domains, the solitons can be not only symmetric, but also antisymmetric and even asymmetric. Formally, the size of an effective domain can be much smaller than the wavelength and, therefore, there exists an applicability limit for the obtained results to describe nonlinear waves in realistic structures.

14.6 Second-order Nonlinear Effects in Metamaterials

Inclusion of elements with non-symmetric current-voltage characteristics such as diodes into the split-ring resonators will result in *a quadratic nonlinear response* of the metamaterial [9]. This quadratic nonlinearity is responsible for the recently analyzed parametric processes such as the second-harmonic generation (SHG) [23, 24] and three-wave mixing [25]. In particular, the first analysis of SHG from a semi-infinite left-handed medium has been briefly presented by Agranovich et al. [23], who employed the nonlinear optics approach. First experiments with SHG in arrays of SRRs were performed in Refs. [26,27].

In this section we consider the problem of SHG during the scattering from a semi-infinite left-handed medium (or a slab of the left-handed material of a finite extent) and demonstrate the possibility of the exact phase-matching, quite specific for the harmonic generation by the backward waves. With this condition, we demonstrate that exact phase matching between a backward

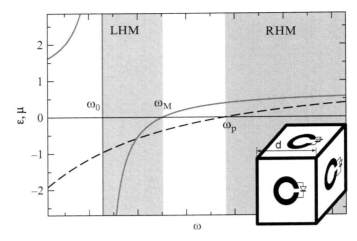

Fig. 14.8. Frequency-dependent magnetic permeability μ (solid) and electric permittivity ε (dashed) of the composite. Two types of the regions (LHM or RHM) where the material is transparent are shaded. For other frequencies it is opaque. Characteristic frequencies ω_0, ω_M, and ω_p are defined in Eqs. (14.4)–(14.6). Inset shows the unit cell of the metamaterial [29]

propagating wave of the fundamental frequency (FF) and the forward propagating wave at the second harmonic (SH) is indeed possible.

Firstly, we will describe our model including both the electric and magnetic responses. Then, we analyze quadratic nonlinearity and the SHG process in metamaterials. Next, we develop the corresponding coupled-mode theory for SHG with backward waves and present the analysis of both lossy and lossless cases of this model. Then, we will present the results of numerical simulations of SHG process a slab of finite-extension.

We consider a three-dimensional composite structure consisting of a cubic lattice of conducting wires and split-ring resonators (SRR), shown schematically in the insert of Fig. 14.8. We assume that the unit-cell size of the structure d is much smaller then the wavelength of the propagating electromagnetic field and, for simplicity, we choose a single-ring geometry of the lattice of SRRs. The results obtained for this case are qualitatively similar to those obtained in more involved cases of double SRRs. This type of microstructured medium is known to possess the basic properties of left-handed metamaterials exhibiting negative refraction in the microwave region.

In the effective-medium approximation, a response of this composite metallic structure can be described by averaged equations allowing one to introduce the effective dielectric permittivity and effective magnetic permeability of the form

$$\varepsilon(\omega) = 1 - \frac{\omega_p^2}{\omega^2}, \tag{14.4}$$

$$\mu(\omega) = 1 + \frac{F\omega}{(\omega_0^2 - \omega^2)}, \tag{14.5}$$

where ω_p is the effective plasma frequency, ω_0 is a resonant frequency of the array of SRRs, F is the form-factor of the lattice, and ω is the angular frequency of the electromagnetic waves. The product of permittivity ε and permeability μ defines the square of the effective refractive index, $n^2 = \varepsilon\mu$, and its sign determines if waves can ($n^2 > 0$) or cannot ($n^2 < 0$) propagate in the medium. Due to the medium dispersion defined by the dependencies (14.4) and (14.5), the wave propagation becomes possible only in certain frequency domains while the waves decay for other frequencies. Metamaterial possesses left-handed properties when both ε and μ become simultaneously negative, and such a frequency domain exists in the model described by Eqs. (14.4) and (14.5) provided $\omega_p > \omega_0$. In this case, the metamaterial is left-handed within the frequency range

$$\omega_0 < \omega < \min\{\omega_p, \omega_M\}, \quad \omega_M = \frac{\omega_0}{\sqrt{1-F}}, \tag{14.6}$$

where ω_p is the plasma frequency introduced in Eq. (14.4).

We assume that $\omega_M < \omega_p$, and in this case we have two frequency ranges where the material is transparent, the range where the material is left-handed (LHM), and the right-handed (RHM) domain for $\omega > \omega_p$, where both permittivity and permeability are positive (shaded domains in Fig. 14.8). For the frequencies outside these two domains, the composite material is opaque.

The composite material becomes nonlinear and it possesses a quadratic nonlinear response when, for example, additional diodes are inserted into the SRRs of the structure [9], as shown schematically in the insert of Fig. 14.8. Quadratic nonlinearity is known to be responsible for various parametric processes in nonlinear media, including the frequency doubling and generation of the second-harmonic field. In dispersive materials, and especially in the metamaterials with the frequency domains with different wave properties, the SHG process can be rather nontrivial because the wave at the fundamental frequency and the second harmonic can fall into *different domains* of the material properties.

The most unusual harmonic generation and other parametric processes are expected when one of the waves (either FF or SH wave) has the frequency for which the metamaterial becomes left-handed. The specific interest to this kind of parametric processes is due to the fact that the waves in the left-handed media are *backward*, i.e., the energy propagates in the direction opposite to that of the wave vector. Both phase-matching condition and nonlinear interaction of the forward and backward waves may become quite nontrivial, as is known from the physics of surface waves in plasmas [28].

In nonlinear quadratic composite metamaterials, interaction of the forward and backward waves of different harmonics takes place when the material is left-handed either for the frequency ω or the double frequency 2ω. Under this condition, there exist two types of the most interesting SHG parametric processes in metamaterials [29].

Case I. The frequency of the FF wave is in the range $\omega_0/2 < \omega < \omega_M/2$ and, therefore, the SH wave is generated with the double frequency in the LHM domain (see Fig. 14.8). For such parameters, the electromagnetic waves at the FF frequency are non-propagating, since $\varepsilon(\omega)\mu(\omega) < 0$. As a result, the field with the frequency ω from this range incident on a semi-infinite left-handed medium will decay exponentially from the surface inside the metamaterial. Taking into account Eqs. (14.4) and (14.5), the depth δ of this skin-layer can be found as

$$\delta = \left(k_\parallel^2 - \varepsilon\mu\frac{\omega^2}{c^2}\right)^{-1/2} < \frac{\lambda}{17}, \qquad (14.7)$$

where $k_\parallel$ is the tangential component of the wavevector of the incident wave, and λ is a free space wavelength. For the SH wave generated in this layer, the metamaterial becomes transparent. In this case, a thin slab of a metamaterial may operate as *a nonlinear left-handed lens* that will provide an image of the source at the second harmonic [30].

Case II. The FF wave is left-handed, whereas the SH wave is right-handed. Such a process is possible when $\omega_p < 2\omega_0$ (see Fig. 14.8). What is truly remarkable here is the possibility of exact phase-matching of the SHG parametric process, in addition to the cases discussed earlier in Ref. [23]. The phase-matching conditions for this parametric process are depicted in the dispersion diagram of Fig. 14.9 for the propagating waves in the metamaterial

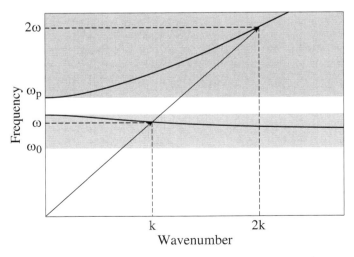

Fig. 14.9. Dispersion of plane waves $k(\omega)$ in the metamaterial. Arrows show the parameters of the FF and SH waves corresponding to the exact spatio-temporal phase matching [29]

where the dispersion of the plane waves is defined by the relation

$$D(\omega, k) = \left[k^2 - \varepsilon(\omega)\mu(\omega)\frac{\omega^2}{c^2}\right] = 0. \qquad (14.8)$$

The exact phase matching takes place when $2\,k(\omega) = k(2\,\omega)$. Different signs of the slopes of the curves at the frequencies ω and $2\,\omega$ indicate that one of the waves is forward, while the other wave is backward.

To study the SHG process in metamaterials we consider a composite structure created by arrays of wires and SRRs. To generate a nonlinear quadratic response of the metamaterial, we assume that each SRR contains a diode, as depicted schematically in the inset of Fig. 14.8. The diode is described by the current-voltage dependence,

$$I = \frac{U}{R_\mathrm{d}}\left(1 + \frac{U}{U_\mathrm{c}}\right), \qquad (14.9)$$

where U_c and R_d are the parameters of the diode, and U is the voltage on the diode. Equation (14.9) is valid provided $U \ll U_\mathrm{c}$, and it represents two terms of the Taylor expansion series of the realistic (and more complex) current-voltage characteristics of the diode.

Following the standard procedure, we consider two components of the electromagnetic field at the fundamental frequency ω and its second harmonic $2\,\omega$, assuming that all other components are not phase matched and therefore they give no substantial contribution into the nonlinear parametric interaction. Subsequently, we write the general coupled-mode equations describing the simultaneous propagation of two harmonics in the dispersive metamaterial as follows,

$$\Delta \boldsymbol{H}_1 + \varepsilon(\omega)\mu(\omega)\frac{\omega^2}{c^2}\boldsymbol{H}_1 = -\sigma_1 \boldsymbol{H}_1^* \boldsymbol{H}_2,$$

$$\Delta \boldsymbol{H}_2 + 4\varepsilon(2\,\omega)\mu(2\,\omega)\frac{\omega^2}{c^2}\boldsymbol{H}_2 = -\sigma_2 \boldsymbol{H}_1^2, \qquad (14.10)$$

where the indices "1", "2" denote the FF and SH fields, respectively, Δ is a Laplacian, and other parameters are defined as follows

$$\sigma_1 = \kappa/2\,R(\omega), \qquad \sigma_2 = \kappa/R^*(\omega),$$

$$\kappa = \frac{6\,\pi\,\left(\pi a^2\right)^3}{d^3 c^5}\left[\frac{\omega_0^4 \omega^2}{U_\mathrm{c} R_\mathrm{d} R(\omega) R(2\omega)}\right], \qquad (14.11)$$

where $R(\omega) = \omega_0^2 \omega^2 + \mathrm{i}\gamma\omega$, the asterisk stands for the complex conjugation, a and d are, respectively, the radius of the SRRs and the period of the metamaterial, and γ is the damping coefficient of the SRR. For simplicity, we assume that both FF and SH waves are of the same polarization, and therefore they can be described by only one component of the magnetic field. In this case, Eqs. (14.10) become scalar. In the derivation of Eqs. (14.10) we take into account the Lorentz-Lorenz relation between the microscopic and macroscopic magnetic fields [31]. Also, it is assumed that the diode resistance R_d is much larger than the impedance of the SRR slit, i.e. $R_\mathrm{d} \gg 1/\omega C$, so that the resonant properties of the composite are preserved.

Using Eqs. (14.10) one can describe SHG processes in metamaterials [29], including unusual concept of opaque nonlinear lens [30]. Moreover, the second order nonlinear response can be further enhanced through appropriate design of double-resonant binary metamaterial [32].

14.7 Conclusions

We have described several nonlinear effects recently predicted for microstructured metamaterials which exhibit left-handed properties and negative refraction. We believe that nonlinear properties of metamaterials can allow for much broader scope of future applications of such materials, including a dynamic control and tunability of the electromagnetic properties of the composite structures, harmonic generation, intensity-dependent switches, and generation of self-localized pulses and beams. We have experimentally demonstrated basic nonlinear effects in metamaterials, including shift of the left-handed frequency range. Due to general physical character of resonant nonlinear phenomena predicted here, similar effects can be observed in plasmonic structures (see, e.g., Ref. [33]), and in future nonlinear optical metamaterials.

Acknowledgements

During last years we have been collaborating with a number of people on the projects involving the theoretical studies of left-handed metamaterials and negative refraction, and we would like to thank all of them and especially those who made major contribution to the results reviewed in this Chapter. In particular, we thank Yuri Kivshar, Alexander Zharov, Nina Zharova, Maxim Gorkunov, Andrey Sukhorukov, Alexander Kozyrev, Costas Soukoulis, Allan Boardman, and Peter Egan. This work has been supported by a Discovery grant of the Australian Research Council.

References

1. V.G. Veselago, Usp. Fiz. Nauk **92**, 517 (1967), in Russian, English translation: Sov. Phys. Usp. **10**, 509 (1968)
2. J.B. Pendry, A.J. Holden, W.J. Stewart, and I. Youngs, Phys. Rev. Lett. **76**, 4773 (1996)
3. J.B. Pendry, A.J. Holden, D.J. Robbins, and W.J. Stewart, IEEE Trans. Microw. Theory Tech. **47**, 2075 (1999)

4. D.R. Smith, W.J. Padilla, D.C. Vier, S.C. Nemat Nasser, and S. Schultz, Phys. Rev. Lett. **84**, 4184 (2000)
5. M. Bayindir, K. Aydin, E. Ozbay, P. Markos, and C.M. Soukoulis, Appl. Phys. Lett. **81**, 120 (2002)
6. C.G. Parazzoli, R.B. Greegor, K. Li, B.E.C. Koltenbah, and M. Tanielian, Phys. Rev. Lett. **90**, 107401 (2003)
7. D. Schurig, J.J. Mock, B.J. Justice, S.A. Cummer, J.B. Pendry, A.F. Starr, and D.R. Smith, Science **314**, 977 (2006)
8. A.A. Zharov, I.V. Shadrivov, and Y.S. Kivshar, Phys. Rev. Lett. **91**, 037401 (2003)
9. M. Lapine, M. Gorkunov, and K.H. Ringhofer, Phys. Rev. E **67**, 065601 (2003)
10. C. Luo, S.G. Johnson, and J.D. Joannopoulos, Appl. Phys. Lett. **81**, 2352 (2002)
11. C. Luo, S.G. Johnson, J.D. Joannopoulos, and J.B. Pendry, Phys. Rev. B **65**, 201104 (2002)
12. M. Gorkunov, M. Lapine, E. Shamonina, and K.H. Ringhofer, Eur. Phys. J. B **28**, 263 (2002)
13. I.V. Shadrivov, A.A. Zharov, N.A. Zharov, and Y.S. Kivshar, Radio Sci. **40**, RS3S90 (2005)
14. A.M. Belyantsev, V.A. Kozlov, and V.I. Piskaryov, Infrared Phys. **21**, 79 (1981)
15. H. Li, A.L. Roytburd, S.P. Alpay, T.D. Tran, L. Salamanca Riba, and R. Ramesh, Appl. Phys. Lett. **78**, 2354 (2001)
16. I.V. Shadrivov, S.K. Morrison, and Y.S. Kivshar, Opt. Express **14**, 9344 (2006)
17. N.A. Zharova, I.V. Shadrivov, A.A. Zharov, and Y.S. Kivshar, Opt. Express **13**, 1291 (2005)
18. K. Zauer and L.M. Gorbunov, Fiz. Plazmy **3**, 1302 (1977), in Russian
19. A.A. Zharov and A.K. Kotov, Fiz. Plazmy **10**, 615 (1984)
20. A.V. Kochetov and A.M. Feigin, Fiz. Plazmy **14**, 716 (1988)
21. Y.S. Kivshar and G.P. Agrawal, *Optical Solitons: From Fibers to Photonic Crystals*, Academic Press, San Diego (2003)
22. I.V. Shadrivov and Y.S. Kivshar, J. Opt. A **7**, S68 (2005)
23. V.M. Agranovich, Y.R. Shen, R.H. Baughman, and A.A. Zakhidov, Phys. Rev. B **69**, 165112 (2004)
24. I.V. Shadrivov, Photonics Nanostruct. **2**, 175 (2004)
25. M. Lapine and M. Gorkunov, Phys. Rev. E **70**, 66601 (2004)
26. M.W. Klein, C. Enkrich, M. Wegener, and S. Linden, Science **313**, 502 (2006)
27. M.W. Klein, M. Wegener, N. Feth, and S. Linden, Opt. Express **15**, 5238 (2007)
28. A.A. Zharov, Fiz. Plazmy **17**, 20 (1991)
29. I.V. Shadrivov, A.A. Zharov, and Y.S. Kivshar, J. Opt. Soc. Am. B **23**, 529 (2006)
30. A.A. Zharov, N.A. Zharova, I.V. Shadrivov, and Y.S. Kivshar, Appl. Phys. Lett. **87**, 091104 (2005)
31. L.D. Landau and E.M. Lifshitz, *Electrodynamics of Continuous Media*, Pergamon Press, Oxford (1963)
32. M.V. Gorkunov, I.V. Shadrivov, and Y.S. Kivshar, Appl. Phys. Lett. **88**, 71912 (2006)
33. R.E. Noskov and A.A. Zharov, Opto-Electronics Review **14**, 217 (2006)

15

Circuit Model of Gain in Metamaterials

Allan D. Boardman[1], Neil King[1], and Yuriy Rapoport[2]

[1] Institute for Materials Research, University of Salford, Salford, M6 5WT, United Kingdom
a.d.boardman@salford.ac.uk, dr.n.king@googlemail.com
[2] Physics Faculty, Taras Shevchenko Kyiv National University, Prospect Glushkov 6, 22 Kyiv, Ukraine
laser@i.kiev.ua

15.1 Introduction

Metamaterials embody exciting prospects for a new generation of novel photonic devices. From their initial emergence as a physical construct in the GHz domain at the start of the 21st century [1–3], they have attracted a significant amount of global interest [4–13] with considerable effort being undertaken to extend their operation into the THz window and even optical regimes [14,15]. However, as they stand, early theoretical indications are that losses will cause potential problems for *all* possible frequencies and, in particular, kill any opportunity [16] for a useful metamaterial operating around and above 30 THz.

Such losses are inevitably closely linked to the resonant behaviour of the metaparticles and is addressed here by the placement of active diodes onto a form of metallic split-ring. The use of diodes to create a nonlinear magnetic response [16] and to create tunability [17] has already been discussed but active diodes [18] not only promise means of reducing losses but they can be deployed to produce an overall gain [19]. This behaviour is readily scalable from GHz to THz and even to nanowire [20] and nanoparticle-based metamaterials [21] operating in the optical frequency window. Nevertheless, it is highlighted here that instabilities could present a serious issue. From an investigation of the dispersion relation for a plane wave, a number of conditions are derived that identify the limits placed upon the system parameters, in order to ensure stable overall gain. Any examination of loss, or gain, must, however, be conducted from the perspective of the entire metamaterial, including the permittivity. Depending on the level of sophistication required in the fabrication technique, split-rings may be engineered with different shapes and deployed in a number of different arrays. The most popular have either a circular, or square shape. The term "split-ring" is treated here as a generic name and is not necessarily indicative of a specific shape.

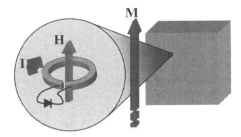

Fig. 15.1. Induced current I within a metamaterial lump with a magnetisation M resulting from a propagating macroscopic magnetic field H

15.2 Negative Resistance Structures

Figure 15.1 is a sketch of the kind of split-ring that could be used as a metaparticle and it shows that a diode is attached across the gap with the aim of using the diode current-voltage characteristic to introduce negative resistance [19]. The purpose of the latter is to introduce some form of amplification into the metamaterial. Figure 15.1 illustrates what is happening at the split-ring metaparticle, in response to a propagating electromagnetic wave that subjects it to a magnetic field H and an induced magnetisation M.

Negative dynamic resistance, or negative *differential* resistance (NDR), relates to a portion of the current-voltage (I, V) characteristic of a device that displays a negative slope. Possible two-port devices that demonstrate such behaviour are the Gunn, or resonant tunnel diodes (RTD) [22–24]. The latter are regarded as one of the fastest devices that it is possible to make with subsequent oscillator circuit frequencies extending into the THz frequency range [18, 22, 24]. Furthermore, they are based upon well established doped semiconductor and molecular beam epitaxy technology. As such, there is tremendous scope for flexibility in RTD design stemming from variations in doping levels, types of semiconductor and layer thickness. Hence, they are the main focus of interest here. A typical (I, V) curve for an RTD is shown in figure 15.2 where no current flows under zero bias. As the voltage increases, the band-structure deforms and it becomes energetically feasible for electrons near the Fermi level E_F to tunnel from occupied states on the left, through the first barrier, into the well and then through the second barrier into unoccupied states on the right. Resonant tunnelling occurs for the applied voltage that provides injected electrons with exactly the same energy as the allowed state within the well [22, 24].

15.3 Diode Inclusions

In order to consider the impact of this diode upon the effective relative permeability of the metamaterial requires, first of all, a consideration of the available dc (I, V) characteristic. If the diode is biased onto a point on the negative

15 Circuit Model of Gain in Metamaterials

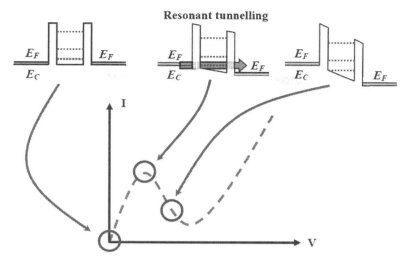

Fig. 15.2. Typical (I, V) characteristic as a result of a deformation of the band structure under an applied voltage. The three snapshots are for key points along the curve [20, 22]. As the voltage increases, occupied electrons states near the Fermi level E_F can tunnel sequentially though the barrier into the well then through the second barrier into unoccupied states on the right. E_C is the conduction band edge and the horizontal dotted lines show possible energy levels in the well

slope of the kind of (I, V) characteristic shown in figure 15.2 then the variation of I and V about this bias point has the following form [23]

$$I = -g_0 V + BV^3 \tag{15.1}$$

where g_0 is the effective linear conductance and B is a constant. An incident electromagnetic plane wave will introduce the necessary ac signals to drive the diode about its bias point. A single plane wave with angular frequency ω can be investigated without loss of generality and the corresponding complex ac voltage on the diode can be written as $V = U_D/2 \exp(-i\omega t) + $ c.c., where t is time, and the corresponding complex ac current is $I = I_D/2 \exp(-i\omega t) + $ c.c. In the Fourier domain, therefore, equation (15.1) becomes

$$I_D = -g_0 U_D + b |U_D|^2 U_D \tag{15.2}$$

where $b = 3B/8$ and it is assumed that everything is operating at the fundamental frequency ω and this can be easily arranged if necessary by including a low-pass filter into the operational circuit [25].

The equivalent circuit of the diode-loaded split-ring is given in figure 15.3 in which C_{BLOCK} is used to prevent any dc biasing currents on the diode affecting the split-ring structure.

In the rest of the circuit, the ac emf source U is created by the oscillating magnetic field carried by a wave passing through the metamaterial and L,

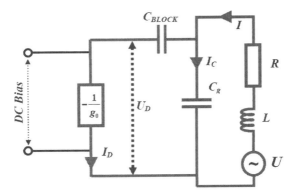

Fig. 15.3. Equivalent circuit of a split-ring resonator (blue) with an attached diode structure (green). The dotted double-headed arrow simply signifies the presence of a separate stable dc biasing circuit

C_g and R are, respectively, the inductance, capacitance and resistance of the split-ring. The exact nature of the biasing circuit is not important to the calculations below so no further discussion will be given here of this feature.

Given that the time-dependence is $\exp(-i\omega t)$, it is clear that

$$U = I(R - i\omega L) + U_D. \tag{15.3}$$

Similarly, the ac current is

$$I = I_D + I_C = -\left(g_0 + i\omega C_g - b|U_D|^2\right)U_D. \tag{15.4}$$

The first-order behaviour of this diode-loaded ring does not need to dwell upon the nonlinear excursion of the current and voltage about the operating point. Building this assumption into the equations gives the following form for the ac current

$$I = \frac{-i\dfrac{U}{L}\omega}{\dfrac{\omega^2 \omega_0^2}{\omega^2 + g_0^2/C_g^2} - \omega^2 - i\omega\left[\Gamma\left(\dfrac{\omega_0^2}{\omega^2 + g_0^2/C_g^2}\right)\dfrac{g_0}{C_g^2}\right]} \tag{15.5}$$

where the definitions

$$\omega_0^2 = \frac{1}{LC_g}, \qquad \Gamma = \frac{R}{L} \tag{15.6}$$

have been adopted. It can be seen from the denominator of (15.5) that the aim of setting the diode to oppose the loss associated with the Γ term has been achieved.

Given this response for a diode-loaded single-ring, the next step is to set-up some form of homogenisation. The latter means that, although it is clear that the metamaterial is being built from discrete metaparticles, it is always assumed that the size of the individual particles must be much less than the operational wavelength. This is called a quasi-static approximation [26] and the outcome is that the material will be described in terms of an effective relative permeability and an effective relative permittivity. It is the permeability that is being investigated here through the addition of diodes and it will be assumed that the dielectric behaviour has not been modified and that the relative permittivity is the standard negative function displayed by a metal at the operational frequency.

Each split-ring is an effective magnetic dipole with a moment equal to IA where A is the effective cross-sectional area of the ring and I is the current. Hence, if there are n rings per unit volume, the total magnetisation of the system is $M = nIA$ and

$$M = \frac{\omega^2 \mu_0 H \frac{nA^2}{L}}{\frac{\omega^2 \omega_0^2}{\omega^2 + g_0^2/C_g^2} - \omega^2 - i\omega\left[\Gamma\left(\frac{\omega_0^2}{\omega^2 + g_0^2/C_g^2}\right)\frac{g_0}{C_g^2}\right]}. \tag{15.7}$$

In practice, H is actually the local field [26] but for clarity this correction is not included here because it does not change the qualitative behaviour of M. Hence, given the fact that the magnetization is related to the magnetic field H and the relative permeability by the formula $\mu = 1 + M/H$, then

$$\mu = 1 + \frac{F'\Omega^2}{\frac{\Omega^2}{\Omega^2 + (g_0/\omega_0 C_g)^2} - \Omega^2 - i\Omega\left[\gamma - \frac{g_0/\omega_0 C_g}{\Omega^2 + (g_0/\omega_0 C_g)^2}\right]}, \tag{15.8}$$

where $F' = nA^2\mu_0/L$ and the normalised quantities are $\Omega = \omega/\omega_0$ and $\gamma = \Gamma/\omega_0$. Note that $g_0/\omega_0 C_g = g_0\sqrt{L}/\sqrt{C_g}$ where the inductance and capacitance have the same dependence upon the characteristic size of the split-ring [27] so that the inclusion of the diode is approximately independent of the ring size. It is interesting that resonant tunnel diodes even down to $2 \times 2\,\mu m^2$ have been studied recently for operation in the 1 THz region [28]. Hence it can be broadly concluded that scalability is applicable to the ideas being exposed here.

For a given set of typical parameters, figures 15.4 show the difference between the passive and active cases. In other words they show quantitatively how the inclusion of diodes can influence the real μ' and imaginary μ'' parts of the relative permeability.

The qualitative behaviour of the important quantity μ' appears to be quite similar whether the system is active or passive. This is an excellent outcome because it means that the resonant behaviour in the relative permeability that underpins the magnetic property of the metamaterial is accessible in both the passive and active operational modes. This leaves the focus of attention upon

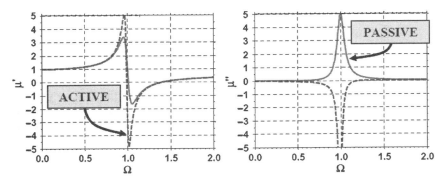

Fig. 15.4. (a) Real and (b) imaginary parts of the relative permeability for $g_0 = 0.15\omega_0 C_g$. The solid red curves are for the passive case whereas the blue dashed curve is the active case

the imaginary part of the relative permeability (μ''), which is the pivotal quantity used to combat loss. It appears that a suitable conductance can always be selected, both to retain the resonant behaviour, and to give a net gain to the system.

15.4 Discussion of Stability

Instability can be classified [29, 30] into absolute and convective, as illustrated in figure 15.5. For absolute instability it can be seen in figure 15.5a that an input excitation is not appreciably propagating along z but is growing substantially in time at every spatial point. Physically, this means that, if the metamaterial is being used in an absolute instability regime, even noise will grow rapidly and swamp the system. Figure 15.5b shows that an excitation can grow in time but it is being swept ("convected") away along z. Conventionally, this is understood to be normal spatial gain, or amplification but the technical term is that this is an example of convective instability. The latter really means that, even though the disturbance can grow while the excitation is traveling, nevertheless, if a particular point in space is selected, it will be found that the excitation is dying away. Hence, the language adopted here refers to convective instability as a regime of stable gain meaning that the catastrophic absolute instability regions have been avoided.

In order to get some quantitative measure of why and how a medium such as a metamaterial that is having its loss diminished by the addition of gain can become unstable is perhaps not immediately clear. The conceptual difficulty derives from the fact that for a loss/gain system that is supporting propagating electromagnetic waves with wave number k and angular frequency ω it is possible that *both* ω and k are complex. Traditionally, it is often assumed that the frequency is real and the wave number is complex, or vice versa. Indeed,

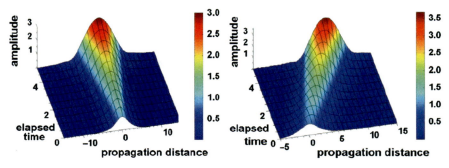

Fig. 15.5. Illustration of the evolution of an input state undergoing (a) absolute instability, (b) convective instability (spatial amplification)

if the wave number is complex and the frequency is real then an examination of the imaginary part of the wave number appears to give an indication of whether a wave is growing or not. The situation, however, is not as simple as this and it is often the case that the frequency is also complex and that some mapping of ω onto the complex k-plane is required in order to work out what stability limits, or otherwise, a material may possess [29]. In fact, a useful technique is to locate the roots of the dispersion equation on the complex k-plane, then choose a real part of the frequency and finally look at the paths traced out by the roots on this complex k-plane as the imaginary part of the frequency is varied. The first requirement for the existence of an absolute instability is that a saddle-point must exist and, hence, a double root is reached, for which $\omega'' > 0$ with the choice $\exp(-i\omega t)$. Basically, this means that an electric field component, for example, which is being evaluated with a contour integration will become infinite because the saddle-point prevents the necessary deformation of the integration path. A vital condition, however, is that the roots of the dispersion equation actually approach the saddle-point from different halves of the complex k-plane [30] as sketched in figure 15.6. This will be nicely illustrated numerically below. As has been stated previously, the physics of this scenario is as follows. If an excitation is created at $z = 0$, the waves will decay away from this point if the initial excitation grows fast enough. It is therefore expected that when the roots of the dispersion equation are tracked on the complex k-plane one of them (k_+) relates to $z > 0$ behaviour and the other (k_-) relates to $z < 0$ behaviour [29]. The question now arises as to what will happen if an excitation is fed into the system at $z = 0$. When this occurs, it should *not* be expected that $k_+ = k_-$ at $z = 0$ yet when this equality does occur it should *not* be expected that an excitation needs to be fed into the $z = 0$ point. In other words, when $k_+ = k_-$ a complex resonance frequency is encountered and no source is required. This is just a physical way of saying that the system suffers from absolute instability, which of course is catastrophic. Any attempt to observe gain in a metamaterial can

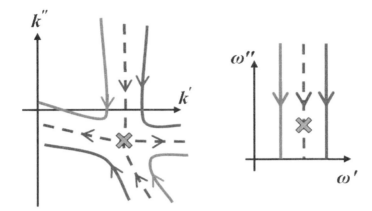

Fig. 15.6. The figure shows the complex k-plane using the complex wave number $k = k' + ik''$ and the complex ω-plane using the complex frequency $\omega = \omega' + i\omega''$. The cross shows the location of the saddle-point and each contour is generated by adopting $\omega' = $ const and then tracking what happens when the imaginary part of the frequency is decreased

be seriously impeded if an unsuitable frequency is selected. A final comment is that a saddle-point occurs where the group velocity is zero.

For the metamaterial under discussion here, the relative permittivity and relative permeability are frequency dependent and the dimensionless dispersion equation for plane wave propagation is simply

$$K^2 = \frac{k^2 c_0^2}{\omega_0^2} = \Omega^2 \varepsilon(\Omega)\mu(\Omega) \tag{15.9}$$

where c_0 is the velocity of light in vacuum. Here, there is a strong dependence of the permittivity $\varepsilon(\Omega)$ and the permeability $\mu(\Omega)$ upon the dimensionless frequency and it is clear that the saddle-point referred to above occurs at $K = 0$, although it must be acknowledged that there is an "essential singularity" [31, 32] for $|K| \to \infty$. The latter needs rather careful consideration because the quasi-static approximation, upon which the metamaterial is based, requires that $Kd \ll 1$ where d is the normalized width of the unit cell in which the metaparticle resides. This point will be returned to later on.

From the dispersion relation, there are three possibilities that yield $K = 0$. The trivial case is when $\Omega = 0$ but the important condition occurs when either complex ε, or μ, goes exactly to zero. Since the relative permittivity in this case has only a loss parameter it is associated with stability, so that the saddle-point is defined by setting the complex relative permeability to zero. This action will define the edge of a frequency window below which it is safe to operate without the onset of absolute instability until the essential singularity region is encountered.

Given that the normalized complex frequency is $\Omega = \Omega' + i\Omega''$, the algebra that leads to the operational frequency window defined as $\Omega'_{\text{POLE}} < \Omega' < \Omega'_{\text{ZERO}}$ is straightforward but somewhat laborious. In fact,

$$\Omega'_{\text{ZERO}} = \frac{\pm\sqrt{4(1-F')\left(1-\gamma\frac{g_0}{C}\right) - \left[(F'-1)\frac{g_0}{C}+\gamma\right]^2}}{2(1-F')}, \qquad (15.10)$$

$$\Omega''_{\text{ZERO}} = \frac{\frac{g_0}{C}(1-F') - \gamma}{2(1-F')}, \qquad (15.11)$$

$$\Omega'_{\text{POLE}} = \frac{\pm\sqrt{4\left(1-\gamma\frac{g_0}{C}\right) - \left(\frac{g_0}{C}-\gamma\right)^2}}{2}, \qquad (15.12)$$

$$\Omega''_{\text{POLE}} = \frac{\frac{g_0}{C} - \gamma}{2}, \qquad (15.13)$$

which shows that the saddle-point condition places the following limitation upon the diode conductance

$$g_0 < \frac{\Gamma C_g}{(1-F')} = g_0^{\text{MAX}}. \qquad (15.14)$$

15.5 Numerical Analysis

For a loss/gain medium it is necessary, in principle, to consider the way in which a complex frequency relates to a complex wave number. Indeed, the way in which the complex Ω-plane maps onto the complex K-plane lies at the heart of a systematic search for stability criteria [29], when attempting to add gain to a metamaterial. Considerable insight can be obtained from numerical analysis and some outcomes of this approach will now be discussed. Figure 15.7, for a passive medium, assumes the dispersion equation given by (15.9) and shows the complex K-plane onto which the movement of the roots of the dispersion equation are tracked as the complex frequency is changed. Note that the normalised complex wave number is $K = K' + iK''$. Specifically, a real part of the frequency (Ω') is selected and the movement of the roots is followed as the imaginary part of the frequency (Ω'') is changed by adopting the values given by the right-hand scale of the figure. It can be seen that, for some values of Ω', in parts of the complex K-plane there is a reasonably rapid colour change induced by the root positions, caused by Ω''. In other parts this is not the case. The figure is labelled to reflect this and identifies regions in terms of the magnitude of Ω'. An examination of the relative permeability will identify it with the $\Omega' \approx 1$ regions and clearly the $\Omega' \ll 1$ is identified with the relative permittivity. It is possible now to build upon this

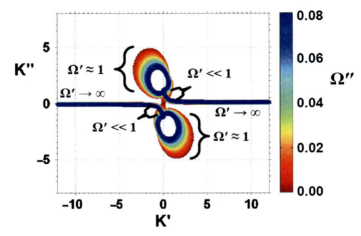

Fig. 15.7. Passive metamaterial. Complex K-plane populated by the roots of the dispersion equation. Real frequency variation: $\Omega' = 0 \to +\infty$. Diode conductance: $g_0 = 0$. Colour scale: measures imaginary part of the frequency. For a given Ω' the roots are tracked for the range $0 \leq \Omega'' \leq 0.08$

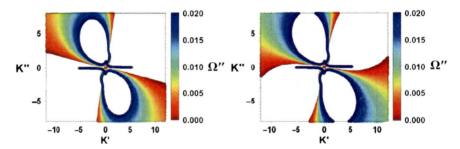

Fig. 15.8. Active metamaterial: root population of complex K-plane. Real frequency variation: $\Omega' = 0 \to +12$. Diode conductance: (**a**) $g_0 = 0.10\omega_0 C_g$ and (**b**) $g_0 = 0.11\omega_0 C_g$. Colour scale: $0 \leq \Omega'' \leq 0.02$

elegant representation. First of all, a comparison between this passive case and the manner in which the diode conductance changes the picture for an active medium can be developed. Secondly, this representation will facilitate a dramatic appearance of the saddle-point that is the harbinger of absolute instability.

Figure 15.8 shows the maps obtained for an active metamaterial and the aim is to demonstrate how the behaviour of the system as the diode conductance changes. The left- and right-hand figures show that, when g_0 changes slightly, an excursion from the upper-half of the complex K-plane to the lower-half is made and vice-versa. The interpretation of these results is as follows. In figure 15.8a the movement of the roots indicate clearly that g_0

15 Circuit Model of Gain in Metamaterials 269

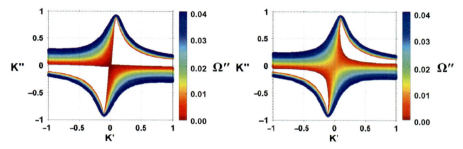

Fig. 15.9. Zooming into the origin shows the edge of the gain window as a saddle-point in the complex K-plane. Real frequency variation: $\Omega' = 0.8 \to +\infty$. Diode conductance: (**a**) $g_0 = 0.20\omega_0 C_g$ and (**b**) $g_0 = 0.22\omega_0 C_g$. Colour scale: $0 \leq \Omega'' \leq 0.04$

has an insufficient influence to create a net amplification. On the other hand, figure 15.8b shows that the diode conductance value is now enough to create a net spatial amplification. Having demonstrated this, it is now interesting to pursue the question of amplification more generally in order to quantify the possible appearance of absolute instability. This means that a search must now be made numerically to discover whether there is actually a saddle-point in the complex K-plane and whether conductance values for the diode can be selected to avoid or encourage its existence.

Figures 15.9 use values of g_0 that show the onset and development of the saddle-point just mentioned. These figures are the magnified "zoomed" representations of what happens in the close vicinity of the origin of figures 15.8. Although the mathematical development given earlier implies that a saddle-point will occur at $K = 0$ the numerical results given in figure 15.9 illustrate vividly that this is, indeed, the case. Hence, this purely numerical approach successfully pinpoints the real part of the frequency at which there is an onset of absolute instability. Naturally, the figures are intended to show the qualitative and quantitative trends rather than the specific numerical values. Referring back to figure 15.3 shows that there is a particular value of Ω' that labels a contour that passes through the saddle-point. From equation (15.10) this frequency is labeled as $\Omega' = \Omega'_{\text{ZERO}}$ and, naturally, this value is part of the data developed for figures 15.9a and 15.9b. In addition to the singularity referred to as the absolute instability point, there is also an essential singularity associated with the $|K| \to \infty$ region [32]. Indeed as equation (15.12) shows, the essential singularity occurs at $\Omega' = \Omega'_{\text{POLE}}$. The outcome of all this stability analysis is that certain frequencies can be labeled as "dangerous" in the sense that operating a gain-driven metamaterial in their vicinity will lead to catastrophic failure. A rough guide is to set the operational window to be $\Omega'_{\text{POLE}} < \Omega' < \Omega'_{\text{ZERO}}$ with the assumption that safe operation should steer clear of the window edges. Figure 15.10 is an illustration of which parts of the complex relative permeability are safe to use given the tendency of the material to engage in unstable behaviour.

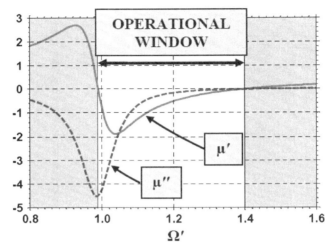

Fig. 15.10. Real and imaginary parts of the complex relative permeability of an active metamaterial plotted for real frequency. Diode conductance: $g_0 = 0.21\omega_0 C_g$

15.6 Summary

This chapter discusses a solution to the problem of how to address loss in metamaterials. It is assumed that the metamaterial is constructed using split-ring metaparticles to provide the complex relative permeability. In combination with the usual conducting wire array, such split-rings contribute the magnetic property that is required to endow a metamaterial with spectacular properties such as the ability to support backward-waves. One of the recognized problems with this type of material is that the very exploitation of a magnetic resonance usually occurs in the region of very high absorption. Naturally, loss of this magnitude, when propagating an electromagnetic wave through the metamaterial, is a serious drawback to any possible application, and something needs to be done to alleviate, or eliminate it. A straightforward theory is provided here, based upon an equivalent circuit, that shows how the conductance of a resonant tunnel diode can be introduced as the important parameter leading to gain. It is demonstrated numerically that gain can offset the loss very well but the chapter also shows that there can be a problem when the resultant metamaterial becomes unstable. A thorough discussion of how to access stability criteria is given and it is demonstrated that this task is quite difficult because it involves the mapping of the complex frequency-plane onto the complex wave number-plane. It is shown that absolute instability is associated with a saddle-point in the complex wave number-plane, which, in turn, involves the investigation of how roots of the dispersion equation move.

In addition to this, it is pointed out that there is also an essential singularity at large wave numbers. Taken together, the singularities that are exposed are used to create an operational window for a typical metamaterial. It should be emphasized that this is very preliminary work and that much more detailed calculations will appear in the future.

References

1. J.B. Pendry, A.J. Holden, D.J. Robbins, and W.J. Stewart, IEEE Trans. Microwave Theory Tech. **47**, 2075 (1999)
2. R.A. Shelby, D.R. Smith, and S. Schultz, Science **292**, 77 (2001)
3. J. Pendry, Physics World **14**, 47, (2001)
4. A.D. Boardman, L. Velasco, N. King, and Y. Rapoport, J. Opt. Soc. Am. B **22**, 1443 (2005)
5. A.D. Boardman, N. King, and L. Velasco, Electromagnetics **25**, 365 (2005)
6. A.D. Boardman, P. Egan, L. Velasco, and N. King, J. Opt. A **7**, S57 (2005)
7. A.D. Boardman, N. King, Y. Rapoport, and L. Velasco, New J. Phys. **7**, 191 (2005)
8. D.R. Smith and N. Kroll, Phys. Rev. Lett. **85**, 2933 (2000)
9. I.V. Shadrivov, A.A. Sukhorukov, Y.S. Kivshar, A.A. Zharov, A.D. Boardman, and P. Egan, Phys. Rev. E **69**, 016617 (2004)
10. R.W. Ziolkowski and E. Heyman, Phys. Rev. E **64**, 055625 (2001)
11. N. Engheta, IEEE Antennas Wireless Propagat. Lett. **1**, 10 (2002)
12. A.A. Houck, J.B. Brock, and I.L. Chuang, Phys. Rev. Lett. **90**, 137401 (2003)
13. A. Grbic and G.V. Eleftheriades, J. Appl. Phys. **92**, 5930 (2002)
14. S. Linden, C. Enkrich, M. Wegener, J.F. Zhou, T. Koschny, and C.M. Soukoutis, Science **306**, 1351 (2004)
15. W.S. Cai, U.K. Chettiar, H.K. Yuan, V.C. de Silva, A.V. Kildishev, V.P. Drachev, and V.M. Shalaev, Opt. Express **15**, 3333 (2007)
16. J.O. Dimmock, Opt. Express **11**, 2397 (2003)
17. M. Lapine, M. Gorkunov, and K.H. Ringhofer, Phys. Rev. E **67**, 065601 (2003)
18. Y. Ookawa, S. Kishimoto, K. Maezawa, and T. Mizutani, IEICE Trans. Electron. **E89C**, 999 (2006)
19. A.D. Boardman, Y.G. Rapoport, N. King, and V.N. Malnev, J. Opt. Soc. Am. B **24**, 11, (2007)
20. A.K. Sarychev, G. Shvets, and V.M. Shalaev, Phys. Rev. E **73**, 036609 (2006)
21. N. Engheta, A. Salandrino and A. Alù, Phys. Rev. Lett. **95**, 095504 (2005)
22. K.K. Ng, *Complete guide to semiconductor devices*, McGraw-Hill, New York (1995)
23. M.A. Lee, B. Easter, and H.A. Bell, *Tunnel Diodes*, Chapman and Hall, London (1967)
24. S.M. Sze, *Semiconductor devices, physics and technology*, John Wiley & Sons, New York (2002)
25. Y.-F. Lau, Q. Xue, C.-H. Chan, and M.-Y. Xia, Micro. Opt. Tech. Lett. **49**, 434 (2006)
26. J. Jackson, *Classical Electrodynamics*, John Wiley & Sons, New York (1999)
27. J. Zhou, T. Koschny, M. Kafesaki, E.N. Economou, J.B. Pendry, and C.M. Soukoulis, Phys. Rev. Lett. **95**, 223902 (2005)

28. N. Orihashi, S. Suzuki and M. Asada, Appl. Phys. Lett. **87**, 233501 (2005)
29. H. Hartnagel, *Semiconductor Plasma Instabilities*, American Elsevier Publishing Co., New York (1969)
30. P.A. Sturrock, Phys. Rev. **112**, 1488 (1958)
31. A. Scott, *Active and Nonlinear Wave Propagation in Electromagnetics*, Wiley Interscience, London (1970)
32. P.M. Morse and H. Feshbach, *Methods of Theoretical Physics*, McGraw-Hill, Boston (1953)

16

Discrete Breathers and Solitons in Metamaterials

George P. Tsironis, Nikos Lazarides, and Maria Eleftheriou

Department of Physics, University of Crete, and Institute of Electronic Structure and Laser, Foundation of Research and Technology Hellas (FORTH), P.O. Box 2208, 71003 Heraklion, Greece
gts@physics.uoc.gr, nl@physics.uoc.gr, marel@physics.uoc.gr

16.1 Introduction

Nonlinear localization is a process that may occur in weakly coupled nonlinear oscillators and leads to the formation of dynamically localized states in an otherwise translationally invariant lattice [1–3]. The main ingredients of nonlinear localization is discreteness, usually stemming from the weak interaction among the oscillators and nonlinearity, arising from the nonlinear nature of the oscillator forces. The dynamical localized states generated in this process are termed discrete breathers (DBs) or intrinsic localized modes (ILMs). These states are collective periodically oscillating modes of the lattice that, at the same time, are localized in a given location of the system. One basic criterion for the formation of DBs in infinite lattices is that their frequency and its sidebands should not coinside with the linearized spectrum of the oscillator lattice [4]. When the interparticle interaction exceeds a certain threshold, DBs become unstable and ultimately disappear. However, if the coupling becomes strong enough, it is possible in some cases to still form localized states that are very extended and have features of nontopological solitons or solitary waves [5]. In the present chapter we will address the generation of localization through nonlinearity both in the discrete, weakly interacting limit, as well in the continuous one where the nonlinear excitations are much larger than the lattice spacing. In all cases we will be focusing on metamaterials made typically of micron sized units that provide desired system electromagnetic properties.

Metamaterials are man made crystals characterized by translational invariance and a unit cell formed by larger than atomic elements. We will be focusing on left handed metamaterials (LHM) characterized by a negative index of refraction; typical sizes of the constituent parts of their unit cell is at the micron level although other sizes exist as well [6–8]. The usual structure of a LHM is that of a periodic array of metallic wires accompanied by periodically placed split ring resonators (SRRs) [9–16]. While the metallic wires are responsible for the negative dielectric permitivity that may be induced in the metamaterial, the SRRs produce negative magnetic permeability; the combination of both turns the index of refraction of the system negative as can be seen by the following argument: Let $\epsilon = \epsilon' \exp[i\phi_1]$ and $\mu = \mu' \exp[i\phi_2]$ the permitivity and permeability respectively of the material, where we assume also an imaginary part denoting very small dissipation. For a positive index material we have both permitivity and permeability positive and ϕ_1, ϕ_2 are taken to be small. The index of refraction is then $n = \sqrt{\epsilon\mu} = \sqrt{\epsilon'\mu'} \exp[i(\phi_1 + \phi_2)/2] \simeq \sqrt{\epsilon'\mu'}$. For a negative index material on the other hand, we have negative permitivity and permeability; this means that $\phi_i = \pi + \theta_i$, $i = 1, 2$ and the angles θ_i are now taken to be small and positive for purely dissipative structures with no gain. As a result $n = \sqrt{\epsilon\mu} = \sqrt{\epsilon'\mu'} \exp(i\pi) \exp[i(\theta_1 + \theta_2)/2] \simeq -\sqrt{\epsilon'\mu'}$, i.e. the index must be negative in this case.

The fact that the index of refraction is negative does not inhibit propagation; this may be seen easily from the dispersion relation of the wave equation in a medium. Seeking plane wave solutions of the wave equations for the electric or magnetic field we arrive at the dispersion relation $k^2 = (\epsilon\mu)/c^2\omega^2$; when both permitivity and permeability are negative, the wave vector k is real, leading to wave propagation in the medium with the negative refractive index. Plane waves give through Maxwell's equations $\boldsymbol{k} = \boldsymbol{E} \times \boldsymbol{B}/\mu$ and thus, for negative μ, the wave vector is exactly opposite to the Poynting vector [17]. Furthermore, if an electromagnetic wave impinges on the interface with a negative index material (NIM), while Snell's law still applies, it results in a negative refraction angle leading to a nonconventional direction of propagation in the metamaterial [17]. These peculiar properties of metamaterials have led to proposals for the formation of flat lenses as well as object cloaking [18].

While both negative ϵ and negative μ are necessary for a NIM, the mechanisms of generating them are quite different. The process for producing negative permitivity is rather simple and it is based on the collective oscillations of the free electron gas that occur at the plasma frequency ω_p. If the electrons at the surface of the metallic wires found in LHMs are driven below their plasma frequency, they respond out of phase with respect to the driving radiation leading to negative ϵ [6]. If the magnetic permeability is not affected at the same time, this would lead to the decay of the wave entering in this medium and typically resulting in wave reflection. Only when μ becomes negative, propagation is restored; this may occur by driving the SRR at frequencies above their resonant frequency. An SRR unit is nothing but a circuit made of an inductance L, capacitance C and resistance R; it is thus characterized by a resonant frequency close to $\omega_0 = 1/\sqrt{LC}$ (for $R \simeq 0$). If a time dependent magnetic field threaded through the SRR loop drives the circuit below ω_0, its response is paramagnetic since the induced current is in phase with the driving field. When, on the other hand, the field has a frequency larger than ω_0, the response becomes diamagnetic and the system may produce an effective negative μ. The SRR unit is central to NIMs since it provides the necessary out of phase magnetic response to the external fields that turns the index of refraction negative [19]. Nonlinearity is introduced in the SRR unit in order to induce tunability in the system.

Nonlinearity in metamaterials may be of two types, either extrinsic or intrinsic. The former may be introduced by embedding the metamaterial lattice in a dielectric with strong nonlinearity that affects primarily the SRR unit [19–21]. When a time dependent electric field is formed accross the gap of the SRR due to alternating charge accumulation, the nonlinearity in the dielectric will affect the resonant frequency of the unit. The new frequency will depend in a complicated way on the magnetic flux threaded through the circuit [20]. Small changes thus in the intensity of the electromagnetic radiation sent to the sample may have dramatic effects in the response of the nonlinear NIM. Nonlinearity may also be introduced intrinsically through a genuine nonlinear mechanism; an example of this is given by the Josephson effect [22]. In the latter the supercurrent may tunnel through a junction in such a way that an external driving field may alter the Copper pair phase nonlinearly. If the SRR is superconducting while a Josephson junction is placed in the gap, the resulting circuit inherits the nonlinearity of the junction leading to an rf-SQUID with possible negative permeability response. A metamaterial made of units of this type may have both negative effective μ but also tunability due to the intrinsic nonlinearity of the Josephson junction.

The present chapter focuses directly on the nonlinear properties of metamaterials generated primarily by extrinsic nonlinearity. We will begin by addressing a nonlinear medium made of weakly interacting SRRs and investigate the onset of nonlinear localization in the form of discrete breathers as well as magnetic solitons in the system. Subsequently we will discuss the more general propagation problem in a left-handed medium embedded in a nonlinear dielectric and show that it proceeds via special type compound electric and magnetic solitons. We will then conclude with a brief summary of the findings.

16.2 Magnetic Breathers

In order to study the effects in metamaterials of the nonliner dependence of the SRR resonant frequency on an external field we will consider a planar one-dimensional (1D) array of N identical SRRs with their axes perpendicular to the plane; each unit is equivalent to an RLC oscillator. An effective magnetic dipole may be induced in each unit either through the time-varying magnetic flux or a time-varying electric field applied parallel to the SRR gap. The mutually inductive magnetic dipole-dipole interaction decays as the cube of the distance and thus we may condider only nearest-neighbor SRR interactions. The SRRs are considered to be nonlinear elements due to the nonlinear Kerr-type dielectric that fills their gap and has dielectric permittivity equal to $\epsilon(|\boldsymbol{E}|^2) = \epsilon_0(\epsilon_\ell + \alpha|\boldsymbol{E}|^2/E_c^2)$, where $\boldsymbol{E}$ is the electric component of the applied EM field, E_c is a characteristic electric field, ϵ_ℓ the linear permittivity, ϵ_0 the permittivity of the vacuum, and $\alpha = \pm 1$ correspond to self-focusing and self-defocusing nonlinearity, respectively [20–22]. Due to the field dependence of the permitivity the SRR gap develops a field-dependent capacitance $C = C(|\boldsymbol{E}|^2) = \epsilon(|\boldsymbol{E}_\mathrm{g}|^2)A/d_\mathrm{g}$ where $A = \pi h^2/4$ is the cross-section area of the SRR wire (assumed circular, with circular cross-section of diameter h), $\boldsymbol{E}_\mathrm{g}$ is the electric field induced along the SRR gap, and d_g is the longitudinal size of the slit. Using the expression $C(U_n) = \mathrm{d}Q_n/\mathrm{d}U_n$ we may find the nonlinear dependence of the charge Q_n stored in the capacitor of the n-th SRR on the applied voltage $U_n \equiv d_\mathrm{g} E_{\mathrm{g},n}$, viz.

$$Q_n = C_\ell \left(1 + \alpha \frac{U_n^2}{3\,\epsilon_\ell U_\mathrm{cr}^2}\right) U_n, \tag{16.1}$$

where $n = 1, 2, \ldots, N$, $C_\ell = \epsilon_0 \epsilon_\ell (A/d_\mathrm{g})$ is the linear capacitance, and $U_\mathrm{cr} = d_\mathrm{g} E_\mathrm{cr}$. The time dependence of the charge Q_n at the n-th unit depends on the internal charge exchange in the unit as well as the inductive coupling to the adjacent elements; this charge dynamics leads to the following set of equations of motion [23, 24]:

$$L \frac{\mathrm{d}^2 Q_n}{\mathrm{d}t^2} + R \frac{\mathrm{d}Q_n}{\mathrm{d}t} + f(Q_n) = M \left(\frac{\mathrm{d}^2 Q_{n-1}}{\mathrm{d}t^2} + \frac{\mathrm{d}^2 Q_{n+1}}{\mathrm{d}t^2}\right) + \mathcal{E}. \tag{16.2}$$

On the lhs of Eq. (16.2) we have the well known dynamics of an RLC circuit that, in the present case involves a nonlinear capacitance through the term $f(Q_n) = U_n$. On the rhs, on the other hand, we have the coupling of neighbouring SRRs with mutual inductance M while $\mathcal{E} = \mathcal{E}(t)$ is the electromotive force applied in each SRR due to the external fields, magnetic and/or electric. In order to simplify Eq. (16.2) we use the relations $\omega_\ell^{-2} = LC_\ell$, $\tau = t\omega_\ell$, $I_\mathrm{cr} = U_\mathrm{cr}\omega_\ell C_\ell$, $Q_\mathrm{cr} = C_\ell U_\mathrm{cr}$, $\mathcal{E} = U_\mathrm{cr}\varepsilon$, $I_n = I_\mathrm{cr} i_n$, $Q_n = Q_\mathrm{cr} q_n$, and find

$$\frac{\mathrm{d}^2}{\mathrm{d}\tau^2}\left[q_n - \lambda\left(q_{n-1} + q_{n+1}\right)\right] + \gamma \frac{\mathrm{d}}{\mathrm{d}\tau} q_n + f(q_n) = \varepsilon(\tau), \tag{16.3}$$

where $\gamma = RC_\ell \omega_\ell$, $\lambda = M/L$ are the loss coefficient and the coupling parameter, respectively.

Analytical inversion of Eq. (16.1) for $u_n = f(q_n)$ with u_n real and $u_n(q_n = 0) = 0$, results to

$$f(q_n) \simeq q_n - \frac{\alpha}{3\,\epsilon_\ell} q_n^3 + 3\left(\frac{\alpha}{3\,\epsilon_\ell}\right)^2 q_n^5 + \mathcal{O}(q_n^7). \tag{16.4}$$

We thus find that the effective on-site potential $V(q_n) = \int_0^{q_n} f(q_n')\,\mathrm{d}q_n'$ is soft for focusing nonlinearity ($\alpha = +1$) and hard for defocusing nonlinearity ($\alpha = -1$).

Upon linearization of Eqs. (16.1) and (16.3) we obtain

$$\frac{\mathrm{d}^2}{\mathrm{d}\tau^2}\left(-\lambda q_{n-1} + q_n - \lambda q_{n+1}\right) + q_n = 0, \tag{16.5}$$

while substituting $q_n = A\cos(kDn - \omega\tau)$ into Eq. (16.5) we find the dispersion relation for the linear lattice modes, viz.

$$\omega_k = [1 - 2\lambda\cos(kD)]^{-1/2}, \tag{16.6}$$

where D is unit cell size and k the wavenumber. Stable localized modes in an infinite sized metamaterial must have frequencies in the zone exterior to the band of the linearized modes.

It is important to have an estimate of the various parameters entering in the reduced Eq. (16.3); we thus consider the SRR systems studied experimentally in Ref. [14] and make simple estimates for the loop self-inductance as well as the coupling λ ignoring the effects of nonlinearity and coupling on the loop resonant frequency. We consider an array of square-shaped SRRs with square cross-section of side length $\ell = 5\,\mu\text{m}$, and $t = w = d_g = 1\,\mu\text{m}$ the SRR depth, width, and slit size, respectively, while the unit cell length is $D = 7\,\mu\text{m}$ [14]. The resulting self-inductance for these parameters is $L \simeq 1.2 \times 10^{-11}\,\text{H}$ [23]. From this value of L we find through the expression $f_r = 1/2\pi\sqrt{LC}$ that a capacitance equal to $C = C_\ell \simeq 7.35 \times 10^{-17}\,\text{F}$ would give a (linear) resonant frequency of $f_r = 6.2\,\text{THz}$. In what regards the mutual coupling of the units we find similarly that for an array of squared SRRs with square cross-section having dimensions as in [14] $\lambda \simeq 0.02$. This mutual coupling coefficient may become one order of magnitude smaller in some other cases [23].

16.2.1 Hamiltonian Discrete Breathers

We focus now on the set of Eqs. (16.3) and consider first the lossless case without an applied field ($\gamma = 0$, $\varepsilon = 0$) assuming further that the power of the emitted dipole radiation is very small. Under these circumstances Eq. (16.3) may be derived from the Hamiltonian

$$\mathcal{H} = \sum_n \left\{ \frac{1}{2}\dot{q}_n^2 + V(q_n) - \lambda\dot{q}_n\dot{q}_{n+1} \right\}. \tag{16.7}$$

We may construct Hamiltonian DBs starting from the anticontinuous limit where all oscillators are uncoupled. In this procedure we fix the initial amplitude of one oscillator (e.g. the one located at $n = n_b$) to a specific value q_b, corresponding to a desired oscillation period T_b that is not resonant to the linearized modes [25]. This trivial breather is then analytically extended through the Newton method to finite couplings λ up to maximal coupling $\lambda_{\max}$ keeping the breather period constant. The resulting localized mode is characterized by the basic frequency $\omega_b = 2\pi/T_b$. The linear stability of these modes in the Hamiltonian SRR system is addressed through the eigenvalues of the monodromy matrix [25]; we find that the modes are stable for small couplings [23].

In Fig. 16.1 we show the time evolution of a typical, linearly stable, magnetic DB as a function of the array site n. We plot specifically the normalized

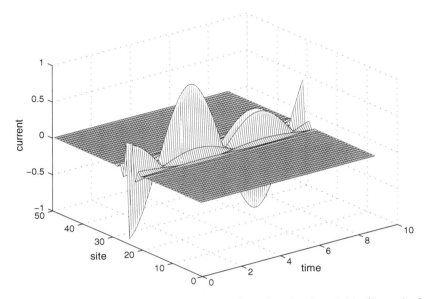

Fig. 16.1. Time evolution of a Hamiltonian breather for $\lambda = 0.04$, dimensionless breather period $T_\text{b} = 6.69$, $\alpha = +1$, and $\epsilon_\ell = 2$. Only few sites of the lattice are engaged in the oscillation

circulating current i_n that is proportional to the magnetic moment of the nth SRR. In the breather region, the lattice performs electromagnetic oscillations that decay spatially very fast in an exponential fashion. Although DBs are very discrete objects, they are seen to be mobile in several cases. Their mobility is studied by perturbing the DBs in the direction of an antisymmetric linearized mode [26]. We find that the Hamiltonian magnetic DBs are generally mobile [23].

16.2.2 Dissipative Discrete Breathers

When the external magnetic flux and/or the applied electric field varies, the system of the nonlinear SRRs is driven externally and, in this case, we may form dissipative breathers. To this effect we start by solving Eqs. (16.1) and (16.3) in the anticontinuous limit [27] using $\varepsilon(\tau) = \varepsilon_0 \sin(\omega \tau)$; the latter represents a sinusoidally varying applied field of amplitude ε_0 and frequency ω. We identify two different amplitude attractors for the single SRR oscillator, with amplitudes q_h and q_ℓ for the high and low amplitude attractor, respectively. We proceed by fixing the initial charge of one of the oscillators (say the one at $n = n_\text{b}$) to q_h while all the others to q_ℓ and the currents i_n are all set to zero. Using as initial condition this configuration, we increase adiabatically the coupling up to a given value $\lambda \neq 0$, leading to dissipative DB formation.

16 Discrete Breathers and Solitons in Metamaterials   279

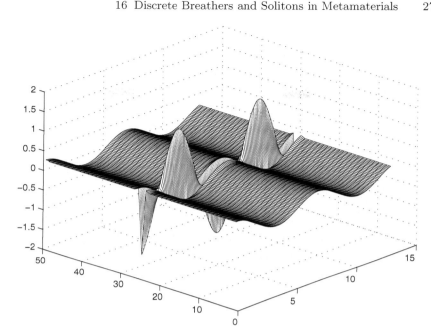

Fig. 16.2. Time evolution of a dissipative breather during approximatelly two periods, for $T_{\rm b} = 6.82$, $\lambda = 0.0045$, $\gamma = 0.01$, $\varepsilon_0 = 0.04$, $\alpha = +1$, and $\epsilon_\ell = 2$. The breather region is left-handed while the rest of the metamaterial is right-handed

Using specifically $q_{\rm h} \simeq 1.6086$ and $q_\ell \simeq 0.28660$ and following the above procedure we find different types of dissipative DBs. One type involves the lattice oscillating close to q_ℓ while the "bright" localized mode has amplitude close to $q_{\rm h}$ as in Fig. 16.2, or the "dark" mode where the amplitudes are reversed. We note that the DB shown in Fig. 16.2 run for over 2×10^4 breather periods without any appreciable change. There is also a domain wall mode where part of the lattice has one of the values and the rest has the other [23]. This state separates the system in different phases and may be potentially interesting in applications.

When the external field is on, the whole electric lattice is locked to the driver and oscillates with the same driver frequency. The presence of nonlinearity, however, allows for the possiblity that different parts of the lattice oscillate at different amplitudes, i.e. different SRRs have different charge values. As a result, one may form, for instance, a single breather such as the one depicted in Fig. 16.2 or more complex ones as in Fig. 16.3 [24]. In these cases we may evaluate the magnetic response of the localized modes by calculating the normalized, site-dependent magnetic moment per period $T_{\rm b}$ for each SRR in the array. We consider the magnetic response of an SRR at a given time instant to be positive (negative), when the driving magnetic flux and the instantaneous current $i_n(\tau)$ have the same (opposite) signs. Without

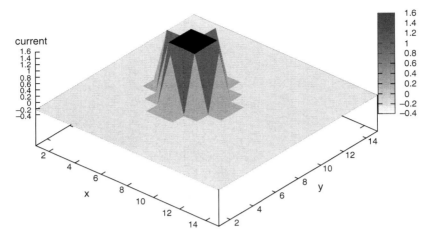

Fig. 16.3. Snapshot of a dissipative multibreather extending over several sites in a two dimensional lattice with equal couplings in the x and y axes, $\lambda_x = \lambda_y = 5\times 10^{-4}$, $\gamma = 0.01$, driver amplitude $\epsilon_0 = 0.04$, $\alpha = +1$ and $T_b = 6.82$. For these parameter values the elevated multibreather region forms a left-handed island in an otherwise right-handed medium. The location and the shape of the breather may be tailored at will

DB excitations, all the SRRs oscillate with the same amplitude and phase, contributing the same magnetization per period to the total. The magnetization however varies with the oscillation amplitude resulting to multiple magnetization states, which is a purely nonlinear effect. For DB solutions the site-dependent magnetization is larger at the central DB site, while it is small and approximately constant away from it. Furthermore, if we select the driving frequency to be below the SRR resonance we may have all the metamaterial in a "right-handed state" while the breather impurity state is a "left-handed state". In other words, nonlinearity enables the formation of small or large "islands" in the metamaterial that have distinct optical properties from the rest of the lattice. The left-handed property of the breather sites is induced through the out of phase motion of the these sites compared to the oscillation of the rest of the system. We note that the presence of a DB enhances substantially the local magnetization.

16.3 Magnetic Solitons

In the previous section we considered the case of discrete localized modes in a weakly coupled SRR lattice. It is also possible that a charge carrier wave forms that is extended in space and has the shape of envelope solitons [28]. We may thus consider an EM wave propagating in the 1D array of N identical SRRs described previously; the linear dispersion relation governing the propagation of magnetoinductive waves is given in Eq. (16.5). We note that the latter has a finite cutoff at $\omega_{\max} = \omega(k=0) = 1/\sqrt{1-2\lambda}$ leading to an optical branch with negative group velocity $v_g \equiv \omega'(k) = -\lambda\omega^3 \sin k$ being negative for all k values within the first Brillouin zone. As a result the wavepacket envelope propagates at the group velocity v_g in a direction opposite to that of the carrier wave propagating at the phase speed $v_{\rm ph} = \omega/k$. The frequency band is therefore bounded by $\omega_{\min} = \omega(k=\pi) = (1+2\lambda)^{-1/2}$ and $\omega_{\max}$ in the physically relevant regime for $\lambda \leq 1/2$.

We may obtain a nonlinear generalization of the dispersion relation by substituting $q_n = \hat{q}\exp[i(kn-\tilde{\omega t})] + $ c.c. in Eq. (16.5) and retaining only first order harmonics. We obtain then the following amplitude dependent expression

$$\omega^2(k;|\hat{q}|^2) = \left(1-\alpha|\hat{q}|^2/\epsilon_1\right)\left(1-2\lambda\cos k\right)^{-1}. \qquad (16.8)$$

Assuming weak dependece on the amplitude and considering a modulated wave frequency ω and wavenumber k close to the carrier values ω_0 and k_0, respectively, we may expand as follows

$$\omega - \omega_0 \approx \left.\frac{\partial\omega}{\partial k}\right|_{k_0}(k-k_0) + \frac{1}{2}\left.\frac{\partial^2\omega}{\partial k^2}\right|_{k_0}(k-k_0)^2$$
$$+ \left.\frac{\partial\omega(k)}{\partial|\hat{q}|^2}\right|_{\hat{q}_0}\left(|\hat{q}|^2-|\hat{q}_0|^2\right), \qquad (16.9)$$

where $\hat{q}_0$ is a reference constant amplitude. Considering slow space and time variables X and T, and thus setting $\omega - \omega_0 \to i\partial/\partial T$ and $k - k_0 \to -i\partial/\partial X$, one readily obtains the nonlinear Schrödinger (NLS) equation

$$i\left(\frac{\partial\psi}{\partial T} + v_g\frac{\partial\psi}{\partial X}\right) + P'\frac{\partial^2\psi}{\partial X^2} + Q'\left(|\psi|^2 - |\psi_0|^2\right)\psi = 0, \qquad (16.10)$$

where we have set $\psi = \hat{q}$ and $\psi_0 = \hat{q}_0$, and defined the dispersion coefficient $P' \equiv \omega''(k)/2 = -\lambda\omega^5(\lambda\cos^2 k + \cos k - 3\lambda)/2$ and the nonlinearity coefficient $Q' = -(\partial\omega/\partial|\psi|^2)|_{\psi_0} = (\alpha\omega/2\,\epsilon_1)(1-\alpha|\psi_0|^2/\epsilon_1)^{-1} \approx \alpha\omega/2\,\epsilon_1$. We have assumed $|\psi_0| \ll 1$, and thus neglected the dependence on $|\psi_0|$ everywhere. Upon a Galilean transformation, viz. $\{X,T\} \to \{X-v_gT,T\} \equiv \{\zeta,\tau\}$, and a phase shift $\psi \to \psi\exp(-iQ'|\psi_0|^2\tau)$, one obtains the usual form of the NLS equation [5]

$$i\frac{\partial\psi}{\partial\tau} + P'\frac{\partial^2\psi}{\partial\zeta^2} + Q'|\psi|^2\psi = 0. \qquad (16.11)$$

The evolution of a modulated wave whose amplitude is described by Eq. (16.11) depends on the sign of the coefficients P' and Q' [5]. Specifically, if $P'Q' < 0$ the wavepacket is modulationally stable dark-type soliton while for $P'Q' > 0$ the wavepacket is modulationally unstable and upon break-up it may reduce to several localized structures in the matetial. We find that P' is negative for low k, while it changes sign at some critical value

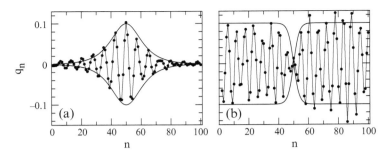

Fig. 16.4. Bright (**a**) and dark (**b**) magnetic solitons. Continuous lines denote the analytical amplitude shape in space, dots are the results of numerical simulation while the lines that link them are a guide to the eye. Parameters used $\lambda = 0.20$, $\epsilon_1 = 2$, $\psi'_0 = 0.1$, $N = 100$, and $\alpha = -1$

$k_{\mathrm{cr}} = \cos^{-1}[(-1 + \sqrt{1 + 12\lambda^2})/2\lambda]$, thus acquiring positive values for $k > k_{\mathrm{cr}}$. On the other hand, the sign of Q' is simply determined by the nature of the nonlinearity, i.e. Q' is positive (negative) for $\alpha = +1\,(-1)$. Therefore, for $\alpha = +1$ the wave is modulationally stable, $P'Q' < 0$ (unstable, $P'Q' > 0$), for $k < k_{\mathrm{cr}}$ ($k > k_{\mathrm{cr}}$), while for $\alpha = -1$ the wave is modulationally unstable, $P'Q' > 0$ (stable, $P'Q' < 0$), for $k < k_{\mathrm{cr}}$ ($k > k_{\mathrm{cr}}$).

The Eq. (16.11) possesses a number of exact solutions; of particular interest are the sech- and tanh-type solutions that correspond to bright and dark solitons respectively [28]. In Fig. 16.4 we present these two types of solutions, both their analytical profile (continuous lines) as well as the shapes determined numerically. The latter were obtained in a regime that is not so close to the "continuous limit", nevertheless we find that the agreement is reasonably good. We thus find that in a system of weakly coupled nonlinear SRRs when the excitations are wide compared to lattice spacing soliton-like objects may form and propagate in the medium. When, on the other hand, the excitation is very narrow discrete breathers are then formed in the metamaterial. In all cases it is possible to have both localized left-handed properties in a right-handed medium, the reverse, or a domain wall separating these two electromagnetically distinct phases.

16.4 Electromagnetic Solitons

So far we analysed primarily the effects of nonlinearity in the magnetic response and propagation properties of metamaterials. If the electric effects are also included we arrive at the more general case of LHM with extrinsic nonlinearity. We focus directly on the continuous case and show that more complex two component solitons may be generated.

The dispersion relations for this compound electromagnetic case are generalizations of the known ones for the linear cases, viz. [20]

$$\epsilon_{\text{eff}}(\omega) = \epsilon_0 \left(\epsilon_D(|E|^2) - \frac{\omega_p^2}{\omega^2} \right), \tag{16.12}$$

$$\mu_{\text{eff}}(\omega) = \mu_0 \left(1 - \frac{F\omega^2}{\omega^2 - \omega_{0\text{NL}}^2(|H|^2)} \right), \tag{16.13}$$

where ω_p is the plasma frequency, F is the filling factor, $\omega_{0\text{NL}} = \omega_{0\text{NL}}(|H|^2)$ is the nonlinear resonant SRR frequency, and $\epsilon_D(|E|^2) = \epsilon_{D0} + \alpha|E|^2$, with α denoting again the strength of nonlinearity. For a linear dielectric $\omega_{0\text{NL}}(|H|^2) \to \omega_0$, where ω_0 is the linear resonant SRR frequency; then Eqs. (16.12)–(16.13) reduce to previously known expressions [6]. The parameters F, ω_p, and ω_0 are related to geometrical and material parameters of the LHM components. Although ϵ_{eff} can be readily put in the form $\epsilon + \epsilon_{\text{NL}}(|E|^2)$, for μ_{eff} this is not an obvious task, since $\mu_{\text{eff}} = \mu_{\text{eff}}(\omega_{0\text{NL}})$, and $\omega_{0\text{NL}}$ depends on $|H|^2$ as [20, 21]

$$\alpha \Omega^2 X^6 |H|^2 = A^2 E_c^2 \left(1 - X^2\right) \left(X^2 - \Omega^2\right)^2, \tag{16.14}$$

where $X = \omega_{0\text{NL}}/\omega_0$, $\Omega = \omega/\omega_0$, E_c is a characteristic (large) electric field, and A is a function of physical and geometrical parameters [20, 21]. Our α is related to the parameters of Eq. (16.14) as $\alpha = \pm 1/E_c^2$. We choose $f_p = \omega_p/2\pi = 10\,\text{GHz}$ and $f_0 = \omega_0/2\pi = 1.45\,\text{GHz}$ leading to a left-handed zone in the range $1.45\,\text{GHz} < f < 1.87\,\text{GHz}$. For relatively small fields when μ_{eff} is truly field dependent, one may approximate the nonlinear expression for the effective permeability as $\mu_{\text{eff}} = \mu + \mu_{\text{NL}}(|H|^2)$, where $\mu_{\text{NL}}(|H|^2) = \beta |H|^2$. Using this approximation we thus turn both the electric and magnetic problem into an effective Kerr nonlinear problem; in what regards the nonlinear coefficient β, we may treat it as fitting parameter.

Using the constitutive relations $\boldsymbol{D} = \epsilon_{\text{eff}} \boldsymbol{E} = \epsilon \boldsymbol{E} + \boldsymbol{P}_{\text{NL}}$ and $\boldsymbol{B} = \mu_{\text{eff}} \boldsymbol{H} = \mu \boldsymbol{H} + \boldsymbol{M}_{\text{NL}}$ where $\boldsymbol{P}_{\text{NL}}$, $\boldsymbol{M}_{\text{NL}}$ are the nonlinear contributions to electric polarization and magnetization respectively, we may use Maxwell's equations and obtain the following general wave vector equations:

$$\nabla^2 \boldsymbol{E} - \mu\epsilon \frac{\partial^2 \boldsymbol{E}}{\partial t^2} - \mu \frac{\partial^2 \boldsymbol{P}_{\text{NL}}}{\partial t^2} - \nabla(\nabla \cdot \boldsymbol{E}) =$$
$$\frac{\partial}{\partial t}\left[(\nabla \mu_{\text{NL}}) \times \boldsymbol{H}\right] + \frac{\partial}{\partial t}\left[\mu_{\text{NL}} \frac{\partial}{\partial t}(\epsilon \boldsymbol{E} + \boldsymbol{P}_{\text{NL}})\right], \tag{16.15}$$

$$\nabla^2 \boldsymbol{H} - \mu\epsilon \frac{\partial^2 \boldsymbol{H}}{\partial t^2} - \epsilon \frac{\partial^2 \boldsymbol{M}_{\text{NL}}}{\partial t^2} - \nabla(\nabla \cdot \boldsymbol{H}) =$$
$$-\frac{\partial}{\partial t}\left[(\nabla \epsilon_{\text{NL}}) \times \boldsymbol{E}\right] + \frac{\partial}{\partial t}\left[\epsilon_{\text{NL}} \frac{\partial}{\partial t}(\mu \boldsymbol{H} + \boldsymbol{M}_{\text{NL}})\right]. \tag{16.16}$$

We consider for simplicity an x–polarized plane wave with frequency ω and wavevecror k propagating along the z-axis, viz. $E(z,t) = q(z,t)\exp[\text{i}(kz - \omega t)]$, $H(z,t) = p(z,t)\exp[\text{i}(kz - \omega t)]$. Assuming that the envelopes $q(z,t)$ and $p(z,t)$ change slowly in z and t allows us to introduce the slow variables $\xi = \varepsilon(z - \omega't)$ and $\tau = \varepsilon^2 t$, where ε is a small parameter, and $\omega' = \partial\omega/\partial k$ is the group velocity of the wave. Substitution of the slow variables into Eqs. (16.15)–(16.16), assuming that $\alpha = \hat{\alpha}\varepsilon^2$, $\beta = \hat{\beta}\varepsilon^2$, and expressing p and q as an asymptotic expansion in terms of a parameter ε, we get various equations in increasing powers of ε [29]. The leading order problem gives the dispersion relation $\omega = ck$, where $c = \sqrt{1/\epsilon\mu}$. At $\mathcal{O}(\varepsilon^1)$, the group velocity is given as $\omega' = kc^2/\omega$. At $\mathcal{O}(\varepsilon^2)$, we obtain [29]

$$\mathrm{i}\frac{\partial q_0}{\partial \tau} + \frac{\omega''}{2}\frac{\partial^2 q_0}{\partial \xi^2} + \frac{\omega c^2}{2}\left(\hat{\alpha}\mu|q_0|^2 + \epsilon\hat{\beta}|p_0|^2\right)q_0 = 0, \qquad (16.17)$$

$$\mathrm{i}\frac{\partial p_0}{\partial \tau} + \frac{\omega''}{2}\frac{\partial^2 p_0}{\partial \xi^2} + \frac{\omega c^2}{2}\left(\epsilon\hat{\beta}|p_0|^2 + \hat{\alpha}\mu|q_0|^2\right)p_0 = 0, \qquad (16.18)$$

where $q_0(\xi,\tau)$, $p_0(\xi,\tau)$ are the zeroth in ε order terms, τ is the slow time, ξ the slow space variable and $\omega'' = (c^2 - \omega'^2)/\omega$. By rescaling τ, ξ and the amplitudes q_0, p_0 according to $\xi = X$, $T = \omega''\tau/2$ and $Q = \sqrt{|\Lambda_\mathrm{q}/\omega''|}q_0$, $P = \sqrt{|\Lambda_\mathrm{p}/\omega''|}p_0$, where $\Lambda_\mathrm{q} = \omega c^2 \mu\hat{\alpha}$ and $\Lambda_\mathrm{p} = \omega c^2 \epsilon\hat{\beta}$, we get

$$\mathrm{i}Q_T + Q_{XX} + \left(\sigma_\mathrm{q}|Q|^2 + \sigma_\mathrm{p}|P|^2\right)Q = 0, \qquad (16.19)$$

$$\mathrm{i}P_T + P_{XX} + \left(\sigma_\mathrm{p}|P|^2 + \sigma_\mathrm{q}|Q|^2\right)P = 0, \qquad (16.20)$$

where $\sigma_{\mathrm{q,p}} \equiv \mathrm{sgn}(\Lambda_{\mathrm{q,p}})$. Eqs. (16.19)–(16.20) is a special case of the fairly general and frequently studied system of coupled NLS equations known to be completely integrable for $\sigma_\mathrm{q} = \sigma_\mathrm{p} = \sigma$ [31]. The sign of the products $\mu\hat{\alpha}$ and $\epsilon\hat{\beta}$ determine the type of nonlinear self-modulation (self-focusing or self-defocusing) effects which will occur. For $\sigma = \pm 1$ both fields experience the same type of nonlinearity.

For $\epsilon, \mu > 0$ and $\hat{\alpha}, \hat{\beta} > 0$ we have $\sigma = +1$, and the system of Eqs. (16.19)–(16.20) accepts solutions of the form [29, 32]

$$Q(X,T) = u(X)\mathrm{e}^{\mathrm{i}\nu_\mathrm{q}^2 T}, \qquad P(X,T) = v(X)\mathrm{e}^{\mathrm{i}\nu_\mathrm{p}^2 T}, \qquad (16.21)$$

where u, v are real functions and ν_q, ν_p are real positive wave parameters. The latter is necessary, if we are interested in solitary waves that exponentially decay as $|X| \to \infty$. Introducing Eq. (16.21) into Eqs. (16.19)–(16.20) we get

$$u_{XX} - \nu_\mathrm{q}^2 u + \left(u^2 + v^2\right)u = 0, \qquad (16.22)$$

$$v_{XX} - \nu_\mathrm{p}^2 v + \left(v^2 + u^2\right)v = 0. \qquad (16.23)$$

For $\nu_{\mathrm{q,p}} = \nu$, Eqs. (16.22)–(16.23) have a one-parameter family of symmetric and single-humped soliton solutions (Fig. 16.5)

$$u(X) = \pm v(X) = \nu\,\mathrm{sech}(\nu X). \qquad (16.24)$$

For $\epsilon, \mu < 0$ and $\hat{\alpha}, \hat{\beta} > 0$ we have $\sigma = -1$, and Eqs. (16.19)–(16.20) accept dark soliton solutions of the form [29, 33]

16 Discrete Breathers and Solitons in Metamaterials   285

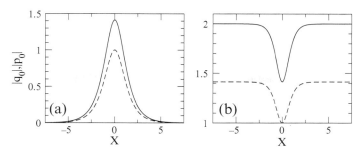

Fig. 16.5. Bright (**a**) and dark (**b**) two component electromagnetic solitons. The continuous line denotes the electric field (q_0) while the dashed one the magnetic field (p_0). The parameters used are (a) $\sigma_p = \sigma_q = +1$ for $\nu = 1$ and $r = \Lambda_q/\Lambda_p = 2$ in arbitrary units while in (b) $\sigma_p = \sigma_q = -1$ for $k = 1$ and $r = \Lambda_q/\Lambda_p = 2$ in arbitrary units; we use appropriate normalizations of the amplitudes

$$Q(X,T) = P(X,T) = k\left[\tanh(kX) - \mathrm{i}\right] \mathrm{e}^{\mathrm{i}(kX - 5k^2 T)}, \qquad (16.25)$$

which are localized dips on a finite-amplitude background wave, as shown in Fig. 16.5. In this very interesting case of LHM, the electric and magnetic fields are coupled together forming a dark compound soliton. Note that the relative amplitudes are controlled by the corresponding nonlinearities and frequency.

In addition to the two component bright and dark solitons, there are also other types of solutions, such as domain wall solitons that separate the left-handed from the right-handed medium. Numerics has shown that the continuous solitons do survive in the discrete lattice provided the coupling of the adjacent elements is not too small. The stability analysis done in Ref. [30] shows that soliton stability depends on the combination of focusing/defocusing nonlinearity as well whether the medium is right-handed or left-handed. The following approximate expression may be used for the stability of the compound solitons [30]

$$K' \simeq \frac{\epsilon_\mathrm{eff}}{\epsilon} + \frac{\mu_\mathrm{eff}}{\mu} - 2. \qquad (16.26)$$

Solitons are stable when $K' < 0$ and, as a result, we find that dark solitons in a left-handed medium are generally stable.

16.5 Summary

Metamaterials provide a natural frame for the investigation as well as practical exploitation of various nonlinear phenomena. The introduction of nonlinearity in the capacitive element of SRRs and the weak interaction among the elements leads directly in the formation of spatially localized structures. When the localization scale is very small the material forms discrete breathers; this may occur in a controled way or in random fashion through modulational instability. Discrete breathers act as elementary, localized antennas in the lattice since their oscillation frequency is typically in the non-propagating zone. Continuous topological solitons, on the other hand, may form when the amplitude of the EM fields change very slowly from element to element leading to a deformation over several unit cells. When only magnetic elements are present in the metamaterial the resulting excitations are magnetic single breathers or multibreathers and NLS-like solitons. When both electric and magnetic elements are present we expect to find more complex, compound electromagnetically localized objects. While the fully discrete case has not been analyzed yet, we know that in the continuous limit we have the formation of two component electromagnetic solitons. The latter are compound objects where both the electric and magnetic fields are localized while they propagate in the metamaterial.

In addition to extrinsic nonlinearity introduced through doping of the dielectric medium of the metamaterial we may also use elements that are intrinsically nonlinear. One way is through the introduction of Josephson junctions in otherwise superconducting rings. In this case the RLC circuit of the SRR is replaced by a resistively and capacitively shunted Josephson junction that is coupled to the self-inductance of the ring. The external driver for this unit is the time-varying magnetic field flux that passes through the loop area. This unit is usually referred to as an rf-SQUID. The resulting, reduced equation of motion for the single unit is

$$\frac{\mathrm{d}^2 f}{\mathrm{d}\tau^2} + \gamma \frac{\mathrm{d}f}{\mathrm{d}t} + \beta \sin(2\pi f) + f = f_{\mathrm{ext}}, \qquad (16.27)$$

where f is the reduced magnetic flux passing through the loop, f_{ext} is the sinusoidally varying external magnetic flux while γ and β are parameters related to the resistance, capacitance and self-inductance of the circuit [22]. We note that the presense of the sin-term in the equation of motion makes the specific system genuinly nonlinear. Simple analysis of the rf-SQUID metamaterial in the decoupled limit shows that negative permeability is possible [22]. When the superconducting units are coupled through their mutual flux, one also finds the possibility for generation of discrete breathers and solitons, similar but not identical to the ones found for extrinsic nonlinearity [34]. Just like in standard optics, the introduction of either extrinsic or intrinsic nonlinearity in metamaterials, enhances the tunabilty of the material and will lead to tailored applications and devices.

References

1. A.J. Sievers and S. Takeno, Phys. Rev. Lett. **61**, 970 (1988)
2. S. Aubry, Physica D **103**, 201 (1997)
3. D.K. Campbell, S. Flach, and Y.S. Kivshar, Physics Today **57**, 43 (2004)
4. R.S. MacKay and S. Aubry, Nonlinearity **7**, 1623 (1994)

5. T. Dauxois and M. Peyrard, *Physics of Solitons*, Cambridge University Press (2005)
6. J.B. Pendry, A.J. Holden, W.J. Stewart, and I. Youngs, Phys. Rev. Lett. **76**, 4773 (1996)
7. J.B. Pendry, Contemp. Phys. **45**, 191 (2004)
8. S.A. Ramakrishna, Rep. Prog. Phys. **68**, 449 (2005), and references therein.
9. D. Smith, W. Padilla, D. Vier, S. Nemat-Nasser, and S. Schultz, Phys. Rev. Lett. **84**, 4184 (2000)
10. R. Shelby, D. Smith, and S. Schultz, Science **292**, 77 (2001)
11. C.G. Parazzoli, R.B. Greegor, K. Li, B.E.C. Koltenbah, and M. Tanielian, Phys. Rev. Lett. **90**, 107401 (2003)
12. J. Zhou, Th. Coschny, M. Kafesaki, E.N. Economou, J.B. Pendry, and C.M. Soukoulis, Phys. Rev. Lett. **95**, 223902 (2005)
13. T.J. Yen, W.J. Padilla, N. Fang, D.C. Vier, D.R. Smith, J.B. Pendry, D.N. Basov, and X. Zhang, Science **303**, 1494 (2004)
14. N. Katsarakis, G. Konstantinidis, A. Kostopoulos, R.S. Penciu, T.F. Gundogdu, M. Kafesaki, and E.N. Economou, Opt. Lett. **30**, 1348 (2005)
15. C. Enkrich, M. Wegener, S. Linden, S. Burger, L. Zschiedrich, F. Schmidt, J.F. Zhou, Th. Coschny, and C.M. Soukoulis, Phys. Rev. Lett. **95**, 203901 (2005)
16. V.M. Shalaev, W. Cai, U.K. Chettiar, H.-K. Yuan, A.K. Sarychev, V.P. Drachev, and A.V. Kildishev, Opt. Lett. **30**, 3356 (2005)
17. V.G. Veselago, Sov. Phys. Usp. **10**, 509 (1968)
18. J.B. Pendry, D. Schuring, and D.R. Smith, Science **312**, 1780 (2006)
19. J.B. Pendry, A.J. Holden, D.J. Robbins, and W.J. Stewart, IEEE Trans. Microwave Theory Tech. **47**, 2075 (1999)
20. A.A. Zharov, I.V. Shadrivov, and Y.S. Kivshar, Phys. Rev. Lett. **91**, 037401 (2003)
21. S. O'Brien, D. McPeake, S.A. Ramakrishna, and J.B. Pendry, Phys. Rev. B **69**, 241101 (2004)
22. N. Lazarides and G.P. Tsironis, Appl. Phys. Lett. **90**, 163501 (2007)
23. N. Lazarides, M. Eleftheriou, and G.P. Tsironis, Phys. Rev. Let. **97**, 157406 (2006)
24. M. Eleftheriou, N. Lazarides, and G.P. Tsironis, Phys. Rev. E **77**, 036608 (2008)
25. J.L. Marin and S. Aubry, Nonlinearity **9**, 1501 (1996)
26. D. Chen, S. Aubry, and G.P. Tsironis, Phys. Rev. Lett. **77**, 4776 (1996)
27. J.L. Marin, F. Falo, P.J. Martinez, and L.M. Floria, Phys. Rev. E **63**, 066603 (2001)
28. I. Kourakis, N. Lazarides, and G.P. Tsironis, Phys. Rev. E, **75**, 067601 (2007)
29. N. Lazarides and G.P. Tsironis, Phys. Rev. E, **71**, 049903 (2005)
30. I. Kourakis and P.K. Shoukla, Phys. Rev. E **72**, 016626 (2005)
31. S.V. Manakov, Sov. Phys. JETP **38** 248 (1974)
32. D.N. Christodoulides and R.I. Joseph, Opt. Lett. **13**, 53 (1988)
33. Y.S. Kivshar and S.K. Turitsyn, Opt. Lett. **18**, 337 (1993)
34. N. Lazarides, G.P. Tsironis, and M. Eleftheriou, Nonlin. Phen. Compl. Syst. **11**, 250 (2008)

Index

Anderson localization, 92
anisotropy, 101
    orientation, 107
    polarization, 108

bandgap, 3, 82, 148, 168
    Bragg reflection, 151, 155
    semi-infinite, 85, 131
bandgap spectrum, 23, 101
bandgap structure, 5, 157
bandgap tunability, 24
basin of attraction, 188
beam
    non-diffracting, 77, 83
beam propagation method, 27
beam steering, 56
    discrete, 28
BEC, *see* Bose-Einstein condensate
Bessel beam, 174
bifurcation, 133
birefringence, 22, 41
bistability, 232
    optical, 218, 228, 234
Bloch momentum, 4, 79, 82
Bloch oscillation, 80, 83, 165, 169, 173, 196, 200
Bloch's theorem, 81, 92, 168
Bose-Einstein condensate, 74, 165, 171, 176, 182, 195, 199
BPM, *see* beam propagation method
Bragg grating, 58
    nonlinear, 56
Bragg reflection, 104, 173, 200

breather, 277
    discrete, 273, 275
    dissipative, 278
    magnetic, 277
    multi-gap, 29
Brillouin zone, 5, 81, 114, 148, 166, 168, 171, 173, 174, 195, 281
Brillouin zone spectroscopy, 95, 104, 107
Brownian motion, 201, 205
BZ, *see* Brillouin zone

c-axis, 74, 101, 105
Casimir-Polder force, 178
chaos, 14, 188
Cherenkov radiation, 43
chirp, 48
cloak of invisibility, 217, 221, 225, 241, 274
CMT, *see* coupled-mode theory
coupled-mode theory, 78, 84
coupling distance, 24
cross-phase modulation, 30

dark irradiance, 103
data storage
    holographic, 114
defect, 12, 130, 146, 156
detuning, 25
diffraction, 4, 7, 56, 79, 145
    anomalous, 5, 7, 79, 87
    conical, 84
    discrete, 13, 24, 29, 147, 154
diffraction length, 154

diffraction relation, 79
diffusion, 149, 186
dispersion, 4, 56, 145, 147, 241
    anomalous, 40, 42, 46, 151
    group velocity, 38, 40, 44
    second order, 57
    third order, 43
dispersion relation, 4, 259, 274, 280, 282
dissipation, 182, 207
DNLS, see Schrödinger equation, nonlinear, discrete
driving, 181, 206
    biharmonic, 185, 210
    multifrequency, 212
    periodic, 202, 206
    quasiperiodic, 185, 188, 207

electromagnetically induced transparency, 55

Fabry-Perot resonator, 232
FB mode, see mode, Floquet-Bloch
Fermi level, 260
fibre
    Bragg, 57, 135
    multi-core, 48
    photonic crystal, 37, 127, 135, 147
        hollow core, 37
        solid core, 37
fine structure constant, 178
four-wave mixing, 39, 40, 44, 88, 149, 228
FWM, see four-wave mixing

Ginzburg-Landau equation, 89
grating
    phase-shifted, 60
Gross-Pitaevskii equation, 166, 171
group velocity, 56, 57, 79, 169, 266, 284
GVD, see dispersion, group velocity

Husimi representation, 190

idler, 41
instability, 264, 269

Jacobian matrix, 222
Josephson effect, 275
Josephson junction, 192, 275

Kramers-Kronig relation, 230
Kronig-Penney model, 81

Landau-Zener tunneling, 31, 83, 166, 170, 173, 174, 196, 200
    optical, 33
laser cooling, 165
laser pulse, 38, 145
    chirped, 48
laser trapping, 165
lattice, 273
    Bessel, 77, 92, 135
    fixed, 112
    flexible, 112
    hexagonal, 63
    hybrid, 112
    multi-periodic, 114
    optical, 22, 165, 182, 195, 199, 206, 208
    optically-induced, 74, 86, 101, 102
    photonic, 102, 157
        nonlinear, 146
    random, 102
LC, see liquid crystal
level of dynamics, 119
LHM, see material, left-handed
light
    polychromatic, 145
light localization, 150
liquid crystal, 21, 149
    nematic, 21

Mach-Zehnder interferometer, 58, 118
material
    left-handed, 218, 241, 248, 273, 282
material response
    nonlinear, 3, 145, 152, 243
Mathieu function, 81
Maxwell's equations, 23, 219, 221, 236, 274, 283
metamaterial, 56, 217, 259, 271, 273
    magnetic, 217
    negative index, 217, 227, 234, 241
mode
    defect, 127, 130
    Floquet-Bloch, 3, 23, 29, 60, 66, 79, 82, 189
    staggered, 155, 169
    surface, 158

modulational instability, 8, 40, 87, 286
  discrete, 8

NIM, *see* metamaterial, negative index
NLC, *see* liquid crystal, nematic
NLS, *see* Schrödinger equation, nonlinear
nonlinearity, 3, 285
  cubic, 8, 11
  defocusing, 7, 8, 10, 13, 31, 75, 85, 87, 129, 135, 138, 147, 150, 154, 243, 248, 276
  focusing, 7, 8, 50, 75, 85, 87, 91, 106, 110, 129, 137, 150, 243, 248, 275, 276
  Kerr, 49, 76, 89, 151, 235, 243
  Kerr-like, 6
  photorefractive, 102
  photovoltaic, 149
  quadratic, 251

OPA, *see* optical parametric amplification
optical bullet, 66
optical parametric amplification, 218, 228, 230

Pauli matrices, 172
PCF, *see* fibre, photonic crystal
Peierls-Nabarro potential, 11, 27, 88
permeability, 219, 235, 241, 243, 248, 263, 270, 274
  nonlinear, 243
permittivity, 219, 224, 235, 241, 243, 259, 274, 275
  negative, 220
  nonlinear, 246
phase matching, 25, 41, 228, 231, 251
phase velocity, 228
photonic crystal, 55, 101, 145, 242
photons
  entangled, 41
photorefractive crystal, 75, 101, 107, 108, 114
plasma frequency, 243, 246, 253, 274
plasmon, 236, 242, 243
Poincaré section, 186
Poynting vector, 228, 230, 233, 274
propagation constant, 25, 29, 82, 131

quasicrystal, 77, 92, 102
quasimomentum, 168, 175, 176, 195, 200

Rabi frequency, 172
Raman generation
  spontaneous, 41
Raman shift, 44, 46, 151
ratchet effect, 181, 188, 201, 205
rectification, 182, 186, 191, 205
reflection
  Bragg, 56, 130
  total internal, 85, 130, 234
refraction, 145
  negative, 56, 218, 241, 242, 252
refractive index, 55
  gradient, 31
  negative, 227, 273
  periodically-modulated, 5, 55, 73, 101, 145
resonance, 209
  Bragg, 66
  Fano, 13
  magnetic, 270
resonator, 234

scattering
  Brillouin
    stimulated, 42
  Raman, 42
Schrödinger equation, 189, 195, 200
  nonlinear, 44, 84, 104, 172, 235, 281
    discrete, 23, 78
second-harmonic generation, 218, 228, 242, 251
self-phase modulation, 39, 151
separatrix, 186
SHG, *see* second-harmonic generation
Sisyphus cooling, 209
slow light, 56
slow light switching, 59
slow light tunneling, 65
Snell's law, 274
solitary wave, 273
soliton, 73, 127, 273
  blocker, 13, 89
  defect, 137, 139
  discrete, 7, 25, 48, 74, 84, 89, 101, 112, 113, 117

dark, 11, 87
dissipative, 89
intra-site, 84
off-site, 84
on-site, 84
staggered, 7
unstaggered, 7
gap, 7, 29, 58, 101, 116, 232, 233
dipole-mode, 118
lattice, 3, 7, 24, 74, 85
higher order, 10
multi-gap, 29
magnetic, 275
quadratic, 88
spatial, 150, 242, 250
surface, 88
temporal, 249
vector, 9, 88
discrete, 9, 14
vortex, 92
off-site, 93
on-site, 92
white light, 150
soliton compression, 48
soliton existence curve, 89
soliton fission, 42
soliton fusion, 13, 118
soliton mobility, 112
solitons
copropagating, 13, 118
counterpropagating, 13, 118
speed of light, 55, 244
split-ring resonator, 220, 231, 236, 241, 242, 246, 259, 270, 273, 275
SPM, *see* self-phase modulation
SRR, *see* split-ring resonator

stability, 264
stability analysis, 285
stability criteria, 267, 270
stabilization, 14
Stark shift, 166, 196
supercontinuum generation, 38, 151
superlattice, 88, 114, 173, 196
superlens effect, 221
superprism effect, 145
superresolution, 218
susceptibility
nonlinear, 39
symmetry, 183
symmetry breaking, 26, 181, 205

Talbot effect, 129
Tamm state, 157
time of flight, 170, 199
TIR, *see* reflection, total internal

Vakhitov-Kolokolov criterion, 90
vortex
optical, 132

WA, *see* waveguide array
Wannier-Stark state, 173, 176
wave vector, 228, 274
wave-number, 66
waveguide, 63
bi-directional, 122
waveguide array, 3, 128, 146
nonlinear, 4, 74
waveguide coupler, 59
waveguiding, 104, 107

XPM, *see* cross-phase modulation